Springer Handbook of Auditory Research

Volume 68

More information about this series at http://www.springer.com/series/2506

The ASA Press

The ASA Press imprint represents a collaboration between the Acoustical Society of America and Springer dedicated to encouraging the publication of important new books in acoustics. Published titles are intended to reflect the full range of research in acoustics. ASA Press books can include all types of books published by Springer and may appear in any appropriate Springer book series.

ASA PRESS

Adrian K. C. Lee • Mark T. Wallace
Allison B. Coffin • Arthur N. Popper
Richard R. Fay
Editors

Multisensory Processes

The Auditory Perspective

ASA PRESS

Springer

Editors
Adrian K. C. Lee
Department of Speech and Hearing Sciences, Institute for Learning and Brain Sciences
University of Washington
Seattle, WA, USA

Mark T. Wallace
Departments of Hearing and Speech Sciences, Psychiatry, Psychology and Vanderbilt Brain Institute
Vanderbilt University
Nashville, TN, USA

Allison B. Coffin
Department of Integrative Physiology and Neuroscience
Washington State University Vancouver
Vancouver, WA, USA

Arthur N. Popper
Department of Biology
University of Maryland
College Park, MD, USA

Richard R. Fay
Department of Psychology
Loyola University Chicago
Chicago, IL, USA

ISSN 0947-2657 ISSN 2197-1897 (electronic)
Springer Handbook of Auditory Research
ISBN 978-3-030-10459-7 ISBN 978-3-030-10461-0 (eBook)
https://doi.org/10.1007/978-3-030-10461-0

Library of Congress Control Number: 2019933097

This Springer imprint is published by the registered company Springer Nature Switzerland AG
The registered company address is: Gewerbestrasse 11, 6330 Cham, Switzerland

Acoustical Society of America

The purpose of the Acoustical Society of America (www.acousticalsociety.org) is to generate, disseminate, and promote the knowledge of acoustics. The Acoustical Society of America (ASA) is recognized as the world's premier international scientific society in acoustics, and counts among its more than 7000 members, professionals in the fields of bioacoustics, engineering, architecture, speech, music, oceanography, signal processing, sound and vibration, and noise control.

Since its first meeting in 1929, the ASA has enjoyed a healthy growth in membership and in stature. The present membership of approximately 7000 includes leaders in acoustics in the United States of America and around the world. The ASA has attracted members from various fields related to sound including engineering, physics, oceanography, life sciences, noise and noise control, architectural acoustics; psychological and physiological acoustics; applied acoustics; music and musical instruments; speech communication; ultrasonics, radiation, and scattering; mechanical vibrations and shock; underwater sound; aeroacoustics; macrosonics; acoustical signal processing; bioacoustics; and many more topics.

To assure adequate attention to these separate fields and to new ones that may develop, the Society establishes technical committees and technical groups charged with keeping abreast of developments and needs of the membership in their specialized fields. This diversity and the opportunity it provides for interchange of knowledge and points of view has become one of the strengths of the Society.

The ASA's publishing program has historically included *The Journal of the Acoustical Society of America, JASA-Express Letters, Proceedings of Meetings on Acoustics*, the magazine *Acoustics Today*, and various books authored by its members across the many topical areas of acoustics. In addition, ASA members are involved in the development of acoustical standards concerned with terminology, measurement procedures, and criteria for determining the effects of noise and vibration.

Series Preface

Springer Handbook of Auditory Research

The following preface is the one that we published in volume 1 of the Springer Handbook of Auditory Research back in 1992. As anyone reading the original preface, or the many users of the series, will note, we have far exceeded our original expectation of eight volumes. Indeed, with books published to date and those in the pipeline, we are now set for over 80 volumes in SHAR, and we are still open to new and exciting ideas for additional books.

We are very proud that there seems to be consensus, at least among our friends and colleagues, that SHAR has become an important and influential part of the auditory literature. While we have worked hard to develop and maintain the quality and value of SHAR, the real value of the books is very much because of the numerous authors who have given their time to write outstanding chapters and our many coeditors who have provided the intellectual leadership to the individual volumes. We have worked with a remarkable and wonderful group of people, many of whom have become great personal friends of both of us. We also continue to work with a spectacular group of editors at Springer. Indeed, several of our past editors have moved on in the publishing world to become senior executives. To our delight, this includes the current president of Springer US, Dr. William Curtis.

But the truth is that the series would and could not be possible without the support of our families, and we want to take this opportunity to dedicate all of the SHAR books, past and future, to them. Our wives, Catherine Fay and Helen Popper, and our children, Michelle Popper Levit, Melissa Popper Levinsohn, Christian Fay, and Amanda Fay Sierra, have been immensely patient as we developed and worked on this series. We thank them and state, without doubt, that this series could not have happened without them. We also dedicate the future of SHAR to our next generation of (potential) auditory researchers—our grandchildren—Ethan and Sophie Levinsohn, Emma Levit, and Nathaniel, Evan, and Stella Fay.

Preface 1992

The Springer Handbook of Auditory Research presents a series of comprehensive and synthetic reviews of the fundamental topics in modern auditory research. The volumes are aimed at all individuals with interests in hearing research including advanced graduate students, postdoctoral researchers, and clinical investigators. The volumes are intended to introduce new investigators to important aspects of hearing science and to help established investigators to better understand the fundamental theories and data in fields of hearing that they may not normally follow closely.

Each volume presents a particular topic comprehensively, and each serves as a synthetic overview and guide to the literature. As such, the chapters present neither exhaustive data reviews nor original research that has not yet appeared in peer-reviewed journals. The volumes focus on topics that have developed a solid data and conceptual foundation rather than on those for which a literature is only beginning to develop. New research areas will be covered on a timely basis in the series as they begin to mature.

Each volume in the series consists of a few substantial chapters on a particular topic. In some cases, the topics will be ones of traditional interest for which there is a substantial body of data and theory, such as auditory neuroanatomy (Vol. 1) and neurophysiology (Vol. 2). Other volumes in the series deal with topics that have begun to mature more recently, such as development, plasticity, and computational models of neural processing. In many cases, the series editors are joined by a co-editor having special expertise in the topic of the volume.

Richard R. Fay, Chicago, IL, USA
Arthur N. Popper, College Park, MD, USA

SHAR logo by Mark B. Weinberg, Potomac, Maryland, used with permission.

Volume Preface

Auditory behavior, perception, and cognition are all shaped by information from other sensory systems. The research examining this multisensory view of auditory function is rapidly expanding and has demonstrated numerous interactions between hearing and other senses at levels of analysis ranging from the single neuron to neuroimaging in human clinical populations. A great deal of evidence now suggests that areas of the brain traditionally considered auditory can be strongly influenced by information from other senses. This volume highlights the multisensory world from an auditory perspective and focuses on the intersection of auditory and visual processing that has a profound impact on communication in everyday social settings. The chapters that make up this volume provide contemporary and comprehensive discussions on an array of concepts related to the behavioral, perceptual, and physiological aspects of audiovisual processing.

Chapter 1 by Lee and Wallace provides a précis of the multisensory world with an overview and road map to the other chapters in the volume. This is followed in Chap. 2 where Alais and Burr begin this journey by describing how redundant cues from the different senses are statistically combined in a so-called optimal manner. Grant and Bernstein (Chap. 3) examine the auditory, visual, and audiovisual factors that influence speech intelligibility. In addition to boosting speech intelligibility in multi-talker environments, visual information can also help listeners to attend the talker-of-interest, helping solve the classic cocktail party problem. Lee, Maddox, and Bizley (Chap. 4) examine whether, and how, auditory and visual information can be grouped perceptually. Then, in Chap. 5, Willet, Groh, and Maddox address the coordinate reference frame problem with a focus on how spatial information is coded in the superior colliculus—a major midbrain hub for multisensory convergence, as well as for other areas along the auditory pathway and throughout the brain.

Subsequently, King, Hammond-Kenny, and Nodal (Chap. 6) take the readers on a deeper exploration of the multisensory neural circuitry along the auditory pathway. Multisensory interactions that influence auditory processing extend far beyond auditory cortex. In Chap. 7, Plakke and Romanski look at how the frontal lobes support the processing of communication signals via the convergence of sensory inputs from many brain regions. Focusing back on auditory cortex, Beauchamp (Chap. 8) elaborates on several of the different neuroimaging approaches that have been used to address mapping of the neural substrates for audiovisual speech processing in the human cortex and highlights the challenges of delimiting functional borders given the individual differences across subjects and the limitations of a method that indirectly indexes neural activity.

Despite these difficulties in accurately parcellating the human temporal cortex, there is abundant evidence both from human and nonhuman primate studies that the temporal lobe is an important site for multisensory processing. Perrodin and Petkov (Chap. 9) provide an overview of the cortical representations of voice and face content in the temporal lobe. The strong connectivity between temporal and prefrontal cortices raises the important question of how information across these brain regions is shared and coordinated. To address this question, Keil and Senkowski (Chap. 10) focus upon neural network dynamics, as reflected in neural oscillations, to describe information processing across different cell assemblies.

It is well known that neuroplasticity is pronounced during development, but there is now a great deal of evidence suggesting significant plastic capacity for the mature brain. Bruns and Röder (Chap. 11) review evidence that spatial, temporal, and speech identification tasks carried out by the auditory modality can also be strongly influenced by cross-modal learning.

The final chapter (Chap. 12) by Baum Miller and Wallace provides a clinical perspective on multisensory influences on auditory processing. In this chapter, the authors review how fundamental changes in both auditory and multisensory processing impact perception in autism spectrum disorder (ASD).

Adrian K. C. Lee, Seattle, WA, USA
Mark T. Wallace, Nashville, TN, USA
Allison B. Coffin, Vancouver, WA, USA
Arthur N. Popper, College Park, MD, USA
Richard R. Fay, Chicago, IL, USA

Contents

Contributors

David Alais School of Psychology, University of Sydney, Sydney, NSW, Australia

Sarah H. Baum Miller Department of Psychology, Institute for Learning and Brain Sciences (I-LABS), University of Washington, Seattle, WA, USA

Michael S. Beauchamp Department of Neurosurgery and Core for Advanced MRI, Baylor College of Medicine, Houston, TX, USA

Joshua G. W. Bernstein National Military Audiology and Speech Pathology Center, Walter Reed National Military Medical Center, Bethesda, MD, USA

Jennifer K. Bizley Ear Institute, University College London, London, UK

Patrick Bruns Biological Psychology and Neuropsychology, University of Hamburg, Hamburg, Germany

David Burr Neuroscience Institute, National Research Council, Pisa, Italy

Department of Neuroscience, University of Florence, Florence, Italy

Ken W. Grant National Military Audiology and Speech Pathology Center, Walter Reed National Military Medical Center, Bethesda, MD, USA

Jennifer M. Groh Department of Neurobiology, Center for Cognitive Neuroscience, Duke University, Durham, NC, USA

Department of Psychology and Neuroscience, Center for Cognitive Neuroscience, Duke University, Durham, NC, USA

Amy Hammond-Kenny Department of Physiology, Anatomy and Genetics, University of Oxford, Oxford, UK

Julian Keil Biological Psychology, Christian-Albrechts-University Kiel, Kiel, Germany

Andrew J. King Department of Physiology, Anatomy and Genetics, University of Oxford, Oxford, UK

Adrian K. C. Lee Department of Speech and Hearing Sciences, Institute for Learning and Brain Sciences (I-LABS), University of Washington, Seattle, WA, USA

Ross K. Maddox Department of Biomedical Engineering, University of Rochester, Rochester, NY, USA

Department of Neuroscience, University of Rochester, Rochester, NY, USA

Del Monte Institute for Neuroscience, University of Rochester, Rochester, NY, USA

Center for Visual Science, University of Rochester, Rochester, NY, USA

Fernando R. Nodal Department of Physiology, Anatomy and Genetics, University of Oxford, Oxford, UK

Catherine Perrodin Institute of Behavioural Neuroscience, University College London, London, UK

Christopher I. Petkov Institute of Neuroscience, Newcastle University Medical School, Newcastle upon Tyne, UK

Bethany Plakke Department of Psychological Sciences, Kansas State University, Manhattan, KS, USA

Brigitte Röder Biological Psychology and Neuropsychology, University of Hamburg, Hamburg, Germany

Lizabeth M. Romanski Department of Neuroscience, University of Rochester School of Medicine, Rochester, NY, USA

Daniel Senkowski Department of Psychiatry and Psychotherapy, Charité—Universitätsmedizin Berlin, Berlin, Germany

Mark T. Wallace Department of Hearing and Speech Sciences, Vanderbilt Brain Institute, Vanderbilt University, Nashville, TN, USA

Department of Psychiatry and Behavioral Sciences, Vanderbilt Brain Institute, Vanderbilt University, Nashville, TN, USA

Department of Psychology, Vanderbilt Brain Institute, Vanderbilt University, Nashville, TN, USA

Shawn M. Willett Department of Neurobiology, Center for Cognitive Neuroscience, Duke University, Durham, NC, USA

Chapter 1
Visual Influence on Auditory Perception

Adrian K. C. Lee and Mark T. Wallace

Abstract Auditory behavior, perception, and cognition are all shaped by information from other sensory systems. The research examining this multisensory view of auditory function is rapidly expanding and has demonstrated numerous interactions between hearing and the other senses at levels of analysis ranging from the single neuron to neuroimaging in human clinical populations. A great deal of evidence now suggests that areas of the brain traditionally considered auditory can be strongly influenced by information from other senses. This chapter highlights the multisensory world from an auditory perspective, in particular, focusing on the intersection of auditory and visual processing that has a profound impact on communication in everyday social settings. It is followed by an introduction of the chapters that make up this volume, which provide contemporary and comprehensive discussions on an array of concepts related to the behavioral, perceptual, and physiological aspects of audiovisual processing.

Keywords Attention · Auditory cortex · Autism spectrum disorder · Cross-modal · Multisensory · Neural typical · Oscillation · Plasticity · Prefrontal cortex · Reference frame · Scene analysis · Sensory integration · Speechreading

---

A. K. C. Lee (✉)
Department of Speech and Hearing Sciences, University of Washington,
Seattle, WA, USA

Institute for Learning and Brain Sciences (I-LABS), University of Washington,
Seattle, WA, USA
e-mail: akclee@uw.edu

M. T. Wallace
Department of Hearing and Speech Sciences, Vanderbilt Brain Institute,
Vanderbilt University, Nashville, TN, USA

Department of Psychiatry and Behavioral Sciences, Vanderbilt Brain Institute,
Vanderbilt University, Nashville, TN, USA

Department of Psychology, Vanderbilt Brain Institute, Vanderbilt University,
Nashville, TN, USA
e-mail: mark.wallace@vanderbilt.edu

A. K. C. Lee et al. (eds.), *Multisensory Processes*, Springer Handbook
of Auditory Research 68, https://doi.org/10.1007/978-3-030-10461-0_1

1.1 Introduction

From the galloping sound of a horse stampede to racing cars zooming past the finish line at a grand prix, the environment is full of events that emit information that is carried as energy and propagated in a variety of forms, such as light and sound. Although individual sensory systems have evolved to transduce this energy into meaningful signals in the nervous system, these individual systems must also work in concert to generate a unified and coherent view of the perceptual world. Additionally, the ability to leverage information across multiple senses can often increase performance, and this construct of multisensory integration undoubtedly impacts survival by conferring a selective advantage. Thus, a gain in the signal attributable to the presence of information in two or more senses can help predators better locate food sources and, conversely, allow potential prey to better avoid or escape their predators.

In humans, face-to-face conversation is a particularly relevant and commonplace multisensory experience. In addition to the meaning derived from the auditory speech signal, visual information from the facial expressions of the interlocutors is also informative for both the content and emotional context of the conversation. If this face-to-face conversation takes place in a crowded restaurant, looking at the talker's lip movement can substantially improve speech intelligibility. Despite the ubiquity of the multisensory experience and the powerful associated behavioral and perceptual benefits, experiments in the laboratory have mostly focused on understanding how unisensory information is used to perform a task. Take, for example, experiments structured to examine the impact of spatial cues in auditory scene analysis as opposed to examining how visual information is combined with this auditory information in the context of the same task. A second related example are experiments evaluating the impact of available visual information alone in the naturalistic environment to help separate talkers in a crowded acoustic scene.

The aim of this volume is to provide a foundation of knowledge about the current state of understanding in regard to multisensory influences on auditory processes, with the goal of inspiring more rapid growth in scientific studies surrounding this topic.

1.1.1 Basic Concepts and Historical Perspectives

From a computational perspective, integrating information across the different senses is nontrivial. Take the face-to-face conversation as an example. Acoustic energy associated with the talker's speech is transmitted as a time-varying pressure-wave signal, whereas visual information is transmitted as electromagnetic radiation (i.e., light waves). Exquisite and dedicated structures perform mechanotranduction in the cochlea and phototransduction in the visual system to turn these different forms of energy into electrical signals. These modality-specific or unisensory signals are first processed by dedicated neural pathways, which are assumed to be largely independent and hierarchically organized (see King, Hammond-Kenny, and Nodal, Chap. 6). In such a traditional and serially oriented view of sensory organization and function, it has been believed that only after these modality-specific computations have been performed can information from the different senses be combined and integrated to carry out multisensory computations.

Seminal work in multisensory systems strived to quantify the product of these multisensory computations and began with a framework in which the effectiveness of multisensory integration was operationally defined as the difference in response evoked by a combination of stimuli from two different modalities when compared with the response evoked by the most effective of its component stimuli (Stein and Meredith 1993). From the perspective of the single neuron, in which many of these multisensory operations were first characterized, the quantification of multisensory integration was summarized as the difference in neuronal firing rates and illustrated that multisensory convergence could give rise to either significant increases in firing (response enhancement) or significant decreases in firing (response depression). Furthermore, this work went on to show that these enhancements and depressions of response were often strongly dependent on the physical features of the stimuli that were combined. Thus, multisensory stimuli that were in close spatial and temporal correspondence generally resulted in response enhancements. Such an organization makes a great deal of sense relative to the physical world because stimuli that are in close proximity in space and time are highly likely to have originated from the same source. Hence, the nervous system can make inferences in common origin by evaluating the spatial and temporal statistics of a given stimulus pairing.

As for other fields of inquiry, the perspective of multisensory processing is often influenced by the field in which each scientist was originally trained. Thus, those coming to the multisensory field from a single-neuron neurophysiological background will focus on changes in neuronal encoding of individual neurons associated with having information present from multiple modalities. Conversely, those coming to the field with a neuroimaging perspective are much more interested in providing a more network-based view into multisensory function. Similarly, those trained in different sensory systems bring a different lens to their study of multisensory function. Whereas vision scientists often focus on spatially based tasks and the impact of adding sounds to these tasks, auditory scientists are generally more interested in questions of how auditory discrimination, such as speech comprehension, is impacted by the presence of visual cues (either concordant or discordant). In this volume, a flavor for these differing approaches and perspectives can be gleaned, but all unified from the viewpoint of better understanding how vision can shape auditory function.

1.2 Volume Roadmap

Auditory and visual information are seamlessly combined to form better perceptual estimates of the multisensory world. Alais and Burr (Chap. 2) begin this journey by describing how redundant cues from the different senses are statistically combined in a so-called "optimal" manner. For example and as alluded to in Sect. 1.1.1, a multisensory event or object is typically one in which the sight and sound energies come from the same place at around the same time. However, these energies are likely not of equal value for the nervous system because the spatial resolution of the visual system is better than that of the auditory system and the temporal resolution of the auditory system is better than that of the visual system. To consider this differential weighting, Alais and Burr make use of a Bayesian statistical framework, here known as maximum likelihood estimation, that combines information based on the uncertainty of the individual cues and thus generates an optimal model of behavioral performance that fits the empirical data exceptionally well. They also trace how this statistical cue-weighting model evolves over development as well as how it is impacted in circumstances of sensory loss.

The benefits of combining auditory and visual information are immediately evident when one tries to communicate in noisy and reverberant environments. Grant and Bernstein (Chap. 3) examine the auditory, visual, and audiovisual factors that influence speech intelligibility, such as which spectral regions of the speech signal are most important for audiovisual speech recognition and what information is complementary or redundant across auditory and visual speech cues. Audiovisual speech intelligibility research can be traced back to the seminal research conducted at Bell Laboratories (a part of the earlier iteration of AT&T) in the last century that addressed how different communication channel qualities can affect speech intelligibility. This line of research has expanded to include different metrics to predict speech intelligibility performance in noisy environments and now includes the significant work focused on listeners with hearing impairment. However, an understudied area is how speech intelligibility can be modeled in active speech communication settings involving face-to-face audiovisual input (i.e., beyond the well-studied auditory target-in-noise scenarios). Grant and Bernstein provide a unique perspective on audiovisual integration, with a distinctive focus coming from the design and evaluation of audio systems and hearing-rehabilitation devices.

In addition to boosting speech intelligibility in multitalker environments, visual information can also help listeners attend to the talker of interest, helping to solve the classic cocktail party problem (Cherry 1953). Lee, Maddox, and Bizley (Chap. 4) examine whether and how auditory and visual information can be grouped perceptually. They also argue that this perceptual binding could help select and focus on the source of interest in the presence of competing sounds. They go on to argue that multisensory grouping phenomena should be strictly delineated from multisensory integration (any process in which information across sensory modalities is combined to make a judgment) to facilitate a deeper understanding of how information is combined across senses. In this chapter, many classic multisensory illusions are revisited, and the authors ask whether there is evidence to unambiguously support the often-presumed perceptual binding cited in the literature. Coming from an auditory-centered perspective, the authors also focus on how visual information can help resolve auditory competition and suggest how future studies can focus on this understudied area of research.

Spatial cues are potent features for visual and auditory scene analyses, and combining spatial information from these two senses could potentially help an observer to better perceive surrounding events, especially in a crowded environment. Computationally, however, there is a significant operational challenge: auditory spatial cues are encoded in a head-centered framework (i.e., based on timing and intensity cues to the two ears), whereas visual spatial cues are initially encoded in an eye-centered framework (i.e., based on the eye-centered location of the visual stimuli available from the retina). Willet, Groh, and Maddox (Chap. 5) address this coordinate reference frame problem with a focus on how spatial information is coded in the superior colliculus, a major midbrain hub for multisensory convergence. In addition to focusing on the physiological properties that help solve these coordinate issues, Willet, Groh, and Maddox also provide evidence from behavioral investigations on how eye movements can affect auditory spatial tasks.

King, Hammond-Kenny, and Nodal (Chap. 6) take readers to a deeper exploration of the multisensory neural circuitry along the auditory pathway. In addition to focusing on visual influences on the auditory cortex and their implications for hearing, they also highlight somatosensory inputs along the auditory pathway because of the tight coupling between the motor aspects of speech production and the associated visual articulation cues. The authors argue that such a multisensory-based perspective will not only improve our understanding of the computational mechanisms of auditory cortical neurons but will also illuminate how perception and behavior can be influenced by multisensory interactions.

Multisensory interactions with an auditory component extend far beyond the auditory cortex. In Chap. 7, Plakke and Romanski look at how the frontal lobes support the processing of communication signals via the convergence of sensory inputs from many brain regions. They focus specifically on the ventrolateral prefrontal cortex, a region known to integrate face and vocal stimuli in nonhuman primates. These authors examine how factors such as the timing and congruence of the auditory and visual information shape how this information is integrated by these prefrontal neurons. Furthermore, the authors go on to review the deactivation studies of the ventrolateral prefrontal cortex that show the central role of this area in the integration of socially relevant face and vocalization information.

Where are the neural substrates for audiovisual speech processing in the human cortex? Beauchamp (Chap. 8) elaborates on several of the different neuroimaging approaches that have been used to address this question. As detailed, somewhat surprisingly, the anatomical and functional mapping studies of the early stages of auditory processing in the temporal cortex reveal this question to be one of ongoing and active debate. Based on evidence from postmortem studies as well as structural and functional magnetic resonance imaging studies, including data from the Human Connectome Project (Van Essen et al. 2013), subdivisions of the human auditory cortex are described. In effect, Chap. 8 highlights the challenges of delimiting functional borders given the individual differences across subjects and the limitations of a method that indirectly indexes neural activity. Beauchamp argues that multisensory processing may be a unifying principle that can further aid the functional parcellation of the human auditory cortex and its surrounding regions, particularly in the context of speech processing.

Despite these difficulties in accurately parcellating the human temporal cortex, there is abundant evidence from both human and nonhuman primate studies that the temporal lobe is an important site for multisensory processing. Perrodin and Petkov (Chap. 9) provide an overview of the cortical representations of voice and face content in the temporal lobe. Based on the results from studies that combine microstimulation and functional magnetic resonance imaging in monkeys, the authors provide insights on effective connectivity between the temporal lobe and the prefrontal cortices and suggest that these sites within the temporal lobe are critical convergence sites for auditory and visual information positioned between sensory-specific cortices and the executive control circuits of the frontal cortex.

The strong connectivity between temporal and prefrontal cortices raises the important question of how information across these brain regions is shared and coordinated. To address this question, Keil and Senkowski (Chap. 10) introduce the concept of neural network dynamics, as reflected in neural oscillations, to describe information processing across different cell assemblies. They argue that such analysis of oscillatory cortical activity provides valuable insight on the network interactions that underlie multisensory processing and, more broadly, any perceptual and cognitive tasks. Based on converging empirical observations, the authors conclude that it is likely that different oscillatory frequencies, reflective of different spatial scales of network assembly, index different facets of multisensory processing.

Can auditory perception be changed as different cross-modal experiences are acquired over time? It is well-known that neuroplasticity is pronounced during development, but there is now a great deal of evidence suggesting significant plastic capacity for the mature brain. Bruns and Röder (Chap. 11) review evidence that spatial, temporal, and speech identification tasks carried out by the auditory modality can all be influenced by cross-modal learning. Both brief as well as longer-term cross-modal exposure can trigger sensory recalibration, but the mechanisms underlying short-term and long-term recalibration appear to be distinct. The authors conclude by reviewing the evidence for the neural mechanisms of such cross-modal learning, which suggest that this learning takes place through the modification of both the cortical and subcortical pathways.

The final chapter provides a clinical perspective on multisensory influences on auditory processing. In Chap. 12, Baum Miller and Wallace review how fundamental changes in both auditory and multisensory processing impact perception in autism spectrum disorder. The authors dive into the behavioral and neural correlates of altered sensory processing and examine instances of both enhanced and diminished sensory function and perception in autism spectrum disorder compared with typical development. Furthermore, they propose that differences in the ability to integrate information across the different senses may link sensory abnormalities with the more canonical autism symptoms (e.g., impairment in social communication). If sensory and multisensory functions form the scaffold on which higher order abilities are built, the authors argue that treatment strategies that target strengthening sensory representations may prove useful in improving social communication.

1.3 Outlook

The perceptual world is not constructed on a strict sense-by-sense basis but rather is experienced as a coherent and integrated multisensory gestalt. Despite the self-evident perspective of the multisensory world, the neuroscience community has been slow to acknowledge the importance of multisensory processing and, consequently, delve into its neural bases. An example of this somewhat biased view comes directly from auditory studies. Given that humans can rely on auditory features (e.g., spatial cues, pitch) to help segregate sounds in a complex acoustical scene, is there a need to study the visual impact on solving this cocktail party problem? It is true that the brain can derive an amazing wealth of information about the perceptual environment using only a single sense. However, integrating information across the different senses often leads to striking improvements in human performance and perception (Calvert et al. 2004; Murray and Wallace 2011). More importantly, sensory integration is the natural *modus operandi* in the everyday environment in which humans live and operate.

With the growing interest and emphasis in multisensory systems, many neurophysiological studies have now sought to describe the brain circuits and encoding features associated with multisensory processing. Much of this work has relied on using animal models, focusing on describing the anatomical convergence that provides the neural substrate for multisensory interactions and then on detailing the neuronal operations carried out by neurons and circuits on multisensory stimulation. However, many of the multisensory encoding principles derived to date have come from work carried out in anesthetized animals, using paradigms in which stimulus characteristics are highly constrained (a necessary prerequisite for beginning to better understand multisensory processes; for reviews, see Stein and Meredith 1993; Stein 2012). However, the field needs to transition to studies in awake and behaving animals, with an emphasis on more naturalistic paradigms. Complementing these animal model studies should be human-imaging studies using similar, if not identical, paradigms, with the goal of bridging across levels of analysis. These human studies should take advantage of the host of approaches currently available, including magnetic resonance imaging and electrocorticography, as well as electro- and magnetoencephalography. Furthermore, building off of the wealth of correlative data that have been gathered, studies need to move more in the direction of causation and employ approaches such as transcranial magnetic stimulation and transcranial direct/alternate current stimulation to activate/deactivate brain regions during task performance. Again, these human studies can and should be complemented by animal model studies that make use of new technologies such as chemo- and optogenetics that represent powerful tools to dissect functional circuits.

As with other sensory systems, a new frontier in multisensory research needs to be in the context of active sensing where there is an acknowledgement that sensation and perception in naturalistic settings are a product of an active interplay between the sensory and motor systems. Take, for example, our eye and head movements, which represent powerful filters that decide what aspects of the multisensory world are sampled from at any given moment. Innovative experimental paradigms need to be established so that the question can be answered: how do observers actively sample the environment through movements of the eyes, head, and body to optimize the information they gather from their multisensory surroundings?

Finally, there is another area of research that has not yet been addressed adequately in the field of multisensory research: How does one build a multisensory environment to optimize human performance? The multisensory scene can be constructed de novo (in a virtual reality setting) or realized by injecting additional sensory information to the natural surrounding (in an augmented reality setting). Consumer electronics have progressed to a point that the differentiating factor for the ultimate user's experience might rest on a better multisensory experience. The ergonomics associated with audiovisual (or other multisensory combinations) experiences to improve human-computer interaction capabilities will be fueled by the needs of the consumers and may represent the next frontier of multisensory behavioral research.

The chapters in this volume focus on the neural circuits related to multisensory integration along the auditory pathway, from the brainstem to the prefrontal cortex as well as the perceptual benefits of leveraging other senses for communication in the complex auditory environment of everyday life. It is hoped that the readers will find this overview of multisensory influences an important contribution to the overall understanding of hearing science and, perhaps more importantly, as an inspiration for new research directions that will continue to improve the understanding of how the behavioral and perceptual representations of the multisensory world within which humans live are assembled.

Compliance with Ethics Requirements Adrian K. C. Lee declares that he has no conflict of interest.

Mark T. Wallace declares that he has no conflict of interest.

References

Calvert, G. A., Spence, C., & Stein, B. E. (Eds.). (2004). *The handbook of multisensory processes*. Cambridge: The MIT Press.

Cherry, E. (1953). Some experiments on the recognition of speech, with one and with two ears. *The Journal of the Acoustical Society of America, 25*(5), 975–979.

Murray, M. M., & Wallace, M. T. (Eds.). (2011). *The neural bases of multisensory processes*. Boca Raton: CRC Press.

Stein, B. E. (Ed.). (2012). *The new handbook of multisensory processing*. Cambridge: The MIT Press.

Stein, B. E., & Meredith, M. A. (1993). *The merging of the senses*. Cambridge: The MIT Press.

Van Essen, D. C., Smith, S. M., Barch, D. M., Behrens, T. E. J., Yacoub, E., Ugurbil, K., & WU-Minn HCP Consortium. (2013). The WU-minn human connectome project: An overview. *NeuroImage, 80*, 62–79.

Chapter 2
Cue Combination Within a Bayesian Framework

David Alais and David Burr

Abstract To interact effectively with the world, the brain must optimize its perception of the objects and events in the environment, many of which are signaled by more than one sense. Optimal perception requires the brain to integrate redundant cues from the different senses as efficiently as possible. One effective model of cue combination is maximum likelihood estimation (MLE), a Bayesian model that deals with the fundamental uncertainty and noise associated with sensory signals and provides a statistically optimal way to integrate them. MLE achieves this through a weighted linear sum of two or more cues in which each cue is weighted inversely to its variance or "uncertainty." This produces an integrated sensory estimate with minimal uncertainty and thus maximized perceptual precision. Many studies show that adults integrate redundant sensory information consistent with MLE predictions. When the MLE model is tested in school-aged children, it is found that predictions for multisensory integration are confirmed in older children (>10 years) but not in younger children. Younger children show unisensory dominance and do not exploit the statistical benefits of multisensory integration, even when their dominant sense is far less precise than the other. This curious finding may result from each sensory system having an inherent specialization, with each specialist sense tuning the other senses, such as vision calibrating audition for space (or audition calibrating vision for time). This cross-sensory tuning would preclude useful combination of two senses until calibration is complete, after which MLE integration provides an excellent model of multisensory cue combination.

D. Alais (✉)
School of Psychology, The University of Sydney, Sydney, NSW, Australia
e-mail: David.Alais@sydney.edu.au

D. Burr
Neuroscience Institute, National Research Council, Pisa, Italy

Department of Neuroscience, University of Florence, Florence, Italy
e-mail: Dave@in.cnr.it

A. K. C. Lee et al. (eds.), *Multisensory Processes*, Springer Handbook of Auditory Research 68, https://doi.org/10.1007/978-3-030-10461-0_2

Keywords Bayesian · Calibration · Cross-modal · Cue combination · Development · Maximum likelihood model · Optimal integration · Sensory deficit · Sensory integration · Sensory noise · Space perception · Time perception · Uncertainty reduction · Weighted sum

2.1 Multisensory Integration and the Problem of Cue Combination

The years since the turn of the twenty-first century have witnessed an explosion of research activity in multisensory processing. Prior to this, most sensory research, whether cognitive or neurophysiological, focused on each modality separately and did not seek to understand multisensory integration (Jones and Powell 1970; Benevento et al. 1977). This reflected the prevailing view of cortical organization that each sensory modality initially processed information independently and that sensory integration or "binding" only occurred at later stages of processing in polysensory association areas of the brain. On this view, the emphasis on unisensory research was sensible and provided a tractable starting point for sensory research when relatively little was known about cortical processing. However, recent findings show the brain's neural architecture contains more connectivity between early unisensory areas than was previously known (Kayser et al. 2008; Murray et al. 2015). The early interaction between unisensory cortices probably reflects the fact that many of the stimuli in the environment are fundamentally multisensory in nature and activate multiple senses, each one encoding a complementary aspect of the stimulus, with the multiple representations also providing redundancy (e.g., of spatial location, timing, intensity). It is appropriate that these sensory signals be combined early so that external stimuli are coherently represented as multisensory objects and events as early as possible, but that raises the question of how to combine different sensory signals efficiently and effectively. This chapter reviews a Bayesian approach to multisensory cue combination known as maximum likelihood estimation (MLE) that provides a statistically optimal model for cue combination and provides a good account of many instances of multisensory integration.

Humans live in a multisensory world where an event in the environment often produces signals in several senses. These multiple signals provide redundant and complementary information about the event, and when they are spatially and temporally correlated (typically, the case for signals originating from a common event), the brain exploits these properties by combining responses across sensory modalities. This is a sensible strategy that brings considerable benefits. First, the redundancy of a multisensory representation provides great flexibility, preventing catastrophic failures of perception if one sense is permanently lost or if environmental conditions render one sense temporarily ineffective (e.g., vision is impaired at night; a critical sound is masked by background noise). Second, the statistical advantages of having two samples of the same stimulus leads to important perceptual benefits, seen in faster reactions times and better discrimination of multisensory stimuli (Alais et al. 2010). This latter aspect of multisensory perception has received a good deal of attention in the last couple of decades, and it is clear that the key to robust and coherent perception is the efficient combination of multiple sources of sensory information (Ernst and Bulthoff 2004). Although the perceptual benefits of multisensory integration are well established, understanding how the brain achieves this integration remains a challenging question in sensory and cognitive neuroscience. Moreover, it is not a trivial problem for the brain to solve because the information to be combined arrives in different primary cortices, is often offset in time, and is mapped (at least initially) in different coordinate systems.

2.2 Cue Combination in a Bayesian Framework

Despite the challenges in doing so, the human perceptual system has an impressive ability to seamlessly integrate the senses into a coherent and reliable perception of the external world. It achieves this despite working with neural signals that are inherently noisy and variable. This variability means that perception is intrinsically a probabilistic process (Fig. 2.1A and B), making interpretations and inferences about the likely nature of external stimuli in a process known as "unconscious inference," as von Helmholtz (1925) termed it. Prior knowledge acquired through experience of the world plays a role in guiding these perceptual inferences, as do the incoming sensory signals. A Bayesian framework (Kersten et al. 2004; Knill and Pouget 2004; Pouget et al. 2013) is perfectly suited to modeling perceptual inference for two reasons. First, it is a mathematical model based on probabilities. Second, its two components, called the prior probability and the likelihood, map perfectly onto the two sources of information for perceptual inference: acquired knowledge of the sensory world (the prior) and incoming noisy sensory signals (the likelihood).

Bayes' theorem states that the posterior probability is proportional to the product of the prior probability and the likelihood (Fig. 2.1C). The prior describes the probability of a stimulus before any stimulus information is received and thus reflects, for example, learning, knowledge, and expectations. The likelihood is the probability of the stimulus given its possible states. As applied to perception and behavior, the prior is thought of as an internal model of the statistics of the environment and the likelihood represents an incoming noisy sensory signal. In the case of multisensory stimuli, there will be signals in two or more modalities and a corresponding likelihood for each component. In the audiovisual example shown in Eq. 2.1, the likelihoods for the auditory and visual stimuli are the first two terms of the numerator and the prior is the third term. Multiplicatively combining these three terms (or four terms for a trimodal stimulus) satisfies Bayes' theorem. If this product is then normalized by the product of the simple probabilities for each component, we obtain Bayes' equality. Equation 2.1 shows Bayes' equality for combining two estimates (here, estimates of spatial location [*S*] from auditory and visual cues, with *P* indicating probability) into an integrated multisensory estimate

$$P(S_{AV} \mid S_A, S_V) = \frac{P(S_A \mid S_{AV})P(S_V \mid S_{AV}) * P(S_{AV})}{P(S_A)P(S_V)} \tag{2.1}$$

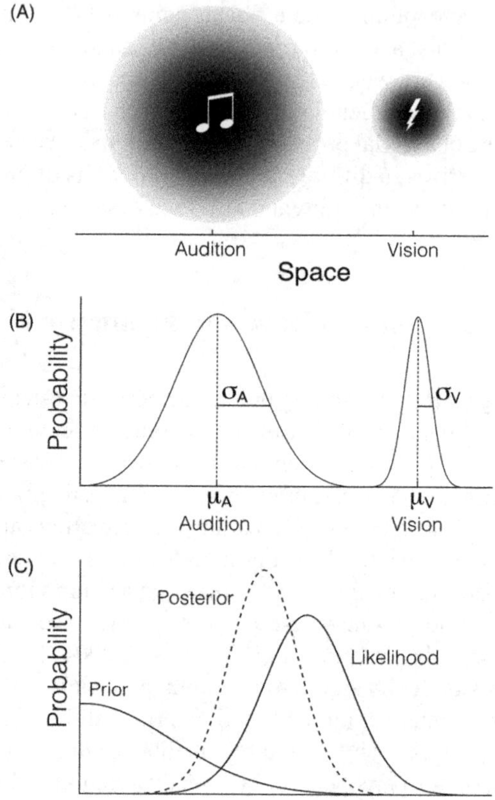

Fig. 2.1 (**A**) The world is crudely sampled through receptive fields of various sizes generating noisy neural signals. Together, these factors degrade the precision of perception. Here the example of spatial location is illustrated, an attribute much more precisely coded in vision than audition. (**B**) The noise accompanying a signal can be modeled by a Gaussian distribution described by two parameters, the mean (μ) and the standard deviation (σ). For spatial location, an auditory estimate is less precise (i.e., higher standard deviation) than a visual one. (**C**) Bayesian theory, being based on probability distributions, provides a convenient way to model the combination of noisy information. Its two components are the prior distribution and the likelihood distribution. Incoming sensory information constitutes the "likelihood," whereas acquired knowledge of the world and its statistics are embodied in the "prior." These can be combined (multiplied) to produce the posterior distribution, an optimal fusion of stored knowledge and current sensory information

One model that has been successful in accounting for many instances of multisensory cue combination is the maximum likelihood estimation (MLE) model. MLE is a simplified Bayesian model that only takes account of the likelihood (it has no prior component, the final term in the numerator of Eq. 2.1). MLE describes how noisy sensory information can be combined from two or more independent sources (e.g., auditory and visual signals). It takes account of the variability of each signal and combines them in a statistically optimal fashion that maximizes the likelihood that the combined response will truly reflect the external stimulus (Fig. 2.2A).

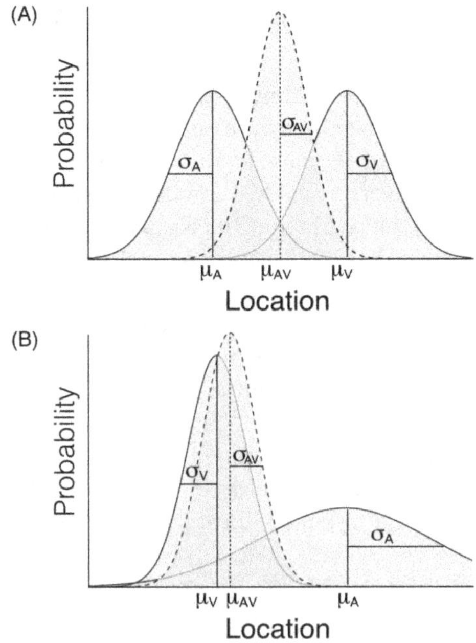

Fig. 2.2 In many cases, perceptual judgments require no access to the stored information and expectations represented by the prior and the Bayesian model simplifies to the likelihood. In multisensory contexts (such as audiovisual localization), each signal will produce a likelihood and combining them produces a product distribution with the highest possible probability, known as the maximum likelihood. Maximizing the likelihood is desirable because it will minimize the distribution's variance, corresponding to maximal perceptual resolution. (**A**) Here the two likelihoods have identical standard deviations but different means. From Eq. 2.3, equal widths (σ) lead to equal component weights, and thus the combined "maximum likelihood" distribution is located at the mean position (see Eq. 2.2), with a narrower width (see Eq. 2.4). (**B**) If component distributions have different widths, their weighting in the product distribution will differ, as per Eqs. 2.2 and 2.3. In effect, the product distribution's location is drawn toward the narrower, and thus the perceptually more reliable, component. Regardless of the relative widths of the component distributions, the product distribution will always be the solution providing the maximum possible probability and thus the minimal standard deviation

In maximizing the likelihood, it minimizes stimulus uncertainty. In essence, MLE is a weighted linear sum that combines two or more signals, each weighted by its reliability. Reliable signals receive a high weight, whereas unreliable signals receive a low weight (Fig. 2.2B). The combination rule is considered statistically optimal because it always provides the result that is most reliable, where "most reliable" means the most probable or least variable solution. In producing the least variable combination, the MLE model effectively minimizes stimulus uncertainty arising from noise in the component signals.

2.3 The Maximum Likelihood Estimation Model

The MLE model is best demonstrated by working through an example of multisensory perception. One of the best-known examples of how the perceptual system deals with redundant spatial signals is the ventriloquist effect (Howard and Templeton 1966). In this effect, provided the auditory and visual stimuli are aligned in time to be synchronous or nearly so (Slutsky and Recanzone 2001), displacing the visual stimulus over modest distances will usually cause the auditory stimulus to be "captured" by the visual event (i.e., perceived as colocalized with the visual stimulus). Being simultaneous and roughly collocated, the signals satisfy the conditions for audiovisual fusion, but how best to fuse them? MLE assumes that the signal in each sensory modality provides an independent estimate about a particular stimulus attribute (here, estimated spatial location, $\hat{s}$) and has a Gaussian-distributed uncertainty. The estimate and its uncertainty are represented by the mean and variance, respectively, of a Gaussian probability distribution. MLE combines the auditory and visual estimates in a weighted linear sum to obtain the estimated bimodal spatial location

$$\hat{s}_{AV} = w_A \hat{s}_A + w_V \hat{s}_V \tag{2.2}$$

where w_A and w_V are the weights allocated to the component modalities. The weights are determined by the relative reliability of each modality's estimate of the stimulus attribute where variance (σ^2) and reliability are inversely related

$$w_A = \frac{1/\sigma^2_A}{1/\sigma^2_A + 1/\sigma^2_V} = \frac{\sigma^2_V}{\sigma^2_A + \sigma^2_V} \tag{2.3}$$

Equation 2.3 shows that the auditory weight and the visual weight are easily obtained by changing the subscript of the numerator. As should be clear from Eq. 2.3, each modality accounts for a proportion of total variance and thus the component weights are relative weights and sum to 1. In short, the more variable a modality is in contributing to the perceptual estimate, the less reliable it is and the less it is weighted in the bimodal percept. The MLE solution is optimal because it provides the combined estimate with the lowest variance, given the available information, and thus provides maximal stimulus precision. Indeed, the combined variance can never be larger than either of the components because of the following relationship

$$\sigma^2_{AV} = \frac{\sigma^2_A \sigma^2_V}{\sigma^2_A + \sigma^2_V} \tag{2.4}$$

From Eq. 2.4, the combined estimate must always have a lower variance than either of the components. The reduction in combined variance (and consequent gain in precision) is maximal when the component variances are equal, reducing variance in that case by a factor of $\sqrt{2}$ (Fig. 2.2A). This benefit reduces if the component variances diverge, and in the limit, very different component variances produce a combined variance that approaches the value of the smaller of the component variances (Fig. 2.2B).

The MLE integration rule therefore makes two key predictions when two signals are combined because it specifies both the mean value of the combined estimate ($\hat{s}_{AV}$) and its variance (σ_{AV}^2). These predictions have been tested and confirmed in a range of different multisensory contexts, showing that multisensory integration closely approximates the MLE model (Clarke and Yuille 1990; Ghahramani and Wolpert 1997; Landy et al. 2011). Examples include audiovisual spatial localization (Alais and Burr 2004) and visual-tactile size estimation (Ernst and Banks 2002). MLE has even been demonstrated in trimodal contexts (Wozny et al. 2008), but it may also occur within a single modality between independent cues (Hillis et al. 2002). The available evidence suggests that MLE integration occurs automatically and does not require that attention to be directed to the component stimuli (Helbig and Ernst 2008). In multisensory contexts, there is evidence that the perceptual estimate of each modality's component cue are not lost when MLE integration occurs, although this appears not to be the case for cues within a single modality where MLE integration is obligatory and the component information is lost (Hillis et al. 2002).

2.4 Maximum Likelihood Estimation: A Flexible Cue Combination Model

The MLE model allows a useful reinterpretation of some earlier ideas in the multisensory literature. One prevalent idea was the "modality appropriateness hypothesis" that stated that conflicts between the modalities were resolved in favor of the most relevant modality (Welch and Warren 1980). In an audiovisual context, the most appropriate modality would be vision for a spatial task and audition for a temporal task. The MLE model supersedes the modality appropriateness hypothesis without resorting to arbitrary notions such as "appropriateness." MLE predicts a dominance of vision over audition for spatial judgments (such as in ventriloquism) because spatial resolution is higher in the visual domain, which means less uncertainty and a higher weighting for vision relative to audition. Conversely, MLE predicts that audition should dominate vision for temporal tasks, such as in auditory driving (Shipley 1964; Recanzone 2003) or for the "double flash" illusion (Shams et al. 2000) because the auditory modality is specialized for temporal processing. Of course, modality appropriateness predicts the same dominances, but it does so within an arbitrary and rigid framework, whereas MLE is flexible and will weight the components in favor of the incoming stimulus with the higher certainty. This flexibility was shown clearly in Alais and Burr's (2004) ventriloquism study (Fig. 2.3) where they demonstrated both conventional ventriloquism and reverse ventriloquism (i.e., auditory capture of visual locations). The reverse ventriloquism

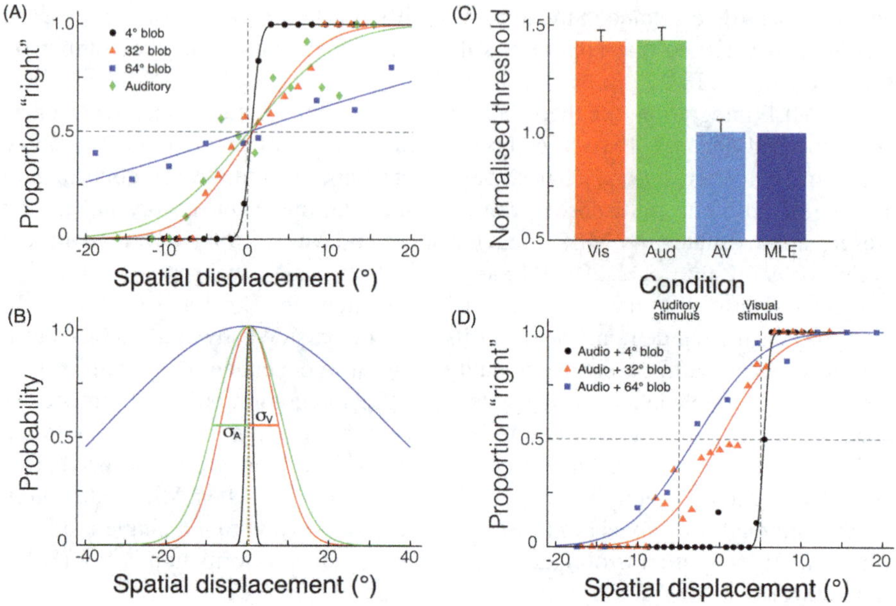

Fig. 2.3 Applying the maximum likelihood estimation model to psychophysics (adapted from Alais and Burr 2004). (**A**) Cumulative Gaussian psychometric functions for localizing an auditory click or Gaussian blobs of various widths ($2\sigma = 4$, 32, or 64°). Functions all pass through ≈0° (all stimuli accurately localized on average) but varied systematically in width. The width is given by the σ term in the cumulative Gaussian equation and defines the discrimination threshold. (**B**) Functions from (**A**) replotted as probability densities to highlight their standard deviations (i.e., σ). The auditory and middle-sized visual stimuli have similar widths and should produce a near-maximal reduction in the combined distribution's width (close to the maximum $\sqrt{2}$ reduction for components of equal widths). (**C**) Audiovisual localization precision (normalized to 1.0) for the collocated auditory and middle-sized visual stimuli was better than for each component separately, indicating increased perceptual precision, and closely matched the maximum likelihood estimation (MLE) prediction. (**D**) Localization of the audiovisual stimulus when the components were separated by ±5° also followed MLE predictions. When the visual component was much better localized than the auditory one (*squares, black curve*), the mean audiovisual position shifted to the visual location (as in Fig. 2.2B). MLE thus accounts for the classic ventriloquist effect. When the auditory stimulus was paired with the poorly localized visual stimulus, audiovisual location was drawn to the (better localized) auditory component (reverse ventriloquism), as MLE predicts

occurred because the visual stimulus was blurred to the point that the auditory signal was more precisely localized (Fig. 2.3A). MLE correctly predicted auditory capture of vision when vision was blurred (Fig. 2.3D), whereas modality appropriateness adheres to a rigid dichotomy of visual spatial dominance and auditory temporal dominance.

MLE is not only a flexible combination rule rather than a rigid assumption of sensory dominances but also takes into account of all the available information. It has been clear since early multisensory studies (Rock and Victor 1964) that one sensory modality rarely dominates completely over another: there is always a residual contribution from the dominated modality. MLE captures this in that the estimate from the less reliable modality is always summed into the combined estimate but is simply downweighted if it has low reliability. It therefore contributes to the combined estimate but with a reduced influence. In this way, MLE provides an intrinsically multisensory estimate, whereas modality appropriateness chooses the most appropriate single modality. The MLE model therefore provides a flexible, quantitative, and principled alternative to the modality appropriateness hypothesis and provides a convenient mathematical framework for combining sensory estimates with their inherent noise and uncertainty.

The MLE model also lends itself readily to psychophysical investigation of multisensory phenomena. This is because simple experiments in which the subject discriminates each of the unisensory components (e.g., which interval is louder, brighter, bigger, more rightward) provide psychometric data that can be modeled with a cumulative Gaussian function to obtain estimates of the mean and its variance, the two parameters needed for Eqs. 2.2–2.4 (see Fig. 2.3A and B). This was the approach adopted in Ernst and Banks's (2002) study of visual-tactile size perception, subsequently applied in Alais and Burr's (2004) audiovisual study of ventriloquism. Both studies found strong evidence for MLE integration. A number of other studies have adopted similar approaches to quantify the variability of sensory estimates and have found results consistent with MLE integration (van Beers et al. 1999; Knill and Saunders 2003; Hillis et al. 2004). The data in Fig. 2.3A show position discrimination results from Alais and Burr (2004). In a two-interval, forced-choice procedure, either a sound source or a Gaussian luminance blob varied in location along the horizon in front of the observer. The subjects had to judge in which interval the stimulus was located further to the right. All stimuli were accurately localized at 0° (directly ahead of the observer) but with a degree of precision that varied with the size of the visual stimuli. As blob size increased over three levels, precision declined. In an analogous experiment varying the location of a sound source, position discrimination data were comparable with the middle-sized visual stimulus. Cumulative Gaussian functions were fitted to the data, which are plotted in Fig. 2.3B as probability densities to highlight the differences in variance.

The MLE model makes the strong prediction that pairing the sound and the middle-sized blob should produce audiovisual discrimination data with significantly higher precision. This is because these two stimuli have roughly equivalent variances and thus should produce an increase in precision that is close to the ideal maximum of $\sqrt{2}$. From Eq. 2.4, the predicted reduction in variance can be calculated and compared against empirical data for discriminating the audiovisual stimulus. As shown by the variances plotted in Fig. 2.3C, discrimination precision for the audiovisual stimulus was indeed significantly lower than for each of the component stimuli and was very close to the value predicted by the MLE model. The test of variance reduction is critical because it provides strong evidence that information was integrated across two sources to produce increased discrimination precision. It rules out alterative possibilities, such as switching between independent information sources, because this would produce a worse performance than the best of the components. It also argues against a probability summation account because this may lead to improved discrimination but by less than a factor or $\sqrt{2}$ (making it imperative to closely match the component variances to distinguish between MLE and probability summation predictions).

The other prediction made by the MLE model concerns the mean of the combined distribution. When the component distributions are centered at different locations, the position of the combined distribution is not simply the average of the two but is a weighted average based on the variability of the components. As shown in Fig. 2.2, the product distribution is drawn to the component with the smaller variance, as predicted by Eqs. 2.2 and 2.3. This aspect of the MLE model is very relevant to multisensory processing because redundant stimulus estimates to be combined across different modalities will often be discrepant despite signaling the same event. This can happen in the temporal domain due to latency differences between the senses or in the spatial domain due to misaligned spatial maps. Provided the signals are roughly spatiotemporally aligned, the brain will attempt to integrate them, but where should the fused stimulus be located? As illustrated in Fig. 2.1A, auditory stimuli will normally be localized with less precision than visual stimuli, meaning that the fused estimate should be drawn toward the (more precise) visual location, according to the MLE model, as shown in Fig. 2.2B. This is the well-known ventriloquist effect. Note that based on the weights in Eq. 2.2 (determined by Eq. 2.3), the MLE model makes a specific quantitative prediction concerning by how much the lesser weighted spatial location should be drawn to the higher weighted location in the fused percept. In this way, it differs from a simple bias to favor one stimulus over the other and from the binary selectivity of the modality appropriateness hypothesis (Welch and Warren 1980) that holds that the most appropriate sense (vision, for a spatial task) will determine perception.

To test if the MLE model could provide an account of the ventriloquist effect, Alais and Burr (2004) compared two conditions. In one, the auditory stimulus and the well-localized visual stimulus (see Fig. 2.3A) were presented simultaneously at horizontally displaced locations and their perceived location was discriminated against the same stimuli both presented at 0° (directly in front of the observer). Location discrimination in this audiovisual condition was compared with another that paired the auditory stimulus with the poorly localized visual stimulus. In the first condition, the spatial discrepancy between the components was resolved by the audiovisual stimulus being localized very near to the visual location. This is the classic ventriloquist effect and is explicable in terms of competing accounts such as simple visual "capture" of auditory location and the modality appropriateness hypothesis (Welch and Warren 1980). However, only the MLE model could account for the second condition. In this case, where the visual stimulus was less reliable than the auditory stimulus, it was the auditory stimulus that dominated audiovisual localization (Fig. 2.3D). This result had never been reported before and is effectively a case of reverse ventriloquism because the location of the visual stimulus was drawn to the location of the auditory stimulus. Importantly, accounts such as modality appropriateness cannot explain such a result, but MLE can; simply, reverse ventriloquism will occur whenever the auditory stimulus is better localized than the visual stimulus (as predicted by Eqs. 2.2 and 2.3).

More recently, interest has turned to the MLE model at the neural level (Rowland et al. 2007; Gu et al. 2008). The study by Gu et al. examined the MLE model using single-neuron recordings from a macaque monkey trained to make heading discriminations in a two-alternative forced-choice task. They measured heading discrimination for vision alone using optic flow stimuli and for vestibular signals alone using a moving platform. The critical condition was the visual-vestibular condition, where conflicting heading directions were introduced from each cue and, as predicted, mean heading direction was drawn to the more reliable component. Confirming the other key prediction of the MLE model, discrimination was better in the visual-vestibular condition (i.e., psychometric functions were steeper, illustrating reduced variance in the bimodal estimate). To bolster the evidence for MLE integration, the authors manipulated the reliability of the visual cue by adding noise to reduce its motion coherence and found that heading discrimination was drawn away from the visual direction toward the vestibular direction, in accordance with MLE predictions of a downweighted visual estimate. Their neural data, recorded while the monkeys performed the behavioral task, showed that spiking rates in single neurons from the dorsal region of the medial superior temporal area were consistent with optimal integration of visual and vestibular cues in heading discrimination.

The evidence for MLE is strong as far as integration of low-level sensory cues is concerned, although to provide an effective explanation for multisensory integration of higher order information, such as speech and semantic information, it may need to be expanded. At this level, other factors exert an influence on multisensory interactions, such as knowledge, expectations, and learning. However, as noted in Sect. 2.2, the MLE model is a simplified Bayesian model in that it does not include a prior, yet these other influences on multisensory integration can be accommodated easily within a Bayesian framework by using a prior probability distribution to account for them. The danger of this approach is that unlike applying the MLE model to low-level sensory cues, which is well constrained and can be well described by psychophysical experiments, priors can be difficult to characterize empirically and there is a risk of invoking them in a post hoc manner to account for unexpected results. Although there is no dispute at a conceptual level about priors embodying learning, experience, and knowledge in a probability distribution that could, in theory, be combined with the likelihood arising from the incoming sensory cues, quantifying and experimentally manipulating priors remains an empirical challenge. Several studies have shown evidence of Bayesian integration involving likelihoods and priors in visual-motor tasks (Kording and Wolpert 2004; Kwon and Knill 2013) and in unisensory tasks involving multiple visual cues (Knill 2007) as well as in the time domain with reproduction of temporal intervals (Jazayeri and Shadlen 2010; Cicchini et al. 2012).

2.5 Maximum Likelihood Estimation Cue Combination in the Time Domain

Multisensory studies finding evidence of MLE integration have used a variety of tasks, including spatial tasks such as judgments of size (Ernst and Banks 2002) or location (Alais and Burr 2004) and visual-motor (Kording and Wolpert 2004) and visual-vestibular (Angelaki et al. 2009) tasks. However, multisensory research assessing MLE in time perception has produced mixed results, some showing that perceptual estimates of elapsed time from a marker event do not obey MLE (Ulrich et al. 2006; Burr et al. 2009; Hartcher-O'Brien and Alais 2011), whereas another report finds that it does (Hartcher-O'Brien et al. 2014). Duration provides a curious case for two reasons. First, elapsed time is not necessarily a sensory representation and may be encoded by central accumulators at a supramodal level. Second, duration estimates cannot be made until the sensory stimulus has ceased so the perceptual judgment must be made on a stored representation, and it may be that these factors preclude MLE integration. Alternatively, there may be procedural differences between these studies that account for the discrepant findings. Studies that failed to find MLE integration in time perception defined elapsed time with brief marker stimuli that could be auditory, visual, or audiovisual at onset and offset of the temporal interval. By using empty intervals (i.e., an interval defined only by onset and offset stimuli), it is not clear whether multisensory integration is expected for the markers or for the elapsed time (which is effectively amodal). Using filled intervals overcomes this problem, and duration perception under these conditions is found to exhibit MLE integration.

In the study of duration discrimination using filled temporal intervals (Hartcher-O'Brien et al. 2014), a sequential two-interval, forced-choice procedure was used to compare a standard and a variable interval, with the intervals both defined by audio, visual, or audiovisual signals. In the audiovisual trials, audio signals with three levels of noise were combined with visual signals with a small temporal conflict to test if the duration mismatch was resolved according to MLE using unisensory signal weights. The finding was that audiovisual duration estimates did exhibit the MLE-predicted weighted average of unisensory estimates with component weights proportional to their reliabilities. This shows that MLE integration is possible on stored duration estimates and suggests that both signal durations and their associated variances needed for appropriate weighting are both available from a stored representation of elapsed time. For further evidence of Bayesian inference in duration perception, the reader is referred to Shi and Burr (2015).

2.6 Changes in Maximum Likelihood Estimation Cue Weightings Over Development

Multisensory perception is often thought to reflect the inherent dominance of one specialized modality over another. Even though recent work inspired by the MLE model shows that the precise weightings of unisensory components can vary flexibly depending on the noise in the incoming signal (e.g., producing reverse ventriloquism when vision is degraded; Alais and Burr 2004), it remains true that in most circumstances vision will dominate for multisensory spatial tasks and audition for temporal tasks. In spatial localization, these multisensory interactions appear to be automatic. For example, when observers need only to localize the auditory component of a pair of simultaneous but spatially displaced audiovisual signals, their judgments still show a bias toward the visual location (Bertelson and Radeau 1981). Other studies too have suggested that ventriloquism occurs automatically (Vroomen et al. 2001), and the same conclusion has been drawn for spatial interactions between touch and vision (Bresciani et al. 2006; Helbig and Ernst 2008) and between touch and audition (Caclin et al. 2002; Guest et al. 2002).

As shown in Sect. 2.5, the MLE model of cue combination accounts well for how information from different senses are combined. However, it is not clear whether this is inherently the case or whether these dominances arise gradually over the span of development. The bulk of the evidence supporting MLE in multisensory integration has been done with adult subjects and does not address the developmental perspective. Sensory systems are not mature at birth but become increasingly refined during development. The brain must take these changes into account and continuously update its mapping between sensory and motor correspondence over the time course of development. This protracted process requires neural reorganization and entails cognitive changes lasting well into early adolescence (Paus 2005). Complicating the matter is that the senses develop at different rates: first touch, followed by vestibular, chemical, and auditory (all beginning to function before birth), and finally vision (Gottlieb 1990). Even though auditory development generally proceeds faster than visual development, perceptual skills within audition continue to develop at different rates, with auditory frequency discrimination (Olsho 1984; Olsho et al. 1988) and temporal discrimination (Trehub et al. 1995) all improving during infancy (Jusczyk et al. 1998). Vision in general develops later than audition, especially visual acuity and contrast sensitivity that continue to improve up until 5–6 years of age (Brown et al. 1987).

These differences in developmental sequences within and between modalities are all potential obstacles for the development of cue integration. Some multisensory processes, such as cross-modal facilitation, cross-modal transfer, and multisensory matching are present to some degree at an early age (Lewkowicz 2000; Streri 2003). Young infants can match signals between different sensory modalities (Dodd 1979; Lewkowicz and Turkewitz 1981) and detect equivalence in the amodal properties of objects across the senses (Rose 1981; Patterson and Werker 2002). For example, they can match faces with voices (Bahrick and Lickliter 2004) and visual and auditory motion signals (Lewkowicz 1992) on the basis of their synchrony. However, the varied time course of sensory development suggests that not all forms of multisensory interaction develop early. For example, multisensory facilitation during a simple audiovisual detection task does not occur until 8 years of age in most children (Barutchu et al. 2009, 2010). Few studies have investigated multisensory integration in school-age children and those that have point to unimodal dominance rather than optimal MLE integration across the senses (McGurk and Power 1980; Hatwell 1987; Misceo et al. 1999).

One study that did test for optimal integration across the senses in school-age children examined visual-haptic integration (Gori et al. 2008). This study tested visual-haptic integration in children aged 5, 6, 8, and 10 years of age and compared their size discrimination against an adult sample. In one experiment, they examined size perception in a paradigm that was essentially the same as that used by Ernst and Banks (2002). Size discrimination thresholds were measured for touch and vision separately to obtain measures of mean and variance for each modality and then the stimuli from both modalities were combined with a small-size conflict to see if the integrated estimate reflected the weights of the individual components. Their results showed that prior to 8 years of age, integration of visual and haptic spatial information was far from optimal. Indeed, directly contradicting the MLE model, in young observers (Fig. 2.4A), they observed that the sense that dominated the multisensory percept was the less precise one: the haptic modality. Haptic information was found to dominate perceived size and its discrimination threshold. Interestingly, however, the weighting of the component signals evolved progressively over development and by 8–10 years of age, visual-haptic integration became statistically optimal and followed MLE predictions, as observed in adults.

In a second analogous experiment, Gori et al. (2008) measured visual-haptic discrimination of orientation in the same age groups. This is another basic spatial task that should favor the visual modality with its specialized orientation-selective neurons in the primary visual cortex (Hubel and Wiesel 1968). Subjects discriminated which one of two bars was rotated more counterclockwise. As with the size discrimination task, thresholds were first measured separately for the visual and haptic modalities and then in a visual-haptic condition with an orientation conflict between the modalities. As with visual-haptic size judgments, the data for 8 year olds were much like the adult data and followed predictions from the MLE model based on the single-modality thresholds. Again, however, the pattern of results for the 5-year-old group was quite different; against the MLE model's predictions, orientation discrimination followed very closely the visual percept rather than incorporating haptic information (Fig. 2.4B). Although both experiments involved visual-haptic spatial tasks, the visual dominance for perceived orientation is the exact opposite to the haptic dominance observed for size discrimination.

In another study, the same group investigated audiovisual integration in both space and time perception across a developmental span covering four age ranges (5–7, 8–9, 10–11, and 13–14 years of age) and compared it to audiovisual integration in adults (Gori et al. 2012). Their goal was to examine the roles of the visual and auditory systems in the development of spatial and temporal audiovisual integration. They used similar tasks to study spatial and temporal perception in which subjects were required to bisect a temporal or a spatial interval. For the temporal bisection task, MLE integration was not observed at all in either in the adult group or any of the four children's age groups. This agrees with another study (Tomassini et al. 2011) showing that multisensory integration is suboptimal for a visual-tactile time reproduction task and with other temporal studies showing auditory dominance over vision rather than optimal integration in adults (Shams et al. 2000;

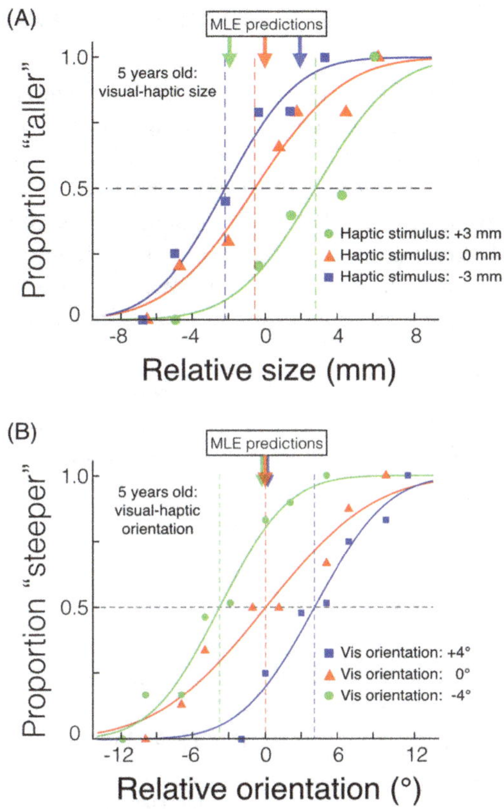

Fig. 2.4 Data showing the absence of MLE integration in children discriminating visual-haptic size and orientation where the haptic stimulus conflicted with the visual one (adapted from Gori et al. 2008). (**A**) Visual-haptic size discrimination: children did not use available visual information to optimize their discrimination and were very strongly dominated by haptic information. This is seen in the locations of the psychometric functions, which were centered at +3 mm when the haptic stimulus was 3 mm larger than vision (*right-hand function, circular data symbols*) and at −3 mm when the haptic stimulus was 3 mm smaller (*left-hand function, square data symbols*). Tellingly, the order of the psychometric functions (*squares, triangles, circles*) was the inverse of the MLE predictions (*indicated by the arrows*). (**B**) Visual-haptic orientation discrimination: children were dominated by visual information for the orientation task and did not use the available haptic information. Showing complete visual dominance, the psychometric functions shifted to +4° when the visual stimulus was 4° clockwise of the haptic stimulus and to −4° when it was 4° counterclockwise of the haptic stimulus

Burr et al. 2009). Alternatively, the lack of optimal temporal integration may have been due to the use of markers to define the start/end of the interval rather than filled intervals (discussed further in Sect. 2.7). For the spatial bisection task, MLE integration was observed only in the adult group, showing that optimal adult-like MLE integration emerges quite late in development for audiovisual temporal tasks, as it does for visual-haptic integration (Gori et al. 2008).

2.7 Cross-Modal Calibration During Development

These studies of multisensory integration over the developmental span (Gori et al. 2008, 2012) show that young children exhibit strong unisensory dominance and that the time course for the development of optimal multisensory integration is rather slow (Fig. 2.5). This is supported by other developmental studies in other sensory domains (Nardini et al. 2008, 2010). With visual-haptic stimuli, haptic information dominates size perception and vision dominates orientation perception. With audiovisual stimuli, audition dominates time perception and vision dominates space perception. The authors account for this developmental change in terms of

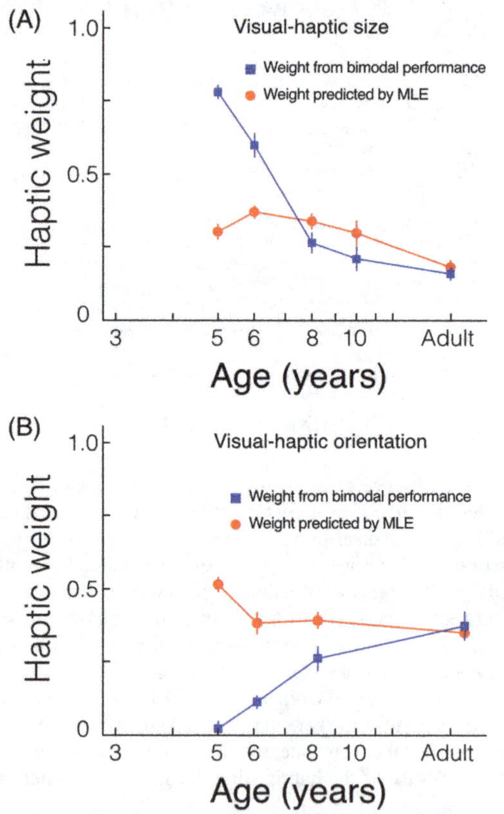

Fig. 2.5 Developmental time course of MLE integration (adapted from Gori et al. 2012). *Circular data symbols* show MLE predictions for the haptic weights when visual and haptic stimuli are combined in a bimodal stimulus. The visual weight is given by the complement (one minus the haptic weight), and both weights are predicted based on Eq. 2.3 using discrimination thresholds (σ) obtained in unimodal experiment (see Fig. 2.3A and B). *Square data symbols* show actual bimodal performance. In both visual-haptic size discrimination (**A**) and visual-haptic orientation discrimination (**B**), there is a large discrepancy at 5 and 6 years of age between bimodal performance and MLE predictions, yet both clearly converge well in adulthood. From about 10 years of age, bimodal visual-haptic performance approximates the statistically optimal MLE performance seen in adults

cross-sensory calibration. The idea is that perceptual systems must be "tuned up" and calibrated during development and that comparing signals across the senses is essential in this process and the more precise modality guides the less specialized one. While one sense is calibrating the other, the sensory signals in those two modalities cannot be usefully combined. The visual dominance for space and the auditory dominance for time could reflect the dominant modality overriding the other modality while it is still developing. This proposal is a reasonable one given that vision is fundamentally spatially specialized and audition is temporally specialized. Many studies in adults show that vision usually dominates audition when spatial locations are in conflict (Warren et al. 1981) and the greater precision of audition (Burr et al. 2009) ensures that it dominates in multisensory temporal tasks (Gebhard and Mowbray 1959; Shams et al. 2000).

Given these established modality specialities, it is reasonable that one particular modality should take the lead in calibrating and "tuning up" the other nonspecialized modalities, with vision tuning both tactile and auditory modalities for spatial tasks and audition tuning vision for temporal tasks. In agreement with this cross-modal calibration proposal, many studies in adults show that the visual system is the most influential in determining the apparent spatial position of auditory stimuli (Pick et al. 1969; Alais and Burr 2004). Only after 12 years of age does visual-auditory integration seem to occur in this spatial task, suggesting a very late development. Audiovisual space integration seems to mature later than visual-haptic spatial integration (which develops after 8–10 years of age; Gori et al. 2008) and visual-auditory temporal integration. This could be related to the time of maturation of the individual sensory systems. Indeed, previous work (Gori et al. 2008) suggested that multisensory integration occurs after the maturation of each unisensory system. The unisensory thresholds for both vision and audition continue to improve over the school years, particularly for the spatial task. For the spatial bisection task, the unisensory thresholds are still not mature at 12 years of age nor is integration optimal at this age. For the temporal task, unisensory thresholds become adult-like after 8–9 years of age, and at this age, the auditory dominance appears. Thus the delay in the development of unisensory systems seems to be related to the delay in the development of optimal sensory integration typically seen in adults.

2.8 Cross-Modal Calibration and Sensory Deficits

The hypothesis that unisensory dominance seen in the early years of development occurs while the dominant modality calibrates other modalities is a generalization of an idea originating with Berkeley's (1709/1963) proposition that touch calibrates vision. More generally, the notion is that the more robust and accurate sense for a particular perceptual task should calibrate the other. This idea raises interesting questions. In particular, what would happen to the nondominant modality if the dominant "calibrating" modality were impaired? A deficit in the more accurate calibrating sense should be detrimental to the system it calibrates. How would visual time perception be impaired in subjects with auditory disabilities? If early unisensory dominance really occurs because cross-modal calibration of the nondominant modality has yet to occur or is incomplete, subjects with visual disabilities should show deficits in auditory spatial tasks because the calibration of space in audition by the visual system will be diminished by the visual impairment. Conversely, subjects with auditory disabilities should show temporal deficits in visual temporal tasks because of the impaired ability of audition to calibrate vision.

Gori et al. (2010, 2014) tested these predictions using stimuli and procedures similar to those used in their other multisensory studies. They established that congenitally blind subjects show severe but selective impairments in haptic discrimination tasks for orientation but not for size discrimination (Gori et al. 2010). Congenitally blind subjects also showed a severe impairment in a task requiring auditory spatial representation, namely auditory space bisection, consistent with the notion that vision is fundamental for space perception (King 2009). On the other hand, thresholds for congenitally blind subjects for simple auditory tasks such as pointing, minimal angle acuity, and temporal bisection were similar to those in control subjects. These findings illustrate the importance of visual spatial representations in establishing and calibrating auditory spatial representations. In another group, it was found that haptically impaired patients showed poor visual size discrimination but not orientation discrimination (Gori et al. 2014). An interesting observation was that in both cases the results were quite different for patients with acquired deficits rather than congenital disabilities, suggesting that cross-sensory calibration at an early age is essential. In addition, blind subjects were not uniformly bad at all auditory tasks but only in the particular spatial bisection task that was designed to tap into a sophisticated map of Euclidean relationships that would require a well-calibrated spatial sense in audition.

In other work pointing to a similar conclusion, Schorr et al. (2005) used the McGurk effect where a visual and an auditory speech signal become perceptually fused into a new phoneme to study bimodal fusion in children born deaf but whose hearing was restored by cochlear implants. Among the group who had implants at an early age (before 30 months), a similar proportion perceived the fused phoneme as normal controls, suggesting that bimodal fusion was occurring. For those who had late implants, however, only one subject showed cross-modal fusion and all the others showed visual dominance. Together, these results highlight the importance of adequate sensory input during early life for the development of multisensory interactions and show that cross-modal fusion is not innate and needs to be learned.

2.9 Summary

To perceive a coherent world, it is necessary to combine signals from the five sensory systems, signals that can be complementary or redundant. In adults, redundant signals from various sensory systems—vision, audition, and touch—are often integrated in an optimal manner following MLE integration and thus lead to an improvement in the bimodal precision relative to the individual unimodal estimates. While much of this work was originally done in adult subjects and showed strong evidence for optimal MLE integration, more recent studies have investigated when and how optimal integration develops in children. A number of studies have shown that multisensory integration is not present at birth but develops over time and optimal integration for some tasks is not attained until about 8 years of age. One of the reasons for this may be that sensory specializations (temporal processing in audition, spatial processing in vision) need to be taught to other nonspecialized senses in a calibration process. Moreover, the continual anatomical and physiological changes occurring during development, such as growing limbs, eye length, and head circumference, mean that a recurrent updating or "recalibration" needs to take place. Until the recalibration process is complete, the two senses cannot be meaningfully combined and the default position is to rely on the specialized sense until optimal integration is possible. This calibration process may occur in different directions between senses, such as touch educating vision for size but vision educating touch for orientation, but in general, the more robust sense for a particular task calibrates the other. Once cross-modal calibration is complete, MLE integration provides an excellent model of multisensory cue combination.

Although this chapter has focused on Bayesian integration of multisensory cues, the principles are general and apply equally to combination of auditory cues. Although less research has been done on Bayesian cue combination in audition than in vision or in cross-modal contexts, a useful overview of Bayesian applications in acoustics has recently appeared (Xiang and Fackler 2015). There are many fundamental research questions remaining to be addressed in Bayesian modeling of auditory processing and psychoacoustics. Among these are, When two cues define a signal, are they combined according to the MLE model or do priors also play a role? How does the variance associated with a given cue get encoded so that cue weightings can be established? Where priors contribute to the Bayesian solution, are they stable internal models of acoustic signal statistics or are they malleable and adaptable? When fusion of two cues takes place, is access to the component cues lost, as occurs in fusion of visual cues (Hillis et al. 2002)? The Bayesian approach has been very effective in modeling visual and multisensory perception and has the potential to provide many insights into auditory perception and psychoacoustics.

Compliance with Ethics Requirements David Alais declares that he has no conflict of interest.

David Burr declares that he has no conflict of interest.

References

Alais, D., & Burr, D. (2004). The ventriloquist effect results from near-optimal bimodal integration. *Current Biology, 14*(3), 257–262.

Alais, D., Newell, F. N., & Mamassian, P. (2010). Multisensory processing in review: From physiology to behaviour. *Seeing and Perceiving, 23*(1), 3–38.

Angelaki, D. E., Gu, Y., & DeAngelis, G. C. (2009). Multisensory integration: Psychophysics, neurophysiology, and computation. *Current Opinion in Neurobiology, 19*(4), 452–458.

Bahrick, L. E., & Lickliter, R. (2004). Infants' perception of rhythm and tempo in unimodal and multimodal stimulation: A developmental test of the intersensory redundancy hypothesis. *Cognitive, Affective, & Behavioral Neuroscience, 4*(2), 137–147.

Barutchu, A., Crewther, D. P., & Crewther, S. G. (2009). The race that precedes coactivation: Development of multisensory facilitation in children. *Developmental Science, 12*(3), 464–473.

Barutchu, A., Danaher, J., Crewther, S. G., Innes-Brown, H., Shivdasani, M. N., & Paolini, A. G. (2010). Audiovisual integration in noise by children and adults. *Journal of Experimental Child Psychology, 105*(1–2), 38–50.

Benevento, L. A., Fallon, J., Davis, B. J., & Rezak, M. (1977). Auditory-visual interaction in single cells in the cortex of the superior temporal sulcus and the orbital frontal cortex of the macaque monkey. *Experimental Neurology, 57*(3), 849–872.

Berkeley, G. (1963). *An essay towards a new theory of vision*. Indianapolis: Bobbs-Merrill. (Original work published 1709).

Bertelson, P., & Radeau, M. (1981). Cross-modal bias and perceptual fusion with auditory-visual spatial discordance. *Perception & Psychophysics, 29*(6), 578–584.

Bresciani, J. P., Dammeier, F., & Ernst, M. O. (2006). Vision and touch are automatically integrated for the perception of sequences of events. *Journal of Vision, 6*(5), 554–564.

Brown, A. M., Dobson, V., & Maier, J. (1987). Visual acuity of human infants at scotopic, mesopic and photopic luminances. *Vision Research, 27*(10), 1845–1858.

Burr, D., Banks, M. S., & Morrone, M. C. (2009). Auditory dominance over vision in the perception of interval duration. *Experimental Brain Research, 198*(1), 49–57.

Caclin, A., Soto-Faraco, S., Kingstone, A., & Spence, C. (2002). Tactile "capture" of audition. *Perception & Psychophysics, 64*(4), 616–630.

Cicchini, G. M., Arrighi, R., Cecchetti, L., Giusti, M., & Burr, D. C. (2012). Optimal encoding of interval timing in expert percussionists. *Journal of Neuroscience, 32*(3), 1056–1060.

Clarke, J. J., & Yuille, A. L. (1990). *Data fusion for sensory information processing*. Boston: Kluwer Academic.

Dodd, B. (1979). Lip reading in infants: Attention to speech presented in- and out-of-synchrony. *Cognitive Psychology, 11*(4), 478–484.

Ernst, M. O., & Banks, M. S. (2002). Humans integrate visual and haptic information in a statistically optimal fashion. *Nature, 415*(6870), 429–433.

Ernst, M. O., & Bulthoff, H. H. (2004). Merging the senses into a robust percept. *Trends in Cognitive Sciences, 8*(4), 162–169.

Gebhard, J. W., & Mowbray, G. H. (1959). On discriminating the rate of visual flicker and auditory flutter. *American Journal of Psychology, 72*, 521–529.

Ghahramani, Z., & Wolpert, D. M. (1997). Modular decomposition in visuomotor learning. *Nature, 386*(6623), 392–395.

Gori, M., Del Viva, M., Sandini, G., & Burr, D. C. (2008). Young children do not integrate visual and haptic form information. *Current Biology, 18*(9), 694–698.

Gori, M., Sandini, G., Martinoli, C., & Burr, D. (2010). Poor haptic orientation discrimination in nonsighted children may reflect disruption of cross-sensory calibration. *Current Biology, 20*(3), 223–225.

Gori, M., Sandini, G., & Burr, D. (2012). Development of visuo-auditory integration in space and time. *Frontiers in Integrative Neuroscience, 6*, 77.

Gori, M., Sandini, G., Martinoli, C., & Burr, D. (2014). Impairment of auditory spatial localization in congenitally blind human subjects. *Brain, 20*, 288–293.

Gottlieb, G. (1990). *Development of species identification in birds: An inquiry into the prenatal determinants of perception*. Chicago: University of Chicago Press.

Gu, Y., Angelaki, D. E., & Deangelis, G. C. (2008). Neural correlates of multisensory cue integration in macaque MSTd. *Nature Neuroscience, 11*(10), 1201–1210.

Guest, S., Catmur, C., Lloyd, D., & Spence, C. (2002). Audiotactile interactions in roughness perception. *Experimental Brain Research, 146*(2), 161–171.

Hartcher-O'Brien, J., & Alais, D. (2011). Temporal ventriloquism in a purely temporal context. *Journal of Experimental Psychology: Human Perception and Performance, 37*(5), 1383–1395.
Hartcher-O'Brien, J., Di Luca, M., & Ernst, M. O. (2014). The duration of uncertain times: Audiovisual information about intervals is integrated in a statistically optimal fashion. *PLoS One, 9*(3), e89339.
Hatwell, Y. (1987). Motor and cognitive functions of the hand in infancy and childhood. *International Journal of Behavioural Development, 10*, 509–526.
Helbig, H. B., & Ernst, M. O. (2008). Visual-haptic cue weighting is independent of modality-specific attention. *Journal of Vision, 8*(1), 21.1–21.16.
Hillis, J. M., Ernst, M. O., Banks, M. S., & Landy, M. S. (2002). Combining sensory information: Mandatory fusion within, but not between, senses. *Science, 298*(5598), 1627–1630.
Hillis, J. M., Watt, S. J., Landy, M. S., & Banks, M. S. (2004). Slant from texture and disparity cues: Optimal cue combination. *Journal of Vision, 4*(12), 967–992.
Howard, I. P., & Templeton, W. B. (1966). *Human spatial orientation*. New York: Wiley.
Hubel, D. H., & Wiesel, T. N. (1968). Receptive fields and functional architecture of monkey striate cortex. *Journal of Physiology, 195*(1), 215–243.
Jazayeri, M., & Shadlen, M. N. (2010). Temporal context calibrates interval timing. *Nature Neuroscience, 13*(8), 1020–1026.
Jones, E. G., & Powell, T. P. (1970). An anatomical study of converging sensory pathways within the cerebral cortex of the monkey. *Brain, 93*(4), 793–820.
Jusczyk, P., Houston, D., & Goodman, M. (1998). Speech perception during the first year. In A. Slater (Ed.), *Perceptual development: Visual, auditory, and speech perception in infancy* (pp. 357–388). Hove: Psychology Press.
Kayser, C., Petkov, C. I., & Logothetis, N. K. (2008). Visual modulation of neurons in auditory cortex. *Cerebral Cortex, 18*(7), 1560–1574.
Kersten, D., Mamassian, P., & Yuille, A. (2004). Object perception as Bayesian inference. *Annual Review of Psychology, 55*, 271–304.
King, A. J. (2009). Visual influences on auditory spatial learning. *Philosophical Transactions of the Royal Society of London. Series B, Biological Sciences, 364*(1515), 331–339.
Knill, D. C. (2007). Learning Bayesian priors for depth perception. *Journal of Vision, 7*(8), 13.
Knill, D. C., & Pouget, A. (2004). The Bayesian brain: The role of uncertainty in neural coding and computation. *Trends in Neurosciences, 27*(12), 712–719.
Knill, D. C., & Saunders, J. A. (2003). Do humans optimally integrate stereo and texture information for judgments of surface slant? *Vision Research, 43*(24), 2539–2558.
Kording, K. P., & Wolpert, D. M. (2004). Bayesian integration in sensorimotor learning. *Nature, 427*(6971), 244–247.
Kwon, O. S., & Knill, D. C. (2013). The brain uses adaptive internal models of scene statistics for sensorimotor estimation and planning. *Proceedings of the National Academy of Sciences of the United States of America, 110*(11), E1064–E1073.
Landy, M. S., Banks, M. S., & Knill, D. C. (2011). Ideal-observer models of cue integration. In K. Tromershauser, K. Körding, & M. S. Landy (Eds.), *Book of sensory cue integration* (pp. 5–30). New York: Oxford University Press.
Lewkowicz, D. J. (1992). Infants' responsiveness to the auditory and visual attributes of a sounding/moving stimulus. *Perception & Psychophysics, 52*(5), 519–528.
Lewkowicz, D. J. (2000). The development of intersensory temporal perception: An epigenetic systems/limitations view. *Psychological Bulletin, 126*(2), 281–308.
Lewkowicz, D. J., & Turkewitz, G. (1981). Intersensory interaction in newborns: Modification of visual preferences following exposure to sound. *Child Development, 52*(3), 827–832.
McGurk, H., & Power, R. P. (1980). Intermodal coordination in young children: Vision and touch. *Developmental Psychology, 16*, 679–680.
Misceo, G. F., Hershberger, W. A., & Mancini, R. L. (1999). Haptic estimates of discordant visual-haptic size vary developmentally. *Perception & Psychophysics, 61*(4), 608–614.
Murray, M. M., Thelen, A., Thut, G., Romei, V., Martuzzi, R., & Matusz, P. J. (2015). The multisensory function of the human primary visual cortex. *Neuropsychologia, 83*, 161–169.

Nardini, M., Jones, P., Bedford, R., & Braddick, O. (2008). Development of cue integration in human navigation. *Current Biology, 18*(9), 689–693.
Nardini, M., Bedford, R., & Mareschal, D. (2010). Fusion of visual cues is not mandatory in children. *Proceedings of the National Academy of Sciences of the United States of America, 107*(39), 17041–17046.
Olsho, L. W. (1984). Infant frequency discrimination as a function of frequency. *Infant Behavior and Development, 7*, 27–35.
Olsho, L. W., Koch, E. G., Carter, E. A., Halpin, C. F., & Spetner, N. B. (1988). Pure-tone sensitivity of human infants. *The Journal of the Acoustical Society of America, 84*(4), 1316–1324.
Patterson, M. L., & Werker, J. F. (2002). Infants' ability to match dynamic phonetic and gender information in the face and voice. *Journal of Experimental Child Psychology, 81*(1), 93–115.
Paus, T. (2005). Mapping brain development and aggression. *Journal of the Canadian Academy of Child and Adolescent Psychiatry, 14*(1), 10–15.
Pick, H. L., Warren, D. H., & Hay, J. (1969). Sensory conflict in judgements of spatial direction. *Perception & Psychophysics, 6*, 203–205.
Pouget, A., Beck, J. M., Ma, W. J., & Latham, P. E. (2013). Probabilistic brains: Known and unknowns. *Nature Neuroscience, 16*, 1170–1178.
Recanzone, G. H. (2003). Auditory influences on visual temporal rate perception. *Journal of Neurophysiology, 89*(2), 1078–1093.
Rock, I., & Victor, J. (1964). Vision and touch: An experimentally created conflict between the two senses. *Science, 143*, 594–596.
Rose, S. A. (1981). Developmental changes in infants' retention of visual stimuli. *Child Development, 52*(1), 227–233.
Rowland, B., Stanford, T., & Stein, B. (2007). A Bayesian model unifies multisensory spatial localization with the physiological properties of the superior colliculus. *Experimental Brain Research, 180*(1), 153–161.
Schorr, E. A., Fox, N. A., van Wassenhove, V., & Knudsen, E. I. (2005). Auditory-visual fusion in speech perception in children with cochlear implants. *Proceedings of the National Academy of Sciences of the United States of America, 102*, 18748–18750.
Shams, L., Kamitani, Y., & Shimojo, S. (2000). Illusions. What you see is what you hear. *Nature, 408*(6814), 788.
Shi, Z., & Burr, D. (2015). Predictive coding of multisensory timing. *Current Opinion in Behavioral Sciences, 8*, 200–206.
Shipley, T. (1964). Auditory flutter-driving of visual flicker. *Science, 145*, 1328–1330.
Slutsky, D. A., & Recanzone, G. H. (2001). Temporal and spatial dependency of the ventriloquism effect. *Neuroreport, 12*(1), 7–10.
Streri, A. (2003). Cross-modal recognition of shape from hand to eyes in human newborns. *Somatosensory and Motor Research, 20*(1), 13–18.
Tomassini, A., Gori, M., Burr, D., Sandini, G., & Morrone, M. C. (2011). Perceived duration of visual and tactile stimuli depends on perceived speed. *Frontiers in Integrative Neuroscience, 5*, 51.
Trehub, S. E., Schneider, B. A., & Henderson, J. L. (1995). Gap detection in infants, children, and adults. *The Journal of the Acoustical Society of America, 98*(5), 2532–2541.
Ulrich, R., Nitschke, J., & Rammsayer, T. (2006). Crossmodal temporal discrimination: Assessing the predictions of a general pacemaker-counter model. *Perception & Psychophysics, 68*(7), 1140–1152.
van Beers, R. J., Sittig, A. C., & Gon, J. J. (1999). Integration of proprioceptive and visual position-information: An experimentally supported model. *Journal of Neurophysiology, 81*(3), 1355–1364.
von Helmholtz, H. (1925). *Treatise on physiological optics* (Vol. 3). New York: Dover.
Vroomen, J., Bertelson, P., & de Gelder, B. (2001). The ventriloquist effect does not depend on the direction of automatic visual attention. *Perception & Psychophysics, 63*(4), 651–659.
Warren, D. H., Welch, R. B., & McCarthy, T. J. (1981). The role of visual-auditory "compellingness" in the ventriloquism effect: Implications for transitivity among the spatial senses. *Perception & Psychophysics, 30*, 557–564.

Welch, R. B., & Warren, D. H. (1980). Immediate perceptual response to intersensory discrepancy. *Psychological Bulletin, 88*(3), 638–667.
Wozny, D. R., Beierholm, U. R., & Shams, L. (2008). Human trimodal perception follows optimal statistical inference. *Journal of Vision, 8*(3), 24.1–24.11.
Xiang, N., & Fackler, C. (2015). Objective Bayesian analysis in acoustics. *Acoustics Today, 11*(2), 54–61.

Chapter 3
Toward a Model of Auditory-Visual Speech Intelligibility

Ken W. Grant and Joshua G. W. Bernstein

Abstract A significant proportion of speech communication occurs when speakers and listeners are within face-to-face proximity of one other. In noisy and reverberant environments with multiple sound sources, auditory-visual (AV) speech communication takes on increased importance because it offers the best chance for successful communication. This chapter reviews AV processing for speech understanding by normal-hearing individuals. Auditory, visual, and AV factors that influence intelligibility, such as the speech spectral regions that are most important for AV speech recognition, complementary and redundant auditory and visual speech information, AV integration efficiency, the time window for auditory (across spectrum) and AV (cross-modality) integration, and the modulation coherence between auditory and visual speech signals are each discussed. The knowledge gained from understanding the benefits and limitations of visual speech information as it applies to AV speech perception is used to propose a signal-based model of AV speech intelligibility. It is hoped that the development and refinement of quantitative models of AV speech intelligibility will increase our understanding of the multimodal processes that function every day to aid speech communication, as well guide advances in future generation hearing aids and cochlear implants for individuals with sensorineural hearing loss.

Keywords Articulation index · Auditory-visual coherence · Hearing loss · Modeling · Place of articulation · Spectrotemporal modulation index · Speech envelope · Speech intelligibility index · Speechreading · Speech transmission index · Temporal asynchrony · Temporal window of integration · Voicing

K. W. Grant (✉) · J. G. W. Bernstein
National Military Audiology and Speech Pathology Center, Walter Reed National Military Medical Center, Bethesda, MD, USA
e-mail: kenneth.w.grant.civ@mail.mil; joshua.g.bernstein.civ@mail.mil

A. K. C. Lee et al. (eds.), *Multisensory Processes*, Springer Handbook of Auditory Research 68, https://doi.org/10.1007/978-3-030-10461-0_3

3.1 Introduction

3.1.1 The Importance of Signal-Based Models of Speech Intelligibility

There can be little doubt of the importance of speech and language skills for cognitive and social development and for the communication of ideas, thoughts, and emotions. For the better part of a century, researchers have been working to develop models of speech perception and language processing, in large part due to the work at AT&T (Bell Laboratories) in the early 1900s. Driven by the development of the telephone and the need for high-quality speech transmission, the research team at Bell Laboratories developed a variety of methods for measuring speech intelligibility and user reactions to the phone. Among the many important discoveries stemming from this work was a characterization of how the signal-to-noise ratio (SNR), loudness, spectral balance, and distortion each impact speech intelligibility. Because of the expensive costs associated with test development and conducting laboratory and field experiments with human listeners, French and Steinberg (1947) and Fletcher and Gault (1950) began to work on methods for predicting the average speech quality of a given communication system as a means of testing new systems before they were put into the field. This work culminated in what became known as the articulation index (AI; American National Standards Institute [ANSI] 1969), which was designed to characterize a device, whether it be a phone, hearing aid, or any sound-transmitting system, based solely on the physical characteristics of the signal output and the environmental noise at the listener's ear.

3.1.2 The Overlooked Problem of Auditory-Visual Speech Intelligibility

Since its development, numerous extensions and simplifications of the AI or alternative metrics based on similar ideas have been proposed to predict speech intelligibility performance in different types of background noise (e.g., steady-state and modulated noise), reverberant environments, and for listeners with hearing impairment (speech intelligibility index [SII], ANSI 1997; speech transmission index [STI], Steeneken and Houtgast 2002). Despite the various iterations of these indices throughout the years, one of the most fundamental facts of human speech communication has been barely examined, namely, that human communication involves auditory-visual (AV) face-to-face input and not just auditory input. It is estimated that well over half of active speech communication takes place in contexts where visual speech cues are available to the listener (Walden et al. 2004). Yet, the prediction of intelligibility for AV speech inputs is woefully underdeveloped. The AI and SII ANSI standards did include a nod to AV speech recognition, but visual cues were treated simply as an additive factor to the basic auditory predictions and failed to understand the intricate manner in which auditory and visual speech cues interact. The goals of this chapter are to illuminate the factors that would necessarily be an important part of any AV speech-intelligibility model and to suggest solutions that are consistent with the original goals of the AI. In addition to being able to accurately predict AV speech intelligibility under a wide variety of noise and reverberation conditions, a practical model should be based on physical measurements of the signal and environment alone to allow for the evaluation of potential benefits of new hearing technologies and algorithms without relying on exhaustive human-subjects testing. (Ideally, any auditory or AV model of speech intelligibility would also consider individual differences in dimensions such as hearing acuity, visual acuity, and cognitive ability; however, accounting for individual differences falls outside the scope of this chapter.) In delineating the factors involved in the development of such a model, this chapter will revisit some of the same issues that had to be addressed during the development of the original auditory-only (A-only) AI. This will include (1) impact of noise and distortion, (2) spectral balance or frequency weighting, (3) integration across spectral channels and across modality, and (4) synchrony between auditory and visual signals.

With few exceptions, listeners are able to improve their speech-recognition performance by combining visual cues (from lipreading; also known as speechreading) and audition (e.g., Sumby and Pollack 1954; Grant et al. 1998). Benefits due to speechreading, especially in reverberant or noisy environments, can be quite substantial for most listeners, often allowing near-perfect comprehension of otherwise unintelligible speech (Grant et al. 1985; Summerfield 1992). Understanding how these large benefits come about is critical because the speech cues that must be relayed to maximize speech understanding in adverse situations are likely to be dramatically different when the listener has access to visual (speechread) cues in addition to acoustic speech information. As discussed, this is the case when considering normal-hearing (NH) listeners in adverse noisy listening environments, hearing-impaired (HI) listeners, or signal-processing strategies for hearing aids and advanced auditory prosthetics such as cochlear implants.

Consider the following scenario (see Fig. 3.1). A speech signal composed of both visual and acoustic information is presented. The listener-observer extracts signal-related segmental (i.e., phonemes and syllables) and suprasegmental (i.e., words and phrases) cues from each modality, integrates these cues, and applies top-down semantic and syntactic constraints in an effort to interpret the message before making a response. The basic components—bottom-up signal-related cue extraction, integration, and top-down linguistic processes—are common to most speech-perception theories (e.g., Liberman et al. 1967; Studdert-Kennedy 1974). The major distinction drawn here from A-only theories of speech perception is that in an AV communication environment, cues from the visual modality must be considered, and the integration of auditory and visual cues, both within and across modalities, must be evaluated (Massaro 1987). From this perspective, consider an individual whose AV recognition of words and sentences is less than perfect. To evaluate the exact nature of the communication problem, it is necessary to determine whether the deficit is due to poor reception of auditory or visual cues, difficulty in integrating

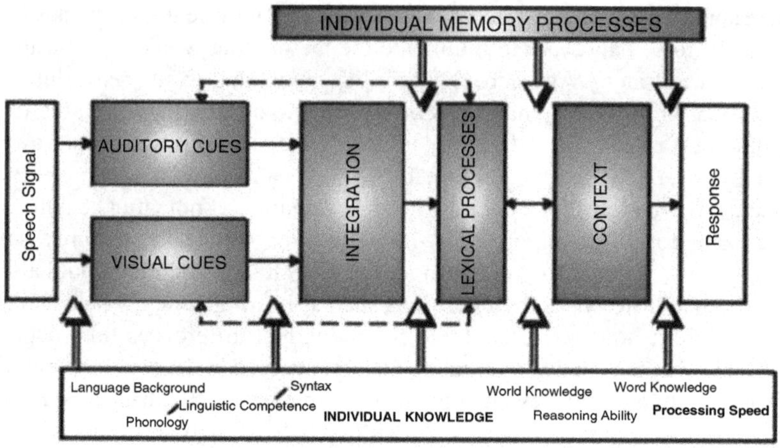

Fig. 3.1 A schematic of the predominant sources of individual variability in auditory-only (A-only) and auditory-visual (AV) speech processing. Processing starts with the common assumption of sensory independence during the early stages of processing. The integration module as a potential source of individual variability uses a model of optimal AV processing

auditory and visual cues, difficulty in applying linguistic and contextual constraints, cognitive limitations such as reduced working-memory capacity or reduced attention, or a combination of these factors. If the problem is determined to be primarily difficulty in receiving auditory or visual cues, signal-processing strategies to enhance the relevant cues and improve the SNR may be used. If, on the other hand, the problem is determined to be difficulty in integrating auditory and visual cues or difficulty in applying top-down language-processing rules, then training and practice techniques may be the better rehabilitation strategy. Simply knowing the individual's AV sentence- or word-recognition performance is not sufficient for determining a plan for rehabilitation.

Based on the simple framework displayed in Fig. 3.1, three questions must be addressed in order to predict speech intelligibility. (1) What are the most important cues for AV speech recognition that can be extracted from acoustic and visual speech signals? (2) How should one measure an individual's ability to integrate auditory and visual cues separate and apart from their ability to recognize syllables, words, and sentences? (3) What are the most important non-signal-related "top-down" processes that contribute to individual variability in AV speech recognition? Because the top-down influences on speech recognition are quite influential, early models of speech intelligibility and most models of AV speech intelligibility and integration limit the types of speech materials used to include mostly nonsense syllables (French and Steinberg 1947; Fletcher 1953). By imposing this limitation on the types of speech signals considered, the focus of the model becomes "bottom-up" and highly dependent on the signal, room, and any equipment (e.g., radio, phone) that resides in the transmission path between the speaker and listener.

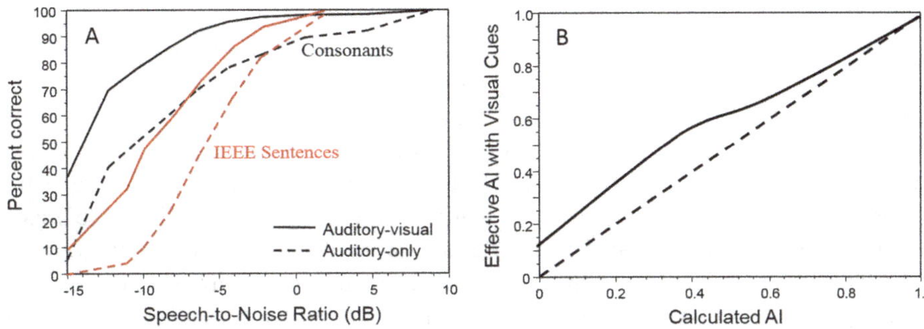

Fig. 3.2 (**A**) Impact of noise on A-only (*dashed lines*) and AV (*solid lines*) speech perception for sentence recognition (*red curves*) and consonant recognition (*black curves*). (**B**) Relationship between calculated articulation index (AI) and "effective" AI when auditory cues are combined with speechreading (from American National Standards Institute [ANSI] 1969)

Figure 3.2A shows a typical outcome for A-only (*dashed lines*) and AV (*solid lines*) speech recognition of low-context sentences (*red curves*) and consonants (*black curves*) for NH young adults (after Grant and Braida 1991; Grant and Walden 1996). For both sets of speech materials, performance was substantially better in the AV condition. At a SNR of −15 dB, the auditory signal was just audible, with performance at zero percent correct for sentences and at chance level for consonants (i.e., 1/18 response options). AV keyword recognition scores at a SNR of −15 dB were roughly 10% correct for sentences. For consonant materials, however, the AV scores at −15 dB SNR were near 40% correct. As will be discussed below in Sect. 3.1.3, this can be explained by the fact that although speechreading alone can barely support word recognition, it can convey very accurate information about certain aspects of speech.

The original ANSI (1969) standard for calculating the AI included an overly simplistic graphical solution to predict the presumed benefit to intelligibility when visual speech cues are present (Fig. 3.2B). In the revised version of the AI known as the SII (ANSI 1997), the effective benefit of visual cues was formalized by a simple two-part equation, essentially mimicking the curve shown in the ANSI (1969) standard. An unmistakable conclusion one can draw from Fig. 3.2B is that the addition of visual cues to speech intelligibility was treated as an effective addition to the AI and that the same AV prediction would be made for a given level of A-only performance regardless of the particular spectral characteristics of the speech signal and noise. In other words, the importance of different spectral regions for A-only intelligibility was assumed to be the same for AV intelligibility.

We now know this assumption to be incorrect. HI listeners show dramatic benefits from speechreading in cases with very little residual auditory function (Erber 1972; Drullman and Smoorenburg 1997). Studies of NH listeners have allowed us to understand this phenomenon. When speechreading is combined with low-frequency, low-intelligibility auditory speech cues, the resulting benefits are enormous. Grant et al. (1985) found that even presenting a sparse acoustic representation of the speech cues located at these low frequencies was sufficient to generate large speechreading benefits on the order of 50 or more percentage points. Adding low-frequency speech signals dramatically sped up the ability to track connected discourse (by repeating back verbatim text read aloud), from 41 words per minute (wpm) for speechreading alone up to 86 wpm for AV speech (tracking rates for full bandwidth speech were 108 wpm). Similarly, Rosen et al. (1981) showed that presenting only the acoustic voice-pitch information provided an 83% improvement in the rate of discourse tracking over speechreading alone. These extremely large increases in the ability to track AV speech when the low-frequency acoustic signals produced zero percent intelligibility by themselves indicate that AV intelligibility does not completely depend on A-only intelligibility as suggested by the AI and SII. Instead, an accurate prediction of AV intelligibility requires an understanding of the information provided by the auditory and visual signals. In particular, Grant and Walden (1996) showed that the addition of visual cues enhances auditory speech perception for low-frequency stimuli much more than for high-frequency stimuli. As will be discussed in Sect. 3.1.3, this is because the visual signal and low-frequency auditory signals provide complementary information. The visual signal facilitates the differentiation of visible speech features generated at the lips (e.g., /ba/ vs. /ga/), whereas the low-frequency auditory signal facilitates the differentiation of invisible speech features generated in the back of the throat or at the larynx (i.e., /ba/ vs. /pa/).

In cases where A-only speech intelligibility is impacted by hearing loss and not just environmental conditions, the importance of speechreading in everyday communication settings increases. Furthermore, when auditory and visual speech cues are integrated successfully, the improvement to speech intelligibility can be so large that the benefit from speechreading can even outweigh the benefit from a hearing aid. Walden et al. (2001) reported consonant-recognition data from 25 adults (mean age 66 years) with an acquired moderate-to-severe high-frequency sensorineural hearing loss. The benefit of visual cues compared with unaided listening was roughly 40 percentage points, whereas the benefit of amplification was only 30 percentage points. (Although this experiment was conducted with older hearing-aid technology, the benefits of amplification for speech understanding in quiet are mostly unaffected by newer technological advances.) A small additional benefit was observed when hearing aids were combined with speechreading, although ceiling effects likely obscured some of the benefits from combining amplification and speechreading. The small difference between aided and unaided AV scores could conceivably contribute to the listener's notion that the hearing aids were not that beneficial under typical AV conditions. In another example where the presence of visual speech might obscure the benefits of newer hearing-aid technologies, directional microphones for improving the SNR are a key feature of almost all modern hearing aids. When evaluated without visual cues, this feature can provide a substantial improvement in SNR (3–5 dB in indoor environments and 4–8 dB in outdoor environments; Killion et al. 1998). However, when evaluated with visual cues, the perceived and objective benefit of directional microphones can be greatly reduced (Wu and Bentler 2010). Thus, even if an advantageous hearing-aid feature is developed that proves to be very useful in an A-only listening situation, it is not guaranteed to be equally beneficial (or even noticed) in an AV listening situation.

In summary, the studies of Grant et al. (1985), Grant and Walden (1996), and Walden et al. (2001) demonstrate two important concepts. First, the advantages for speech understanding provided by integrating auditory and visual speech cues are determined by a complex interaction among auditory and visual speech information as well as several important top-down influences. This means that AV speech-reception performance cannot be predicted from A-only speech-reception performance without an understanding of the information relayed by each modality and some assessment of information redundancy and complementarity. Second, the effects of hearing loss and hearing aids might be very different under AV and A-only conditions. The most commonly used hearing-aid fitting algorithms are based on maximizing model-predicted A-only speech intelligibility (e.g., Byrne et al. 2001). The fact that AV speech perception is likely to depend on hearing loss and hearing-aid features differently than A-only speech perception highlights the need for an AV model of speech intelligibility.

Because of the importance of visual cues for speech communication and the fact that speechreading and auditory cues interact in a nonadditive manner, studies measuring the contribution of these cues to speech perception and theories of AV speech perception have become more common in the literature (see Summerfield 1987; Massaro 1998 for reviews). Furthermore, despite the obvious importance of speech communication for maintaining the health and fitness of elderly persons, little is known about the combined effects of hearing loss, visual acuity, and aging on AV speech recognition, making the task of developing an AV version of the AI that much more difficult. However, for the purposes of this chapter and in the spirit of the original AI, the first step of accurately modeling AV speech recognition for a NH population is the current focus, leaving aside for now the more complex questions related to sensory impairment and individual variability (hearing loss, aging, visual acuity, and cognitive decline).

3.1.3 Speech-Feature Complementarity and the Relative Importance of Different Spectral Regions

How can an acoustic signal that generates near zero intelligibility on its own so dramatically improve speechreading performance? An important clue to understanding this synergy comes from research that has carefully analyzed the pattern of particular errors that listeners make in speech-recognition tests (Grant and Walden 1996). These analyses show that some of most reliable information relayed by speechreading are surface features of the lips and tip of the tongue that help to differentiate between certain consonants. For example, by speechreading alone, it is very easy to tell the difference between /bɑ/, /gɑ/, and /dɑ/, even though these tokens would often be confused in the case of A-only speech processing in a noisy situation or by listeners with hearing loss. In contrast, speechreading provides very little information regarding speech contrasts generated at the larynx. For example, visual representations of /bɑ/, /pɑ/, and/mɑ/ are often confused with one another. Although not usually enough to support high levels of intelligibility, being able to accurately recognize these visual categories of contrast greatly reduces the number of possible choices when making a response. When combined with the right kind of complementary acoustic information, the integration of auditory and visual speech cues can lead to very high levels of speech intelligibility.

To illustrate the complementary nature of auditory and visual speech cues, it is useful to discuss the relative contributions of the speech signals in terms of their articulatory and phonetic distinctive features (voicing [e.g., /bɑ/ vs. /pɑ/], manner of articulation [e.g., /bɑ/ vs. /mɑ/] and place of articulation [e.g., /bɑ/vs. /gɑ/]). Briefly, *place of articulation* refers to the location within the vocal tract where the airflow has been maximally restricted. For example, the primary place of constriction for the consonant /mɑ/ is the lips. Place-of-articulation cues are often clearly visible in terms of lip and tongue-tip position. Acoustically, these dynamic high-frequency speech cues associated with the second and third formant transitions are considered to be fragile and easily corrupted by noise or hearing loss (Kewley-Port 1983; Reetz and Jongman 2011). *Voicing* cues mainly reflect the presence or absence of waveform periodicity or vocal-fold vibration. Taking place in the larynx, these cues are not visibly apparent. Acoustically, voicing is well represented in the low frequencies of speech and is marked by attributes such as voice-onset time and the trajectory of the first formant immediately following the consonant release (Reetz and Jongman 2011). *Manner of articulation* refers to the way the speech articulators interact when producing speech. For example, for the consonant /s/, the tip of the tongue forms a narrow constriction with the alveolar ridge (gum line) just behind the teeth. The result of this constriction is a turbulent airflow that serves as the primary source of the sound, making /s/ a fricative. These three broad phonetic and articulatory features are not orthogonal, although each sound in English can be uniquely identified by a combination of place, manner, and voicing (e.g., /bɑ/ is classified as a voiced, bilabial, plosive; /sɑ/ is classified as a voiceless, alveolar, fricative).

Figure 3.3 illustrates how auditory and visual information interact across these three types of consonant feature. Each panel shows the percentage correct in identifying a particular consonant feature under A-only and AV conditions (Grant and Walden 1996). Figure 3.3C shows that place-of-articulation information is readily available to the speechreader, is not affected by noise, and does not need auditory place cues to reach ceiling performance. In contrast, voicing information (Fig. 3.3A) is determined entirely by auditory cues with very little contribution from the visual speech signal. Figure 3.3B shows the results for manner of articulation and, at first glance, suggests that visual information is helpful for making consonantal manner determinations and combines with auditory cues as they become available with improving SNR. However, further analysis (not shown) suggests that this is due to the high degree of redundancy between place and manner cues for consonant identification. In other words, the score observed for manner information by speechreading alone is what one would predict by chance given 100% correct transmission-of-place information. Thus, for these consonant materials, speechreading contributes almost exclusively to the reception of place information.

Grant and Walden (1996) also provided insight into how the complementarity of speech features (Fig. 3.3) translates into a complex interaction between speechreading

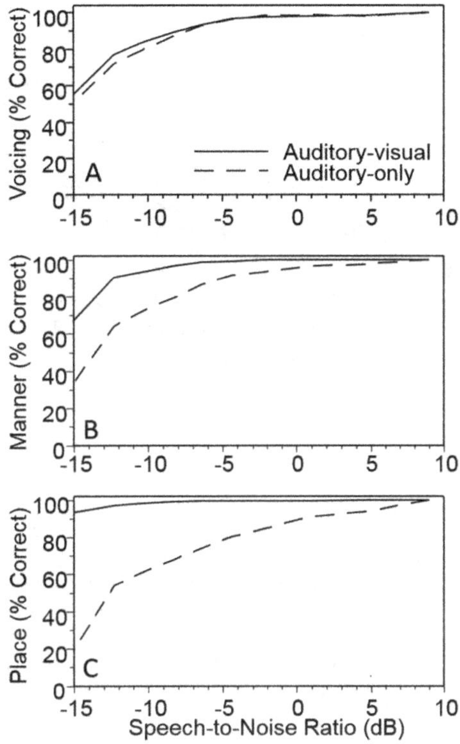

Fig. 3.3 A-only and AV feature transmission for consonant identification (Grant and Walden 1996). The information contained in the visual signal is derived by comparing A-only and AV performance for each feature. Visual cues contribute almost zero information regarding voicing (**A**), some manner information (**B**), and near perfect place-of-articulation information (**C**)

benefit and the spectral content of the speech signal. The AI makes the basic assumption that better A-only speech-reception performance will also result in better AV performance (Fig. 3.2B). In contrast, when Grant and Walden examined the speechreading benefit for filtered bands of speech, they found that the AV speech scores did not increase monotonically with A-only performance. Instead, speechreading benefit varied substantially depending on the filter bandwidth and center frequency, even for frequency bands that generated equal A-only performance. Twelve bandpass-filter conditions were chosen to carefully control the A-only AI prediction while varying the bandwidth and center frequency of the filter. Figure 3.4A shows the results, with the A-only conditions (solid bars) arranged in ascending order based on percentage- correct consonant-identification scores. The AV speech-reception scores were only weakly correlated with A-only performance, demonstrating clear nonmonotonicity between A-only and AV speech recognition. The relationship between AV benefit and spectral region is clearly exemplified in the comparison between filter conditions 1 and 6. Whereas filter condition 6 (containing high-frequency speech information) yielded a substantially higher A-only speech score, AV performance was substantially better in condition 1 (containing only low-frequency speech information). This pattern was observed repeatedly across filter-band conditions (e.g., compare conditions 7 and 9 and conditions 10 and 12). (It should be noted that this same pattern of results holds whether the difference between AV and

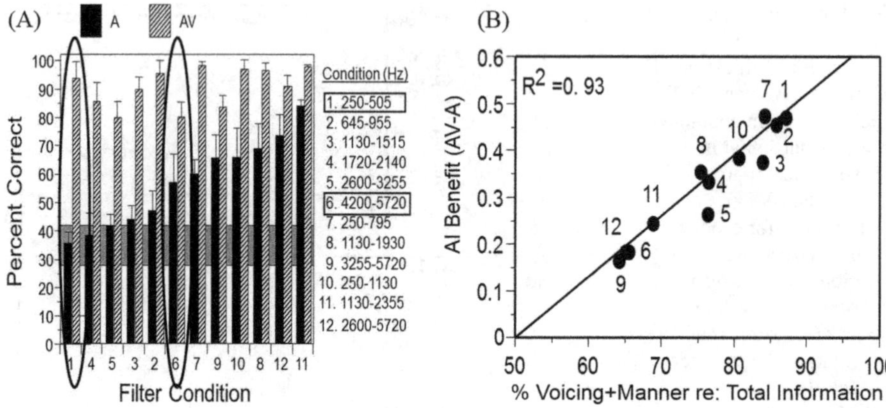

Fig. 3.4 (**A**) Consonant recognition scores for A-only and AV filtered speech. *Horizontal gray band* between 30 and 40% correct reflects the range of speechreading-only scores. Ellipses highlight two conditions (filter bands 1 and 6) representing a narrow low-frequency and a high-frequency band, respectively. Note that although A-only performance is significantly greater for band 6 than for band 1, the AV score for high-frequency band 6 is much less than that for band 1, demonstrating nonmonotonicity between A-only and AV performance. (**B**) Visual benefit as predicted by the proportion of voicing plus manner-of-articulation information relative to the total amount of transmitted information for the 12 bandpass-filtered conditions tested. The greatest AV benefit occurs for filtered speech with a high concentration of low-frequency energy and a high relative transmission rate for voicing and manner information. From Grant and Walden (1996)

A-only speech scores are measured in terms of percentage correct or as relative benefit, taking into account how close to ceiling performance the A-only score might be; Sumby and Pollack 1954).

The results in Fig. 3.4A show that AV intelligibility was greater when the audible acoustic speech spectrum was dominated by low-frequency energy than when it was dominated by high-frequency energy. This suggests that the frequencies that are most important for speech understanding are very different under A-only conditions (mid-to-high frequencies; ANSI 1969) than under AV conditions (low frequencies). To investigate why low-frequency auditory information is so highly complementary with visual speechreading cues, Grant and Walden (1996) examined the relationship between an information analysis of consonant features (Miller and Nicely 1955) and the frequency dependence of the speechreading benefit (Fig. 3.4B). This analysis clearly showed that the magnitude of the AI benefit due to the addition of visual cues was strongly correlated with the amount of voicing and manner of information transmitted for a given frequency band. Low-frequency conditions (e.g., bands 1, 2, 3, 7, and 10) transmitted a great deal of voicing and manner information relative to the total amount of information contained in each band and generated the largest AI benefit. The reverse was true for high-frequency conditions (e.g., bands 6, 9, and 12). Comparable analyses of the visual-only (V-only) condition confirmed that the low-frequency auditory bands were essentially complementary with speechreading while the high-frequency bands were mostly redundant with speechreading. In other words, the reason that visual speech cues provided such a large benefit when the auditory signal was limited to low frequencies is because voicing (and manner) information available at low frequencies was highly complementary to the place information provided by the visual signal. In contrast, the visual signal provided much less additional benefit for high-frequency stimuli because both modalities contributed largely redundant place information.

3.1.4 Auditory-Visual Integration Efficiency

The integration of the auditory and visual modalities of speech information requires a neural process that combines the two inputs and hence could be susceptible to individual differences in integration efficiency (see Fig. 3.1). In fact, it is often assumed that if a particular stimulus condition demonstrates a large visual benefit to speech intelligibility, then the listener must have been able to integrate auditory and visual information with a high degree of efficiency (Sommers et al. 2005). However, as just demonstrated, the processes involved in integrating auditory and visual information efficiently and the amount of AV benefit obtained compared to A-only or V-only intelligibility are distinctly different processes. As shown in Fig. 3.4, the amount of AV benefit observed is much more closely related to the spectral region of the acoustic speech signal than to the A-only or V-only recognition scores. Thus, the fact that one acoustic condition shows a much larger visual benefit than another could be because it provides access to very different auditory information and not necessarily because there is a problem integrating information across modalities. Stated another way, the fact that filter-band condition 6 demonstrated far less benefit than filter-band condition 1 (Fig. 3.4A) does not mean that AV integration was less efficient for filter-band 6. The question of integration efficiency can be specifically examined using a class of models of AV integration for consonant identification developed by Massaro (1987) and Braida (1991). These models take as input confusion matrices that describe the speech information contained in separate auditory and visual speech signals (or for separate frequency bands of auditory speech signals). They then make an AV prediction based on the mutual information contained in the V-only and A-only conditions. Grant et al. (2007) applied the modeling approach of Braida (1991), defining integration efficiency in terms of the ratio between the model prediction and the actual AV performance (or performance for combinations of auditory frequency bands). NH listeners were found to have nearly perfect integration efficiency both within and across modalities. HI listeners were found to have slightly reduced efficiency (although not significantly so) for combining auditory and visual speech information but significantly reduced efficiency for combining auditory speech information across frequency bands. Similarly, Tye-Murray et al. (2007) found that HI adults do not exhibit a reduced ability to integrate auditory and visual speech information relative to their age-matched, NH counterparts. Thus, HI listeners demonstrate greater difficulty integrating acoustic bands across the spectrum than they do integrating auditory and visual cues.

3.1.5 Auditory-Visual Asynchrony

Although the AV integration models of Massaro (1987) and Braida (1991) (and more generally of Alais and Burr, Chap. 2) can successfully account for the role of feature complementarity and redundancy in predicting AV speech intelligibility using probabilistic approaches such as fuzzy logic, multidimensional scaling, or maximum likelihood estimation, they each share an important shortcoming that prevents their wider application in the tradition of AI, SII, or STI models. To apply any of these models to the problem of AV integration, the uncertainty contributed by each separate modality regarding the identity of a given speech token or speech feature must be determined. In tests of nonsense-syllable identification, confusion matrices for A-only and V-only (at a minimum) must be obtained before any predictions of bimodal processing can take place (Massaro 1987; Braida 1991). Because these models as applied to speech identification require an abstraction of the auditory and visual speech information to phoneme labels before they are integrated, they cannot achieve what the AI methodology can accomplish by making speech-intelligibility predictions based on the physical properties of the speech signals alone.

Some clues for how one might accomplish the goal of a signal-based prediction of AV speech perception come from a number of studies that have examined how the temporal relationship between auditory and visual speech signals affects AV integration (see Fig. 3.5A). Studies have shown that AV integration does not require precise temporal alignment between A-only and V-only stimuli (e.g., McGrath and Summerfield 1985; Massaro et al. 1996). However, these studies also demonstrated

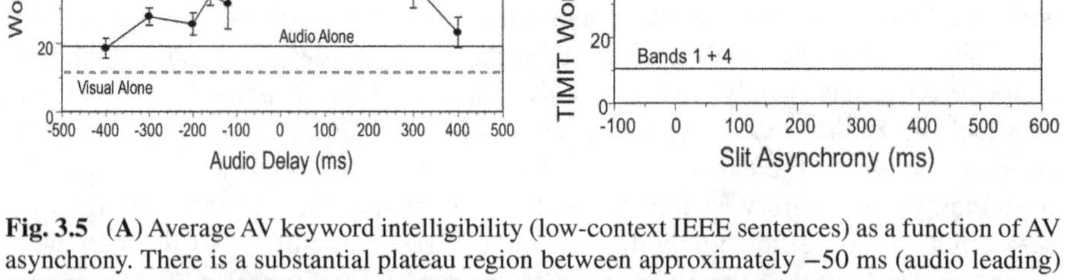

Fig. 3.5 (**A**) Average AV keyword intelligibility (low-context IEEE sentences) as a function of AV asynchrony. There is a substantial plateau region between approximately −50 ms (audio leading) to +200 ms (audio lagging) where intelligibility scores are high relative to the A-alone (*horizontal solid line*) or V-alone (*horizontal dashed line*) conditions. Error bars are ±1 SD. (**B**) Average A-only sentence intelligibility (Texas Instruments/Massachusetts Institute of Technology [TIMIT] sentences; Garofolo et al. 1990, 1993) for synchronous and asynchronous presentations of one-third octave, widely spaced auditory spectral slits. Unlike the AV condition, peak word-recognition performance in the A-only condition occurs when the different bandpass-filtered signals are presented synchronously and intelligibility falls off precipitously when any asynchrony is introduced across the spectral bands. From Grant et al. (2004)

that the temporal windows of integration (TWI) over which AV interactions can successfully occur are very asymmetric, with much greater tolerance found for visual-leading than for visual-lagging conditions. For naturally produced "congruent" speech, where the speaker's articulations and speech sounds are matched, auditory-lagging misalignments of up to 200 ms are easily tolerated, whereas visual-lagging misalignments greater than 20 ms lead to a breakdown in AV integration (Grant et al. 2004; Shahin et al. 2017). The asymmetry of the TWI favoring audio delays is consistent with the idea that for most speech utterances, the movement of the mouth begins before any sound is emitted. It has also been suggested that because visual speech information is available to the listener before the acoustic speech signal, it has the potential to facilitate language processing (e.g., lexical access) by allowing initial lexical pruning to proceed before any speech is heard (van Wassenhove et al. 2005, 2007). The fact that AV integration takes place over limited and multiple time windows suggests that bimodal speech processing is based on neural computations occurring at an earlier stage than a speech feature-based analysis.

In contrast to the long asymmetric temporal windows associated with AV integration, the TWI for combining information across acoustic frequency bands is both symmetric and narrow (see Fig. 3.5B). One interpretation of these data is that there are multiple time intervals over which speech is decoded in the auditory system. These include short-range analysis windows (1–40 ms), possibly reflecting various aspects of phonetic detail at the articulatory feature level (e.g., voicing); midrange analysis windows (40–120 ms), possibly reflecting segmental processing; and long-range analysis windows (beyond 120 ms), possibly reflecting the importance of prosodic cues, such as stress accent and syllable number, in the perception of running speech. The differences observed for cross-spectral (within modality) and cross-modal integration are important considerations for models of intelligibility as they highlight the different timescales associated with processing fine structure (formant transitions), syllabicity, and intonation. The different time frames may also implicate cortical asymmetries whereby left auditory areas process primarily short temporal integration windows while the right hemisphere processes information from long integration windows (Poeppel 2003). Yet the fact that the auditory and visual signals must be at least somewhat temporally coherent (Fig. 3.5A) suggests that a model of AV speech perception based on the coherence of auditory and visual signals might better represent the underlying process of AV integration than a feature-based or intelligibility-based approach.

3.1.6 Perception of Auditory-Visual Coherence and the Enhancement of the Auditory Speech Envelope

Another clue for how the auditory and visual speech signals might temporally interact comes from a set of speech-detection experiments conducted by Grant and Seitz (2000) and Grant (2001). The goal of these experiments was to determine whether movements of the lips perceived during speechreading could be used to improve the masked detection thresholds of congruent auditory signals. The basic idea used a variant of the comodulation masking-release paradigm (Hall et al. 1984), but in this coherence-protection paradigm (Gordon 1997, 2000), the audio speech target and visible movements of the lips were comodulated while the masker (e.g., speech-shaped noise) was uncorrelated with the target speech signal. The fact that the movements of the lips were coherent with the audio speech envelopes should have helped to protect the target speech from being masked and therefore improve detection thresholds.

From a strictly psychophysical perspective, it is reasonable to assume that the correlation between lip movements and acoustic envelope would be useful in detecting speech in noise and, further, that the greatest synergistic effects would be seen for sentences with the highest correlations. This is exactly what was found in studies by Grant and Seitz (2000) and Grant (2001). These studies showed a significant masking release for detecting spoken sentences (1–3 dB depending on the particular sentence) when simultaneous and congruent visual speech information was provided along with the wideband acoustic speech signal (Fig. 3.6, AV_{WB}). Incongruent speech (not shown) had no effect and resulted in the same threshold as the A-only condition. Finally, knowing prior to each trial (by orthography; AV_O) the content of the specific sentence to be detected had a mild positive influence (roughly 0.5 dB masking release) and was independent of which particular sentence was presented.

Critically, Grant (2001) showed that the degree of AV masking protection was related to the degree to which the auditory and visual signal envelopes were correlated. Figure 3.7 shows the time-intensity waveform, amplitude envelopes, and area of mouth opening for the sentence "Watch the log float in the wide river" (similar relationships can be seen for almost any AV sentence with only minor variations in the results). The traces in Fig. 3.7A represent the envelope extracted from wideband (WB) speech and from the speech filtered into three different spectral bands repre-

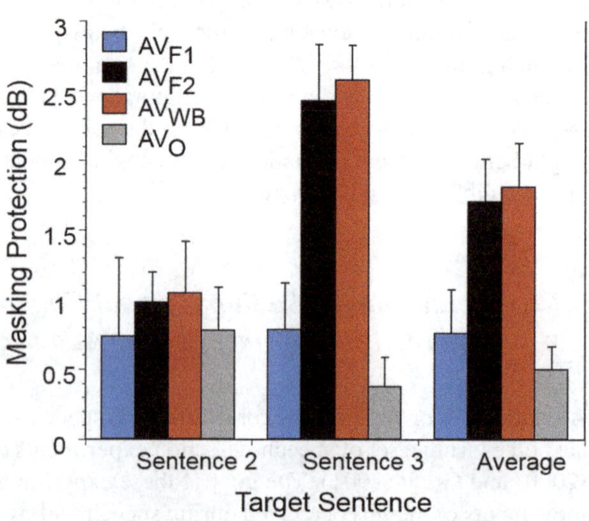

Fig. 3.6 Difference in A-only and AV masked detection thresholds (masking protection) for spoken filtered sentences (Grant 2001). AV_{F1}, AV visual presentation of speech filtered between 100 and 800 Hz; AV_{F2}, AV presentation of speech filtered between 800 and 2200 Hz; AV_{WB}, AV presentation of wideband speech (100–8500 Hz); AV_O, auditory presentation of wideband speech preceded by visual orthography. Error bars show +1 SD

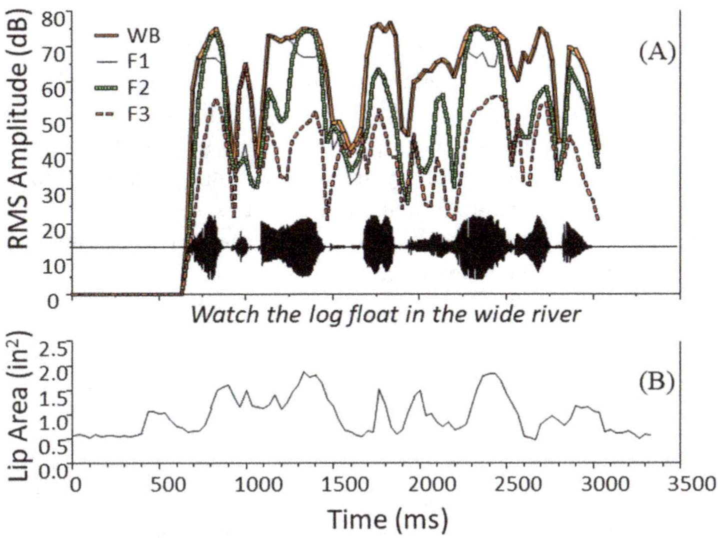

Fig. 3.7 (**A**) Waveform and amplitude envelopes extracted from wideband (WB) speech and from bandpass-filtered speech with filters centered at the F1 (100–800 Hz), F2 (800–2200 Hz), and F3 (2200–8500 Hz) formant regions. RMS, root-mean-square. (**B**) Amplitude envelope of the kinematic lip movements over time during speech production. The correlation between acoustic envelope and visual movement (area of mouth opening) was greatest for the envelope in the F2 region (0.65) and weakest in the F1 region (0.49). From Grant (2001)

senting the first (F1), second (F2), and third (F3) speech formants. These envelopes were clearly comodulated with the area of mouth opening extracted from the video image (Fig. 3.7B). However, the degree of correlation was largest for the acoustic envelopes derived from the higher frequency regions (F2 and F3) than for the F1 envelope. Grant (2001) found that WB speech or speech filtered into the F2 (800–2200 Hz) region also produced substantially more masking protection (about 2 dB on average) than for speech filtered into the F1 (100–800 Hz) region (less than 1 dB; Fig. 3.6, AV_{F2}, AV_{F1}). Thus, as long as the speech signal contained energy in the F2 region associated with place of articulation, the addition of synchronized visual information from the face of the speaker provided significant masking protection and lower detection thresholds. Overall, these results showed that listeners used the visible modulations of the lips and jaw during speechreading to make auditory detection easier by informing them about the probable spectrotemporal structure of a near-threshold acoustic speech signal, especially with peak energy in the F2 frequency range.

The temporal coherence of the acoustic and visual signals and the fact that the brain can make use of this temporal coherence to more easily detect the acoustic signal offer the possibility of analyzing the acoustic and visual signals within a single common mechanism of time-intensity envelope processing (see Lee, Maddox, and Bizley, Chap. 4). The fact that the modulation envelopes for speech of mid- to high-frequency auditory channels and the slowly time-varying visual kinematics of the speaker's face (e.g., area of mouth opening over time) are strongly correlated with one another provides a mechanism for combining the auditory and visual inputs directly at the physical signal level without requiring lengthy and costly behavioral experimentation. Section 3.2 describes efforts toward the development of a signal-based model of AV speech perception that makes predictions based on (1) the coherence between the auditory and visual signals over long temporal windows of integration, (2) greater AV benefit relative to A-only speech recognition at poorer SNRs, and (3) greater correlation between visual kinematics and the acoustic envelopes in the higher speech frequencies.

3.2 Modeling Auditory-Visual Speech Intelligibility

A model of AV speech perception based on the temporal coherence of the auditory and visual modalities necessarily requires an analysis of the temporal modulations of speech across the frequency spectrum. In this regard, the model would share many characteristics of the STI (Steeneken and Houtgast 2002), a model that takes into account the degree of modulation degradation as a result of noise, reverberation, or hearing loss. By considering the dynamics of the visual speech signal as additional modulation channels that can be used to reduce some of the deleterious effects of noise and reverberation, this approach can be easily expanded to include the influence of speechreading on speech intelligibility.

Grant et al. (2008, 2013) described a signal-based AV speech-intelligibility model that considered both auditory and visual dynamic inputs, combining them at the level of the speech envelopes to generate a prediction of AV speech intelligibility in noise. The basic premise was that the brain can use the visual input signals to help reconstruct the temporal modulations inherent in the "clean" auditory signal (minus noise or reverberation) based on *a priori* knowledge of the relationship between facial kinematics and the temporal envelopes of the audio speech signal. This approach was inspired by the engineering applications of Girin et al. (2001) and Berthommier (2004) showing that a video signal of the talker's face could be used to enhance a noise-corrupted audio speech signal.

Grant et al. (2008, 2013) used a biologically inspired auditory spectrotemporal modulation index (STMI) model (Elhilali et al. 2003) to make A-only and AV speech-intelligibility predictions. Like the STI, the STMI bases its predictions on analysis of the critical modulations present in the speech signal. However, the STMI includes an additional dimension, spectral modulation, which is critical to the prediction of the effects of spectral smearing caused, for example, by the reduced frequency selectivity associated with hearing loss (Bernstein et al. 2013). The model (Fig. 3.8) consisted of three main stages: (1) a "peripheral" stage that processed the acoustic waveform into frequency bands and derived the envelope in each band, (2) a "cortical" stage that processed the resulting envelopes to derive the modulation spectra, and (3) an "evaluation" phase that compared the resulting spectrotemporal modulation profile of speech presented in noise with the profile associated with

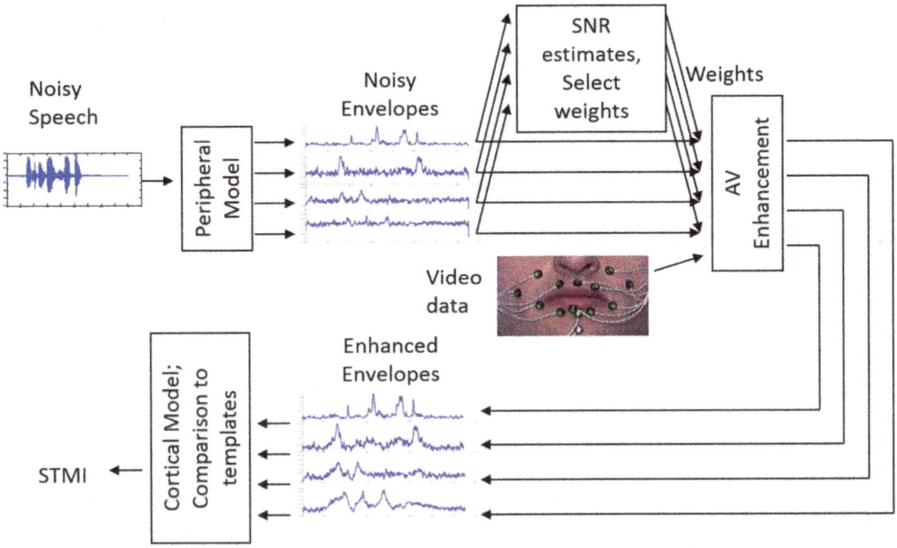

Fig. 3.8 Schematic of the expanded AV-spectrotemporal modulation index (STMI) model showing the inclusion of visual speech-movement envelopes to enhance the outputs of each auditory channel. The enhanced AV envelope channels were then processed by the cortical model and compared with "clean" speech templates to make the final AV speech-intelligibility estimate. *SNR* signal-to-noise ratio

clean speech ("comparison to speech templates" in Fig. 3.8). To extend the STMI model to include visual modulation channels, the model also included an "AV enhancement" component that cleaned up the noisy acoustic speech envelopes based on *a priori* knowledge about the relationship between the auditory and visual stimuli.

An example of the output of the peripheral stage of the model for an acoustic speech signal presented in speech-shaped noise is shown in Fig. 3.9A. Each individual curve represents a different SNR condition. As the SNR increased, the correlation between the envelope of the speech-plus-noise signal and the clean (speech-in-quiet) signal in each spectral band became greater, ultimately reaching a correlation coefficient of 1.0 (no noise or reverberation). These correlations were reflected in the output of the STMI model: with increasing SNR, as the spectral and temporal modulations of the speech-plus-noise envelopes began to resemble the modulations in the "clean" speech envelope, the model predicted an increase in speech intelligibility (Fig. 3.10). To model AV interaction, the visual enhancement was carried out based on dynamic measurements of the two-dimensional positions of 14 reference points on the talker's face made using an OPTOTRAK camera (Fig. 3.8, video data). The 28 resulting visual waveforms (*x*- and *y*-coordinates for each transmitter), along with the speech-in-noise envelopes from each frequency channel (cochlear filter), were input as predictor variables into a linear-regression model to predict the clean-speech envelope in each of 136 peripheral frequency bands.

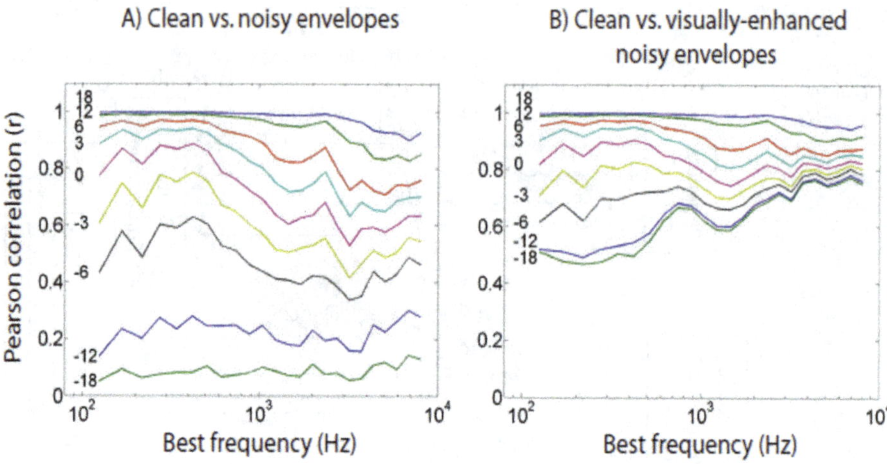

Fig. 3.9 (**A**) Correlation between clean and noisy acoustic speech envelopes for 136 peripheral auditory channels with center frequencies between 125 and 8000 Hz. The speech materials consisted of spoken IEEE sentences. The parameter is the SNR for the A-only speech signal. (**B**) Same as (**A**) except that the speech envelopes were enhanced using visual speech kinematics derived from 14 optical sensors positioned around the lips, cheeks, and chin of the speaker. From Grant et al. (2013)

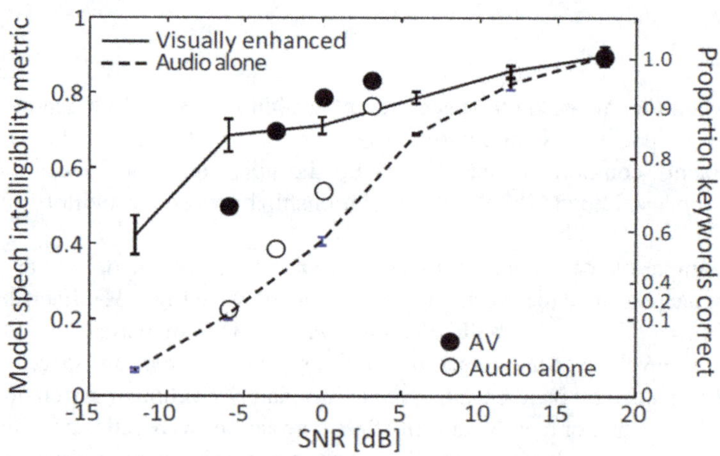

Fig. 3.10 Predicted AV (*solid line*) and A-only (*dashed line*) intelligibility based on the visually enhanced STMI model. *Circles*, intelligibility data measured in normal-hearing listeners Error bars are ±1 SD from model estimates for a list of IEEE sentences processed at each SNR tested

Figure 3.9B shows the correlations between the enhanced speech envelopes (based on the speech-in-noise envelope for each channel and the visual inputs) and the clean-speech envelopes. As in Fig. 3.9A, the correlations generally increased with increasing SNR because the speech-in-noise envelopes became more like the clean-speech envelopes. However, for the AV model, the correlation with the SNR was considerably higher, especially for low-SNR conditions, than in the A-only case (Fig. 3.9A). This is because the visual speech motion information also contributed to the prediction of the speech envelope. In fact, the AV correlations never decreased below the A-only values obtained for an SNR of approximately −6 dB. At very low SNRs (i.e., −12 and −18 dB), the speech-in-noise acoustic envelopes contained virtually no target speech information, and the prediction was based purely on the visual inputs. Thus, the predicted speech intelligibility was never poorer than that based on the visual channels alone.

By comparing the two panels in Fig. 3.9, it can be seen that the model accounted for the frequency dependence of the AV enhancement similar to what has been observed in perceptual studies (e.g., Grant and Walden 1996). At low frequencies, there was a relatively small difference between the correlations for A-only speech (Fig. 3.9A) and the correlations for AV speech (Fig. 3.9B), meaning that the model showed relatively little visual enhancement to the auditory envelopes when the low-frequency auditory information was corrupted. This is because the visual signal was relatively uninformative (complementary information) about acoustic speech information in this frequency region. In contrast, at high frequencies where the visual signal was predictive of the auditory envelope (redundant information), the visual signal more dramatically enhanced the resulting correlation, meaning that the model showed a large enhancement when high-frequency auditory information was corrupted.

Once the noisy-speech envelopes were enhanced using the temporal dynamics of the visual speech signal to more closely resemble the clean auditory speech envelopes, the cortical and evaluation stages of the model were carried out just as if the envelopes had been generated in the purely acoustic domain but now predicted a higher level of speech intelligibility because the peripheral envelopes more closely resembled clean speech. Figure 3.10 plots the model-predicted speech-intelligibility scores (solid and dashed curves) against the speech-intelligibility scores for sentence materials presented to NH adults (closed and open circles) in speech-shaped noise. The model captured the increase in intelligibility provided by the visual signal as well as the diminishing visual benefit associated with higher SNRs.

The key advantage of this signal-based approach to modeling AV speech intelligibility is that it could successfully account for important aspects of AV speech perception (cf. Sect. 3.1) that traditional models cannot achieve. Although Fig. 3.10 shows that this model captured some of the key features of the relationship between AV benefit and SNR, this is not the same as demonstrating that the model represents an improvement in the ability to predict AV speech intelligibility. In fact, the AI and SII models also predict a contribution of the visual component decreasing with SNR (Fig. 3.2). What this model accomplished beyond the traditional models is (1) the ability to predict AV speech intelligibility based on physical measurements of the speech and noise signal (like the AI, SII, and STI) without requiring a feature-based analysis of auditory- and visual-cue redundancy or an information analysis of A-only and V-only consonant confusions; and (2) an ability to account for spectral interactions when predicting AV speech perception (Fig. 3.9). The model also has the potential to account for AV synchrony effects, although that was not investigated here. Specifically, the imperviousness of the AV benefit to temporal misalignment (Fig. 3.5) could be modeled by computing a cross-correlation and choosing the delay in each channel that produces the maximum cross-correlation, while adhering to the characteristics of the AV temporal integration window.

3.3 Future Challenges

3.3.1 Complex Auditory Backgrounds

All the AV speech-intelligibility phenomena and modeling (cf. Sects. 3.1 and 3.2) deal with the simple case of NH listeners presented with speech in stationary background noise or filtered speech. In everyday environments, listening situations are much more complex, involving, for example, speech maskers, modulated noise, and spatial separation between target and masker. Although standard speech-intelligibility models (e.g., AI, SII, STI) do not generally address these complex factors, even in A-only situations, substantial research has taken place to understand how these factors influence speech perception in everyday environments. As a result, steps have been taken to incorporate some of these effects into models of auditory speech perception. For example, Rhebergen and Versfeld (2005) and Rhebergen et al. (2006) modified the SII to allow for predictions of speech intelligibility in modulated-noise backgrounds.

Despite the advances made in understanding the complex factors that influence A-only speech perception, relatively little is known about how visual cues interact with spatial cues, variability in masker type, or hearing loss. There have been a handful of studies investigating some of these interactions. For example, Helfer and Freyman (2005) have shown that visual cues can provide an important grouping cue for auditory-scene analysis in multitalker settings, with AV coherence providing the listener with information to perceptually segregate the speech produced by the target talker of interest from a concurrent interfering talker. Bernstein and Grant (2009) found little interaction between hearing loss and the influence of visual cues for speech perception in complex backgrounds. Together, these results suggest that the effects of hearing loss and visual benefit can be modeled independently, but the interaction between the availability of visual information and the perceptual separation of concurrent talkers is likely more complex.

3.3.2 Individual Differences: Hearing Acuity, Visual Acuity, and Integration Efficiency

Several attempts have been made to model the effects of hearing impairment on speech intelligibility (e.g., Bernstein et al. 2013; Bruce 2017). In most of these attempts, only the effects of reduced audibility have been modeled. Individual differences in spectral and temporal resolution, central auditory processing, and cognitive processing (e.g., working memory, speed of processing, attention allocation), each known to be important for speech intelligibility and understanding, remain a significant challenge (but see Bernstein et al. 2013).

Another area of AV speech perception that would need to be incorporated into any comprehensive model involves degradation in the visual domain due to vision loss (Hardick et al. 1970). Although typical age-related vision loss does not eliminate the visual speech-intelligibility benefit (Hickson et al. 2004), blurred vision can reduce the effect (Legault et al. 2010). Evidence from earlier studies suggests that speechreading performance significantly declines with age, especially for those over 70 years old (Shoop and Binnie 1979; Middelweerd and Plomp 1987). Although the reasons for this decline are not fully understood, it has been suggested that reductions in peripheral visual acuity and motion perception associated with aging may play a role. Unfortunately, there are very few studies that have examined the relationship between overall speechreading ability, individual differences in the transmission of visual place-of-articulation information, and visual acuity. Therefore, if the goal is to predict AV speech intelligibility as well as individual differences in AV processing due to hearing and vision loss, basic tests of auditory and visual function will have to be incorporated into the modeling efforts.

Finally, there is the possibility that some individuals are better able than others to integrate auditory and visual information. As discussed in Sect. 3.1.4, although many of the differences in AV benefit observed by HI listeners can be ascribed to an impoverished auditory signal, there was at least some evidence that certain individuals might also have had a particular deficit in the ability to integrate speech information from the two modalities (Grant et al. 2007). To the extent that individual variability in integration efficiency exists, this factor would also need to be included in an individual-specific model of AV speech perception.

3.4 Summary

Signal-based models of speech perception are critically important for the design and evaluation of audio systems and hearing-rehabilitation devices. Models such as the AI, SII, and STI have undergone decades of development and scrutiny and are mostly successful in predicting average speech intelligibility for acoustic signals under a variety of conditions. Yet more than half of speech communication takes place in face-to-face situation where the listener is looking at the talker and has access to visual speech cues (Walden et al. 2004). It is clear that the simplistic approach in the manner in which that these models predict AV speech intelligibility, assuming that the speechreading benefit to auditory speech intelligibility can be modeled as a simple additive factor, is incorrect. Thus, these extant models are inadequate for predicting AV speech intelligibility for a given audio input signal, transducer, and hearing loss.

Section 3.1 reviewed several important phenomena associated with AV speech perception that highlight the complex interaction between these modalities that any model would need to take into account. In particular, it was shown that the amount of speechreading benefit depends dramatically on the spectral content of the speech signal (Grant et al. 1985; Grant and Walden 1996). This interaction can be understood in terms of the complementary or redundant nature of the speech features provided by the visual and acoustic speech cues (Grant and Walden 1996). Although extant models of speech-feature integration proposed by Massaro (1987) and Braida (1991) do a good job predicting AV speech recognition for nonsense syllables, they cannot predict sentence or connected discourse performance and require significant time and effort to obtain unimodal perceptual confusion-matrix data. Other important aspects of AV speech perception that the simple additive models cannot account for include a limited tolerance to temporal asynchrony within a range of −20 ms (audio leading) to +200 ms (audio lagging) (Grant et al. 2004; Shahin et al. 2017) and the possibility of individual variability in AV integration efficiency (Grant et al. 2007).

Section 3.2 described a signal-based modeling approach to predicting AV speech perception. One of the greatest obstacles to developing a model of AV speech perception has been the centuries-old tradition of treating sensory modalities as independent receivers of information, combined at an abstract linguistic level. However, physiological data showing the existence of multimodal neurons that only fire when certain temporal constraints across inputs from different sensory modalities are met suggest a different story. In fact, listeners are sensitive to coherence in the modulation of the acoustic envelope and the temporal dynamics of lip movements (Grant and Seitz 2000; Grant 2001), providing a clue for how AV speech performance might be predicted from the physical properties of the visual and acoustic signals. In the model, visual speech motion was used to help reconstruct and enhance corrupted auditory speech-envelope information from different frequency channels into envelopes that more closely resemble those from clean speech. This approach was shown to be consistent with experimental evidence that the visual signal is best able to stand in for corrupted acoustic speech information in the mid-to-high speech frequencies associated with F2 and F3 transitions (place-of-articulation information). Although work remains to integrate other important known aspects of AV speech processing (e.g., tolerance to asynchrony, individual variation in visual or hearing acuity, and integration efficiency), this approach represents an important step toward the development of a signal-based AV speech-perception model in the spirit of the AI, SII, and STI.

Compliance with Ethics Requirements Kenneth W. Grant declares that he has no conflict of interest.

Joshua G. W. Bernstein declares that he has no conflict of interest.

The views expressed in this chapter are those of the authors and do not reflect the official policy of the Department of Army/Navy/Air Force, Department of Defense, or US Government.

References

American National Standards Institute (ANSI). (1969). *American National Standard Methods for the calculation of the articulation index. ANSI S3.5-1969*. New York: American National Standards Institute.

American National Standards Institute (ANSI). (1997). *American National Standard Methods for calculation of the speech intelligibility index. ANSI S3.5–1997*. New York: American National Standards Institute.

Bernstein, J. G. W., & Grant, K. W. (2009). Audio and audiovisual speech intelligibility in fluctuating maskers by normal-hearing and hearing-impaired listeners. *The Journal of the Acoustical Society of America, 125*, 3358–3372.

Bernstein, J. G. W., Summers, V., Grassi, E., & Grant, K. W. (2013). Auditory models of suprathreshold distortion and speech intelligibility in persons with impaired hearing. *Journal of the American Academy of Audiology, 24*, 307–328.

Berthommier, F. (2004). A phonetically neutral model of the low-level audio-visual interaction. *Speech Communication, 44*(1), 31–41.

Braida, L. D. (1991). Crossmodal integration in the identification of consonant segments. *Quarterly Journal of Experimental Psychology, 43*, 647–677.

Bruce, I. (2017). Physiologically based predictors of speech intelligibility. *Acoustics Today, 13*(1), 28–35.

Byrne, D., Dillon, H., Ching, T., Katsch, R., & Keidser, G. (2001). NAL-NL1 procedure for fitting nonlinear hearing aids: Characteristics and comparisons with other procedures. *Journal of the American Academy of Audiology, 31*, 37–51.

Drullman, R., & Smoorenburg, G. F. (1997). Audio-visual perception of compressed speech by profoundly hearing-impaired subjects. *Audiology, 36*(3), 165–177.

Elhilali, M., Chi, T., & Shamma, S. A. (2003). A spectro-temporal modulation index (STMI) for assessment of speech intelligibility. *Speech Communication, 41*(2), 331–348.

Erber, N. (1972). Auditory, visual, and auditory-visual recognition of consonants by children with normal and impaired hearing. *Journal of Speech, Language, and Hearing Research, 15*(2), 413–422.

Fletcher, H. (1953). *Speech and hearing in communication*. New York: Van Nostrand.

Fletcher, H., & Gault, R. H. (1950). The perception of speech and its relation to telephony. *The Journal of the Acoustical Society of America, 22*, 89–150.

French, N. R., & Steinberg, J. C. (1947). Factors governing the intelligibility of speech sounds. *The Journal of the Acoustical Society of America, 19*, 90–119.

Garofolo, J. S., Lamel, L. F., Fisher, W. M., Fiscus, J. G., et al. (1990). *DARPA, TIMIT acoustic-phonetic continuous speech corpus CD-ROM*. Gaithersburg, MD: National Institute of Standards and Technology, US Department of Commerce.

Garofolo, J. S., Lamel, L. F., Fisher, W. M., Fiscus, J. G., & Pallett, D. S. (1993). *DARPA TIMIT acoustic-phonetic continous speech corpus CD-ROM. NIST speech disc 1-1.1*. Gaithersburg, MD: National Institute of Standards and Technology, US Department of Commerce.

Girin, L., Schwartz, J. L., & Feng, G. (2001). Audio-visual enhancement of speech in noise. *The Journal of the Acoustical Society of America, 109*(6), 3007–3020.

Gordon, P. C. (1997). Coherence masking protection in speech sounds: The role of formant synchrony. *Perception & Psychophysics, 59*, 232–242.

Gordon, P. C. (2000). Masking protection in the perception of auditory objects. *Speech Communication, 30*, 197–206.

Grant, K. W. (2001). The effect of speechreading on masked detection thresholds for filtered speech. *The Journal of the Acoustical Society of America, 109*, 2272–2275.

Grant, K. W., Ardell, L. H., Kuhl, P. K., & Sparks, D. W. (1985). The contribution of fundamental frequency, amplitude envelope, and voicing duration cues to speechreading in normal-hearing subjects. *The Journal of the Acoustical Society of America, 77*, 671–677.

Grant, K. W., Bernstein, J. G. W., & Grassi, E. (2008). Modeling auditory and auditory-visual speech intelligibility: Challenges and possible solutions. *Proceedings of the International Symposium on Auditory and Audiological Research, 1*, 47–58.

Grant, K. W., Bernstein, J. G. W., & Summers, V. (2013). Predicting speech intelligibility by individual hearing-impaired listeners: The path forward. *Journal of the American Academy of Audiology, 24*, 329–336.

Grant, K. W., & Braida, L. D. (1991). Evaluating the articulation index for audiovisual input. *The Journal of the Acoustical Society of America, 89*, 2952–2960.

Grant, K. W., Greenberg, S., Poeppel, D., & van Wassenhove, V. (2004). Effects of spectro-temporal asynchrony in auditory and auditory-visual speech processing. *Seminars in Hearing, 25*, 241–255.

Grant, K. W., & Seitz, P. F. (2000). The use of visible speech cues for improving auditory detection of spoken sentences. *The Journal of the Acoustical Society of America, 108*, 1197–1208.

Grant, K. W., Tufts, J. B., & Greenberg, S. (2007). Integration efficiency for speech perception within and across sensory modalities. *The Journal of the Acoustical Society of America, 121*, 1164–1176.

Grant, K. W., & Walden, B. E. (1996). Evaluating the articulation index for auditory-visual consonant recognition. *The Journal of the Acoustical Society of America, 100*, 2415–2424.

Grant, K. W., Walden, B. E., & Seitz, P. F. (1998). Auditory-visual speech recognition by hearing-impaired subjects: Consonant recognition, sentence recognition, and auditory-visual integration. *The Journal of the Acoustical Society of America, 103*, 2677–2690.

Hall, J. W., Haggard, M. P., & Fernandes, M. A. (1984). Detection in noise by spectro-temporal pattern analysis. *The Journal of the Acoustical Society of America, 76*, 50–56.

Hardick, E. J., Oyer, H. J., & Irion, P. E. (1970). Lipreading performance as related to measurements of vision. *Journal of Speech and Hearing Research, 13*, 92–100.

Helfer, K. S., & Freyman, R. L. (2005). The role of visual speech cues in reducing energetic and informational masking. *The Journal of the Acoustical Society of America, 117*(2), 842–849.

Hickson, L., Hollins, M., Lind, C., Worrall, L. E., & Lovie-Kitchin, J. (2004). Auditory-visual speech perception in older people: The effect of visual acuity. *Australian and New Zealand Journal of Audiology, 26*, 3–11.

Kewley-Port, D. (1983). Time-varying features as correlates of place of articulation in stop consonants. *The Journal of the Acoustical Society of America, 73*(1), 322–335.

Killion, M., Schulein, R., Christensen, L., Fabry, D., Revit, L., Niquette, P., & Chung, K. (1998). Real-world performance of an ITE directional microphone. *The Hearing Journal, 51*, 24–39.

Legault, I., Gagné, J. P., Rhoualem, W., & Anderson-Gosselin, P. (2010). The effects of blurred vision on auditory-visual speech perception in younger and older adults. *International Journal of Audiology, 49*(12), 904–911.

Liberman, A. M., Cooper, F. S., Shankweiler, D. P., & Studdert-Kennedy, M. (1967). Perception of the speech code. *Psychological Review, 74*(6), 431–461.

Massaro, D. W. (1987). *Speech perception by ear and eye: A paradigm for psychological inquiry*. Hillsdale, NJ: Lawrence Erlbaum Associates.

Massaro, D. W. (1998). *Perceiving talking faces: From speech perception to a behavioral principle*. Cambridge, MA: MIT Press.

Massaro, D. W., Cohen, M. M., & Smeele, P. M. (1996). Perception of asynchronous and conflicting visual and auditory speech. *The Journal of the Acoustical Society of America, 100*(3), 1777–1786.

McGrath, M., & Summerfield, Q. (1985). Intermodal timing relations and audio-visual speech recognition by normal-hearing adults. *The Journal of the Acoustical Society of America, 77*(2), 678–685.

Middelweerd, M. J., & Plomp, R. (1987). The effect of speechreading on the speech-reception threshold of sentences in noise. *The Journal of the Acoustical Society of America, 82*(6), 2145–2147.

Miller, G. A., & Nicely, P. E. (1955). An analysis of perceptual confusions among some English consonants. *The Journal of the Acoustical Society of America, 27*(2), 338–352.
Poeppel, D. (2003). The analysis of speech in different temporal integration windows: Cerebral lateralization as 'asymmetric sampling in time. *Speech Communication, 41*(1), 245–255.
Reetz, H., & Jongman, A. (2011). *Phonetics: Transcription, production, acoustics, and perception.* Chichester, West Sussex: Wiley-Blackwell.
Rhebergen, K. S., & Versfeld, N. J. (2005). A speech intelligibility index-based approach to predict the speech reception threshold for sentences in fluctuating noise for normal-hearing listeners. *The Journal of the Acoustical Society of America, 117*(4), 2181–2192.
Rhebergen, K. S., Versfeld, N. J., & Dreschler, W. A. (2006). Extended speech intelligibility index for the prediction of the speech reception threshold in fluctuating noise. *The Journal of the Acoustical Society of America, 120*(6), 3988–3997.
Rosen, S. M., Fourcin, A. J., & Moore, B. C. J. (1981). Voice pitch as an aid to lipreading. *Nature, 291*(5811), 150–152.
Shahin, A. J., Shen, S., & Kerlin, J. R. (2017). Tolerance for audiovisual asynchrony is enhanced by the spectrotemporal fidelity of the speaker's mouth movements and speech. *Language, Cognition and Neuroscience, 32*(9), 1102–1118.
Shoop, C., & Binnie, C. A. (1979). The effects of age upon the visual perception of speech. *Scandinavian Audiology, 8*(1), 3–8.
Sommers, M. S., Tye-Murray, N., & Spehar, B. (2005). Auditory-visual speech perception and auditory-visual enhancement in normal-hearing younger and older adults. *Ear and Hearing, 26*(3), 263–275.
Steeneken, H. J., & Houtgast, T. (2002). Validation of the revised STI_r method. *Speech Communication, 38*(3), 413–425.
Studdert-Kennedy, M. (1974). The perception of speech. *Current Trends in Linguistics, 12*, 2349–2385.
Sumby, W. H., & Pollack, I. (1954). Visual contribution to speech intelligibility in noise. *The Journal of the Acoustical Society of America, 26*, 212–215.
Summerfield, Q. (1987). Some preliminaries to a comprehensive account of audio-visual speech perception. In B. Dodd & R. Campbell (Eds.), *Hearing by eye: The psychology of lip-reading* (pp. 3–52). Hillsdale NJ: Lawrence Erlbaum Associates.
Summerfield, Q. (1992). Lipreading and audio-visual speech perception. *Philosophical Transactions of the Royal Society of London B, Biological Sciences, 335*(1273), 71–78.
Tye-Murray, N., Sommers, M. S., & Spehar, B. (2007). Audiovisual integration and lipreading abilities of older adults with normal and impaired hearing. *Ear and Hearing, 28*(5), 656–668.
van Wassenhove, V., Grant, K. W., & Poeppel, D. (2005). Visual speech speeds up the neural processing of auditory speech. *Proceedings of the National Academy of Sciences of the United States of America, 102*, 1181–1186.
van Wassenhove, V., Grant, K. W., & Poeppel, D. (2007). Temporal window of integration in auditory-visual speech perception. *Neuropsychologia, 45*, 598–607.
Walden, B. E., Grant, K. W., & Cord, M. T. (2001). Effects of amplification and speechreading on consonant recognition by persons with impaired hearing. *Ear and Hearing, 22*(4), 333–341.
Walden, B. E., Surr, R. K., Cord, M. T., & Dyrlund, O. (2004). Predicting hearing aid microphone preference in everyday listening. *Journal of the American Academy of Audiology, 15*(5), 365–396.
Wu, Y. H., & Bentler, R. A. (2010). Impact of visual cues on directional benefit and preference: Part I—Laboratory tests. *Ear and Hearing, 31*(1), 22–34.

Chapter 4
An Object-Based Interpretation of Audiovisual Processing

Adrian K. C. Lee, Ross K. Maddox, and Jennifer K. Bizley

Abstract Visual cues help listeners follow conversation in a complex acoustic environment. Many audiovisual research studies focus on how sensory cues are combined to optimize perception, either in terms of minimizing the uncertainty in the sensory estimate or maximizing intelligibility, particularly in speech understanding. From an auditory perception perspective, a fundamental question that has not been fully addressed is how visual information aids the ability to select and focus on one auditory object in the presence of competing sounds in a busy auditory scene. In this chapter, audiovisual integration is presented from an object-based attention viewpoint. In particular, it is argued that a stricter delineation of the concepts of multisensory integration versus binding would facilitate a deeper understanding of the nature of how information is combined across senses. Furthermore, using an object-based theoretical framework to distinguish binding as a distinct form of multisensory integration generates testable hypotheses with behavioral predictions that can account for different aspects of multisensory interactions. In this chapter, classic multisensory illusion paradigms are revisited and discussed in the context of multisensory binding. The chapter also describes

---

A. K. C. Lee (✉)
Department of Speech and Hearing Sciences, University of Washington, Seattle, WA, USA

Institute for Learning and Brain Sciences (I-LABS), University of Washington, Seattle, WA, USA
e-mail: akclee@uw.edu

R. K. Maddox
Department of Biomedical Engineering, University of Rochester, Rochester, NY, USA

Department of Neuroscience, University of Rochester, Rochester, NY, USA

Del Monte Institute for Neuroscience, University of Rochester, Rochester, NY, USA

Center for Visual Science, University of Rochester, Rochester, NY, USA
e-mail: ross.maddox@rochester.edu

J. K. Bizley
Ear Institute, University College London, London, UK
e-mail: j.bizley@ucl.ac.uk

A. K. C. Lee et al. (eds.), *Multisensory Processes*, Springer Handbook of Auditory Research 68, https://doi.org/10.1007/978-3-030-10461-0_4

multisensory experiments that focus on addressing how visual stimuli help listeners parse complex auditory scenes. Finally, it concludes with a discussion of the potential mechanisms by which audiovisual processing might resolve competition between concurrent sounds in order to solve the cocktail party problem.

Keywords Binding · Cross-modal · McGurk illusion · Multisensory · Object-based attention · Scene analysis · Sensory integration · Sound-induced flash illusion · Ventriloquism

4.1 Introduction

There are many different perspectives on how auditory and visual information can be combined to influence our perception of the world. Chapter 2 by Alais and Burr focuses on how the cues in each of these sensory modalities could be optimally combined to maximize perceptual precision through the lens of Bayesian modeling. From a communication perspective, Grant and Bernstein in Chap. 3 describe how speech intelligibility could be altered by the presence of visual information, especially in situations where the auditory signal is embedded in masking noise. The focus of this chapter takes on yet another perspective: does visual information help segregate sounds in a cacophony, and if so, how?

4.1.1 Multisensory Cocktail Party: Disambiguating Sound Mixtures Using Visual Cues

Most normal-hearing listeners can recognize what one person is saying when others are speaking at the same time. This is the classic "cocktail party problem" as defined by Cherry more than six decades ago (Cherry 1953; Middlebrooks et al. 2017). Current state-of-the-art machine-learning algorithms still struggle with this auditory scene analysis problem, yet the brain exploits the statistics of natural sound to accomplish this task comparatively easily. Psychoacoustic studies in the past decades have shown that sound elements are likely to be grouped together to form an auditory stream when they are harmonically related to each other (Culling and Stone 2017), temporally coherent with one another (Shamma et al. 2011), or share common spatial cues across time (Maddox and Shinn-Cunningham 2012). All of these past studies examined how the brain solves the cocktail party problem using auditory cues alone (see Lee 2017 for a summary of the human neuroimaging efforts to understand the listening brain). It is noteworthy to point out that Cherry in his original paper (1953) highlighted "lip reading" as a potential component of the cocktail party problem's solution. Even though visual cues are usually present and can potentially be used to separate sound mixtures, this process is considerably less explored in behavioral listening experiments. Because visual cues in conversation are often intimately linked to speech reading—using a talker's lip and articulator movements and other facial expressions to better understand conversation (see Grant and Bernstein, Chap. 3)—many studies have focused on characterizing the increase in speech intelligibility when a listener can see a talker's face. This chapter, instead, focuses on how visual cues help a listener to better segregate and select a sound of interest from a mixture.

4.1.2 Object-Based Attention

Auditory and visual information propagate in different physical forms and reach the brain via coding in different sensory epithelia, but features across these sensory modalities are seamlessly bound to create a coherent percept. Binding stimulus features from a common source is not a unique problem across sensory modalities; within a modality, independently encoded perceptual features (e.g., color, shape, and orientation in vision; pitch, timbre, and spatial cues in audition) must also be combined to form a single perceptual object. These perceptual objects are the "units" on which attention operates, both in vision (Desimone and Duncan 1995) and audition (Shinn-Cunningham et al. 2017). Given that there is a central limitation in the amount of information the brain can process, attention helps to determine what object(s) the brain analyzes to make sense of a complex scene.

As elaborated in Sect. 4.2, audiovisual binding can be viewed through the lens of object-based attention, extending theories that have been well developed in each sensory modality. For example, watching the bowing action of the first violin in an orchestra can help a listener pick out the string melody. Conversely, it is more difficult to follow the violin melody if one instead watches the timpanist's mallets. This example illustrates two fundamental aspects of object-based attention, namely, (1) attending to one feature of an object (visual motion of the bow) automatically enhances another feature of the same object (the auditory melody produced by the bow) and (2) there is a cost associated with dividing attention across objects (listening to the violin melody while watching the timpanist). An *a priori* problem the brain must solve is determining which sets of features belong to which object.

4.1.3 The Auditory Perspective

From an auditory perspective, one of the most important questions to address is how and to what extent visual information can help someone listen, especially in a crowded environment. In this chapter, an object-based perspective of multisensory processing will be presented to account for the visual benefits in auditory perception, especially when the auditory signal of interest is embedded in other competing sounds. In particular, multisensory integration will be defined as any integration of information across sensory domains, whereas the term multisensory binding will be reserved for the binding of auditory and visual information into a single multisensory object. We propose that such a multisensory object is more easily segregable from competing stimuli, which allows us to distinguish binding as a distinct form of multisensory integration. Whereas resolving competition between sensory objects is a basic problem the brain solves readily in the natural environment, laboratory procedures that employ competing stimuli are rare, with most focusing on characterizing the interactions between two individual (or, rarely, streams of) auditory and visual stimuli.

The most common way to study multisensory interaction is with multisensory illusions: experimenters present multisensory stimuli with a conflict between the modalities and measure the change in perception compared with presenting either of these stimuli in a single modality. Reports of these illusory percepts have been often discussed as a demonstration of multisensory integration and multisensory binding, without consideration of their differences. Among the most widely used paradigms, the first is the ventriloquist illusion whereby the location of a sound is "captured" by the visual stimulus (Howard and Templeton 1966). The second is the sound-induced flash illusion (Shams et al. 2000) in which the number of visual flashes reported is influenced by the number of rapidly presented auditory stimuli. Finally, the McGurk illusion (McGurk and MacDonald 1976) in which visual mouth movements for a given syllable are paired with an incongruent auditory syllable, resulting in the percept of a third syllable (e.g., a visual /ga/ and an auditory /ba/ can result in a /da/ percept). The findings from experiments eliciting these illusions are discussed in the context of multisensory binding in Sect. 4.3.

Furthermore, in contrast to the illusion paradigms outlined above that generally employ brief auditory and visual stimuli, this chapter highlights the benefits of using longer, dynamic stimuli in multisensory experiments. It is hoped that such experiments will provide a new perspective on how visual information can help listeners solve the cocktail party problem.

4.2 Visual, Auditory, and Auditory-Visual Objects

A red car moving to the left, a trumpet riff in a jazz band, these are examples of visual and auditory objects that are not difficult to conceptualize. Intuitively, a visual or auditory object is a perceptual entity that is subjectively perceived to originate from one physical source. Yet a precise (and concise) definition of what makes a visual (Feldman 2003) or auditory (Bizley and Cohen 2013) object is difficult to pin down, perhaps due to sensitivities to specific stimulus factors and context (as described in Sect. 4.2.1.2) that contribute to how objects are formed. Nevertheless, it is generally accepted that visual and auditory attention operates on objects (Desimone and Duncan 1995; Shinn-Cunningham et al. 2017) and that the attentional network is often assumed to be supramodal (i.e., not specific to one sensory modality; Knudsen 2007).

Two classic studies elegantly demonstrate the fundamental attributes of object-based attention introduced above. In a functional magnetic resonance imaging study, O'Craven et al. (1999) showed visual stimuli consisting of a face transparently superimposed on a house, with one moving and the other stationary, and asked the subjects to attend to either the house, the face, or the motion. They showed that attending to one feature of an object (e.g., motion) enhanced not only the neural representation of that feature (i.e., motion) but also of the other feature of the same object (i.e., the house, if that was moving) compared with features of the other object (i.e., the face).

In a psychophysical study, Blaser et al. (2000) asked subjects to track and make judgments about a Gabor patch—a staple stimulus in vision studies consisting of a sinusoidal alternation in space between high and low luminance, also known as gratings, smoothed by a 2-D Gaussian window—that dynamically changed its features (viz. orientation, spatial frequency, and color saturation) in the presence of another competing Gabor patch at the same location but with its features changed according to different temporal trajectories. Not only could observers track one Gabor patch in the presence of the other, they also reported that the target Gabor patch was more salient than the competing one, not dissimilar to figure-ground segmentation. Furthermore, when observers were asked to make two judgments on feature perturbations introduced to specific features of these stimuli, they performed worse when these perturbations were divided *across* the two Gabor patches (e.g., reporting a color perturbation in one Gabor patch and the orientation perturbation of the other) compared with when they were *within* the same Gabor patch (e.g., reporting the color and orientation perturbation of the same Gabor patch).

Figure 4.1A provides a sketched illustration of the benefits bestowed on reporting features from one object compared with across two objects, modified from the original study by Behrmann et al. (1998). The displays are made up of two overlapping rectangles, with a set of "cuttings" (or features) that looks as if one or

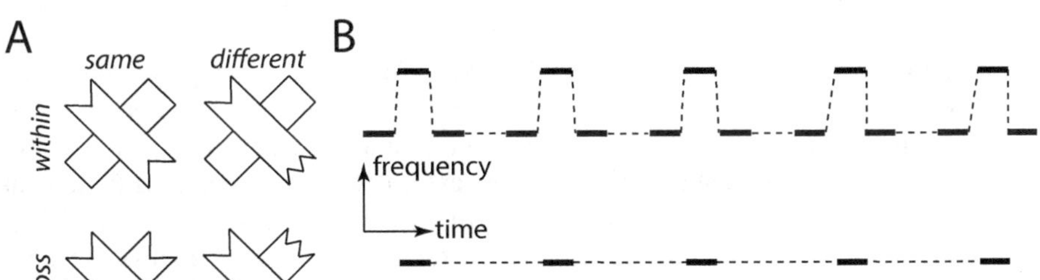

Fig. 4.1 Visual and auditory examples illustrating object-based attention. (**A**) Visual example inspired by Behrmann et al. (1998) showing two objects intersecting in an X configuration. Subjects are asked to perform a same/different judgment task (whether the "cuttings" at the end of the rectangle(s) are the *same* (*left column*) or *different* (*right column*). This task was performed with a faster reaction time when these cuttings/features appeared *within* the same object (*top row*) compared with when they were spread *across* two objects (*bottom row*) despite having a bigger spatial separation for the displays in the *bottom row*. (**B**) Auditory example from Cusack and Roberts (2000). Subjects were asked to detect a change in the isochronous (i.e., equally paced) rhythm. Deviation from the isochronous rhythm was much easier to detect when tones were grouped as one object *(top)* compared with when they were segregated as two objects (*bottom*)

two triangles have been cut away from one of the four possible edges of the X figure. These cuttings appear either at the ends within the same rectangle (Fig. 4.1A, *top row*) or across two rectangles (Fig. 4.1A, *bottom row*). The features had either the same (Fig. 4.1A, *left column*) or different (Fig. 4.1A, *right column*) number of cuttings. Consistent with object-based attention, subjects could perform this same/different psychophysics task faster and without loss of accuracy when the "cuttings" appeared on the same object (Fig. 4.1A, *top row*) compared with features spread across two objects (Fig. 4.1A, *bottom row*).

These concepts translate intuitively to the auditory domain. For example, selectively listening to a female talker in the presence of a male talker will enhance all the features of the female's voice (e.g., prosody and phonation). The ability to judge temporal relationships across two sounds is impaired when those sound elements belong to two separate streams (Fig. 4.1B, *bottom row*) instead of a single one (Cusack and Roberts 2000). Indeed, the ability to identify deviations from an isochronous (i.e., equally paced) rhythm within the same auditory stream but not across two separate streams is often leveraged in psychoacoustics paradigms to objectively measure stream segregation (Micheyl and Oxenham 2010).

The notion that attending to a feature belonging to one object will enhance the other features of the object leads to a difficult question: how are these features bound to form an object in the first place? This feature-binding problem still vexes both psychophysicists and neurophysiologists. One of many hypotheses is the temporal coherence theory of object formation (Shamma et al. 2011). This theory is based on the observation that the sound features (e.g., pitch, timbre, loudness, spatial location) associated with one source would be present whenever the source is active and absent when it is silent; features within the same source will be modulated coherently through time. Furthermore, different sound sources (and their associated features) will fluctuate according to their own time courses, which are independent of those of other sources. However, it is still unclear whether and how the coherence between neural populations is computed on a neural level. Nevertheless, temporal coherence across neural populations as a way to solve the binding problem can also be extended to other sensory domains where there is a natural temporal fluctuation of objects in the scene, but a caveat must be applied for the case of static visual scenes (which an observer is well able to segment into its constituent objects; Treisman 1998). The temporal coherence model could account for how observers process dynamic visual scenes (Alais et al. 1998; Blake and Lee 2005) or even multisensory stimuli.

An audiovisual object can functionally be defined as "a perceptual construct which occurs when a constellation of features are bound within the brain" (Bizley et al. 2016a). Most naturally occurring audiovisual objects have auditory and visual features that evolve coherently in time, dictated by the physical nature of the source. For example, mouth shape must covary with the dynamic acoustics of speech because it is the physical configuration of the speech articulators that determines the sounds being produced (Chandrasekaran et al. 2009). If one is watching the trumpet player generating the riff in the jazz band, it is likely that one will see the player move the instrument and body in rhythm with the playing, providing temporal coherence between the visual and auditory scenes.

4.2.1 Integration Versus Binding

Although many multisensory investigators use the terms "integration" and "binding" interchangeably (Stein and Stanford 2008), defining and distinguishing these two terms can provide clarity when attempting to distinguish changes in sensory representation from other interactions, such as combining independent sensory information at the decision-making stage; such clarity will be important in behavioral as well as neurophysiological studies (Bizley et al. 2016a). Specifically, multisensory integration can be defined as any process in which information across sensory modalities is combined to make a perceptual judgment, whereas multisensory binding should be reserved to describe a specific form of integration in which perceptual features are grouped into a unified multisensory object. In other words, binding is a form of integration; however, integrating information at the decision stage, for example, is not a form of binding.

Here is an example to illustrate the intuitive concept of multisensory integration and the steps required to substantiate multisensory binding leading to a unified multisensory percept. At a 100-meter dash, the starting pistol is an important audiovisual stimulus that marks the beginning of the race. The runners in the race are concerned with *when* the gun is fired so they could potentially make a decision based on the timing of the sound of the gun combined with the timing of the light flash to determine when to jump off the starting block. This is an example of multisensory integration. Nevertheless, in practice, because auditory stimuli generally provide more precise temporal information, runners would rely on hearing the sound of the gun rather than seeing the flash to start running.[1] If, instead, the task of the observer (most likely someone in the audience, not the runners themselves) were to locate *where* the pistol was, one could weigh and combine the location estimates of the visual flash and the sound of the gun. This would be another example of multisensory integration. In practice, someone in the audience who couldn't see *where* the pistol was before it was fired would be cued to its exact location by the visual flash because visual stimuli provide much better spatial information. As discussed by Alais and Burr in Chap. 2, weighing evidence from each sensory system by their reliability—specifically, temporal in audition and spatial in vision—to reach a decision is an example of how multisensory integration is shaped by current behavioral demands.

What would multisensory binding mean from the above example? It is unlikely that an observer would ever perceive the sound of the gunshot and the motion of the athlete as features of a unified, multisensory object, even though one could integrate this multisensory information to decide when the 100-meter dash started. An observer would more likely associate the sound and flash of the gun as sensory events that "go together," possibly forming an audiovisual object; after all, these two pieces of sensory information originate from the same location at the same time.

[1] This example is best understood if acoustic propagation delay is ignored; modern track and field competitions use a loudspeaker mounted on each starting block, making that a practical reality.

What factors influence an observer to report that different sensory events "go together" and how can experimenters test whether perceptual features across sensory modalities truly are bound together into a multisensory object? Sections 4.2.1.1 and 4.2.1.2 address factors that influence multisensory integration and ultimately how to test for multisensory binding (see Sect. 4.2.2).

4.2.1.1 Unity Assumption

In the multisensory literature, a hypothesis known as the "unity assumption" posits a process in which an observer considers whether various unisensory stimuli originate from the same object or event (Welch and Warren 1980; Chen and Spence 2017). The degree to which observers infer these unisensory inputs as belonging together can be influenced by stimulus statistics, such as spatial and temporal coincidence, and other top-down influences, such as prior knowledge, context, and expectations. Conceptually, the "unity assumption" provides an intuitive way to probe multisensory binding; based on one's belief, is there evidence that different sensory information should be grouped together to form a cohesive object? However, empirical evidence to support the unity effect is contentious, due, in part, to a confluence of factors listed above. Furthermore, it remains unclear whether the unity assumption requires conscious belief of the observer or just an implicit assessment that the multisensory inputs belong together. Instead, many studies in the past few decades have focused on the individual factors that influence this unity assumption.

4.2.1.2 Stimulus Factors Guiding Multisensory Integration

Based on the findings of electrophysiological studies at the neuronal level in the deep layers of the superior colliculus—a convergence zone of multisensory inputs—three stimulus factors are thought to influence multisensory integration. The first two factors are concerned with whether sensory inputs are spatially and temporally proximal. Typically, multisensory stimuli that are close in space and time would lead to the largest enhancement in neuronal response (Stein and Stanford 2008) and these guiding principles are often referred to as the spatial and temporal rule, respectively. The third factor, inverse effectiveness, postulates that the cross-modal effect is maximal when at least one of the unisensory inputs is only weakly informative when presented on its own.

On the surface, behavioral studies seem to demonstrate that these neural observations extend well to the perceptual domain. For example, in agreement with the inverse effectiveness principle, visual cues are most useful when auditory targets are embedded in environments at low signal-to-noise ratios. There are also many studies that show behavioral facilitations when stimuli are presented close together in time and space (see Wallace and Stevenson 2014 for a review). However, on closer inspection, the effects of spatial colocation and temporal coincidence across modalities can be both subtle and highly task dependent at the psychophysical level.

Spatial Colocation

According to the spatial rule, multisensory integration is maximal when stimuli from different sensory modalities are presented in the same spatial location, i.e., spatial coincidence facilitates multisensory integration. From a neuronal perspective, particularly in relation to the orienting role of the superior colliculus with respect to multisensory integration, this spatial rule makes intuitive sense; each multisensory neuron has multiple excitatory receptive fields and maximal neuronal gain would occur when these receptive fields align spatially (see Willett, Groh, and Maddox, Chap. 5, about the issue of reference frame alignment). However, evidence from behavioral studies suggests that spatial colocation has more of a consistent effect on tasks involving spatial attention or tasks in which space is somehow relevant to the participant's task compared with other nonspatial tasks (Spence 2013). For example, Harrington and Peck (1998) found that human saccadic reaction time was faster when bimodal auditory and visual stimuli were presented together compared with when they were presented alone in each modality, suggesting that there was an enhancement in multisensory integration. Furthermore, they found that saccadic latency increased as spatial distance between the auditory and visual targets increased, supporting the idea that behavioral enhancement is maximal when there is spatial correspondence across sensory modalities. This behavioral benefit extends to spatial cueing studies in which subjects are cued to attend covertly (i.e., while maintaining a central gaze direction) to a particular location. In general, subjects respond more rapidly and more accurately when the cue and target are presented from the same rather than from opposite sides of fixation irrespective of whether the cue and target are the same modality (Spence and McDonald 2004) These results can be interpreted either in terms of the spatial rule or that there is a robust link in cross-modal spatial attention (Spence and Driver 2004).

However, when subjects perform a nonspatial task in which they have to either identify the target stimuli and/or report the temporal content, spatial colocation seems to become unimportant for multisensory integration. For example, in a visual shape discrimination task, performance of the subjects improved when a sound was presented simultaneously along with the visual stimulus, but this improvement was present regardless of whether the location of the sound matched that of the visual stimulus (Doyle and Snowden 2001). Spatial colocation also generally seems not to play a significant role in modulating multisensory integration in many of the classic audiovisual illusion paradigms such as the McGurk effect (e.g., Colin et al. 2001) and flash-beep illusion (e.g., Innes-Brown and Crewther 2009; Kumpik et al. 2014). Although there are examples of nonspatial tasks where integration is modulated by spatial colocation (e.g., Bizley et al. 2012), many of these exceptions required that subjects deploy spatial attention to resolve stimulus competition.

Temporal Coincidence

Similar to the spatial rule, the temporal rule was derived from the temporal tuning functions of individual superior colliculus neurons (Meredith et al. 1987); the gain of a multisensory unit is maximal when the stimulus onset asynchrony is minimal (i.e., close to temporally coincident). Behaviorally, multisensory enhancement has been shown by an increase of stimulus sensitivity when accompanied by a task-irrelevant, but synchronous, stimulus presented in another sensory modality. In one study, auditory signals were better detected when accompanied by a synchronous, although task-irrelevant, light flash (Lovelace et al. 2003). Analogously, in another study, visual sensitivity was only enhanced when the accompanied sound was presented simultaneously and not when the acoustic stimulus was presented 500 ms preceding the visual stimulus. Synchrony between a nonspatialized tone "pip" can also make a visual target "pop" out in cluttered displays. In the "pip and pop" paradigm (Van der Burg et al. 2008), subjects are tasked to search for a visual target (e.g., defined by an orientation) when an array of visual elements is flickered repeatedly and asynchronously with respect to one another. This is often a difficult visual search task because of the clutter of surrounding elements. However, when a tone pip is presented in synchrony with an abrupt temporal change of the visual target, subjectively this makes the target pop out and the visual search becomes quick and efficient, even if the auditory signal is not spatialized and provides no information about where to search in the visual scene.

Is perfect temporal alignment a prerequisite for these multisensory enhancements? So long as cross-modal stimulus pairs are presented in close temporal proximity, audiovisual integration can accelerate reaction time (Colonius and Diederich 2004) as well as improve speech perception (see Grant and Bernstein in Chap. 3). Furthermore, when subjects are asked to make subjective simultaneity judgments of an auditory-visual stimulus pair that is presented with various stimulus onset asynchronies, they are likely to report that these stimuli are simultaneous, even with delays of a hundred milliseconds or more (Wallace and Stevenson 2014). This leads to the concept of a temporal window of integration (also known as temporal binding window and see Baum Miller, and Wallace in Chap. 12 using this concept to probe multisensory dysregulation in different developmental disorders). On a descriptive level, this time window describes probabilistically whether information from different sensory modalities will be integrated (Colonius and Diederich 2010). This temporal window differs in width depending on the stimuli, with it being narrowest for simple flash-beep stimuli and widest for complex multisensory speech stimuli (Wallace and Stevenson 2014). Estimation of the width of these temporal windows also varies markedly across subjects and its variability can be linked to individuals' susceptibility to audiovisual illusions (Stevenson et al. 2012).

Context Influencing Multisensory Integration

When multisensory stimuli have congruent low-level cues, as when each sensory stimulus is either spatially or temporally proximal, many studies have observed behavioral benefits, mostly attributed to the process of multisensory integration.

But is there a cross-modal benefit when multisensory stimuli are congruent at a higher level of cognitive representations; for example, does showing a picture of a dog influence one's perception of a barking sound compared with showing a picture of a guitar? Many behavioral and electrophysiological studies have shown some form of behavioral enhancement or modulated brain activity by this type of semantic congruency (Laurienti et al. 2004; Thelen et al. 2015). These studies generally postulate that semantic congruency can lead to a bound multisensory object due to the learned associations between the individual sensory elements of a single event based on the unity assumption argument. However, even proponents of the unity assumption argument point out that most studies of the effect of semantic congruency on multisensory integration have used unrealistic stimuli lacking ecological validity. Furthermore, the semantic congruency effect seems to be maximal when the auditory stimulus precedes the visual by a few hundred milliseconds as opposed to when they are presented simultaneously (Chen and Spence 2013). Thus, rather than attributing the congruency effect to binding, a more parsimonious explanation is simply semantic priming of one stimulus by the other; for example, hearing a barking sound primes one to react to a picture of a dog (Chen and Spence 2017).

Context can also rapidly modulate multisensory integration on a trial-by-trial basis. For example, the McGurk illusion is reduced in subjects who were first exposed to repeated presentations of incongruent visual lip movement and speech sounds (Nahorna et al. 2015). This contextual influence is surprisingly rapid; an incongruent audiovisual presentation of one syllable suffices to produce a maximum reduction of this McGurk illusion.

4.2.2 Strong Test of Multisensory Binding and Multisensory Objecthood

The evidence presented in Sect. 4.2.1 illustrates the difficulty in teasing apart the way in which information from multiple senses interacts. Distinguishing binding from integration experimentally is nontrivial. Returning to the example of the 100-meter dash, if one is listening for the judge's gunshot in a noisy stadium, it may be easier to achieve with eyes open (i.e., with cross-modal input) rather than closed (i.e., without cross-modal input). Is this caused by the visual and auditory events being perceived as a unified object? It may be, but it is also possible that visual input simply biases the observer toward reporting hearing the shot. Experimenters often use signal detection theory to decouple the change in detection sensitivity (that comes from a perceptual change) from a shift in decision bias. However, if the observer uses one criterion for the gunshot with eyes opened and another criterion with eyes closed, a signal detection analysis may incorrectly conclude zero bias (because the bias shifts in equal amounts in opposite directions for the two conditions) and an increase in sensitivity and thus an underlying change in the sensory representation. This error occurs because the decision model used in standard applications of signal detection theory assumes a fixed, unchanging criterion across conditions (Green and Swets 1966; Durlach and Braida 1969).

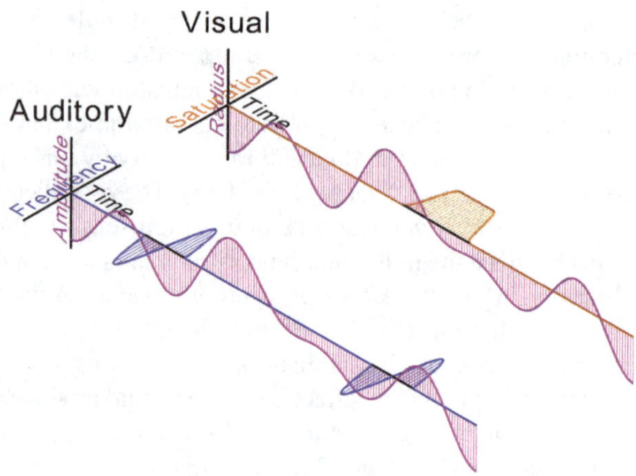

Fig. 4.2 Auditory and visual stimuli with evolving features over time (in the style of those used by Maddox et al. 2015). In this example, the auditory amplitude and visual radius change coherently (*pink*), which facilitates binding into a cross-modal object. The task is based on deviations in the auditory frequency (*blue*) and visual saturation (*orange*), which are orthogonal features to those that facilitate binding (amplitude and radius). In other words, the auditory amplitude (visual radius) provides no task-relevant information about the changes in visual saturation (sound frequency). Thus, improved perception of these orthogonal feature deviants when the amplitude and radius change coherently (versus when they change independently) demonstrates that this coherence leads to binding

To circumvent the experimental confound of bias versus enhancement through binding, Bizley et al. (2016a) suggested that binding can be identified behaviorally by observing cross-modal effects on a stimulus feature that is *orthogonal* to the features that create the binding. In other words, if a subject is presented with an audiovisual stimulus with temporally coherent changes in sound and light intensity, we might expect that these two stimuli would be bound. To demonstrate this, subjects should perform a perceptual judgment on some other feature such as pitch or saturation that changes independently of the intensity (see Fig. 4.2). If the multisensory binding features are task irrelevant (i.e., they provide no information that could aid in a decision about the task-relevant feature), they cannot (meaningfully) influence the decision criterion, and any measured changes in behavior can be assumed to result not from a simple criterion shift but from changes in sensory representation.

4.2.3 *Models of Audiovisual Integration and the Role of Attention*

Behaviorally, multisensory integration of auditory and visual stimuli clearly makes an impact on decision making, but two questions remain to be answered regarding the neural basis of such multisensory processing: (1) where is multisensory information integrated and (2) does attention play a role in multisensory processing?

Theoretically, there are two models that represent the extremes of a spectrum of ways multisensory processing could be realized. In one extreme, the late integration model postulates that sensory information is processed separately (e.g., in the sensory cortices) and those unisensory sources of evidence are integrated at a later stage (e.g., in higher order cortical areas). In this framework, auditory and visual information can be weighted through unimodal attention at the integration stage. Alternatively, the early integration model postulates that multisensory integration begins early at the unisensory cortices (or before, in subcortical areas) with cross-modal inputs modifying the representations of incoming stimuli. In this early integration framework, integrated sensory information across modalities contributes to the accumulation of sensory evidence, and decision making in higher order cortical areas is thus based on an already multisensory representation (Bizley et al. 2016b). This bottom-up framework suggests that early multisensory binding can occur even independently of attention (Atilgan et al. 2018), even though selective attention can act to further shape and define this representation.

Whether multisensory information is integrated early in the sensory cortices and/or at later stages by combining independent unisensory information may depend on the specific task. However, the early integration model provides the necessary neural substrate for multisensory binding and the formation of a multisensory object. Hence, the early integration model provides a theoretical conceptualization of how multisensory binding could be realized. Because attention operates at the level of objects, if attention was applied to this multisensory representation, this would imply that all cross-modal features associated with the multisensory object would also be enhanced.

Although there is substantial physiological (Bizley et al. 2007; Lakatos et al. 2007) and anatomical (Bizley et al. 2007; Falchier et al. 2010) evidence to demonstrate that multisensory processing occurs in primary and nonprimary sensory cortices, behavioral paradigms (Raposo et al. 2012) and neuroimaging (Rohe and Noppeney 2015) have provided evidence in favor of integration occurring in higher brain areas. Generally speaking, the neural basis for different kinds of multisensory integration remains underexplored. Therefore, when performing behavioral experiments, it is important to conceptually separate multisensory binding from general multisensory integration so that the findings can better inform neurophysiologists on discovering the different neural architectures that support multisensory processing. Even though this distinction is not often made (with some exceptions; e.g., Odgaard et al. 2004), previous work can be reinterpreted in this framework (see Sect. 4.3).

4.3 Reinterpreting Classic Audiovisual Illusions: Binding or Multisensory Integration?

Many multisensory behavioral studies focus on illusions that place cross-sensory information in conflict to understand how the brain normally integrates sensory information. Often, the effects are so perceptually salient that researchers assume not only that information has been integrated across the sensory modalities concerned but that it has also been bound to form a strong cohesive multisensory object.

In other cases, authors have used the terms integration and binding interchangeably. In this section, three well-known multisensory paradigms are examined to see whether there is evidence that these illusions pass the strong test of multisensory binding as previously outlined in Sect. 4.2.2.

4.3.1 Ventriloquism

In the ventriloquism illusion, the observer's perception of the location of a sound source is "captured" (or if not fully captured, biased) by a visual stimulus. However, does this illusion demonstrate binding of the auditory and visual signals into a multisensory object, as often stated? It has been demonstrated that observers combine the visual and auditory location cues in an optimal Bayesian manner. In fact, the ventriloquism effect can be reversed when the visual stimuli used are so blurred that their spatial estimate is less reliable than that of the auditory cues (Alais and Burr 2004). If observers are asked to provide a location estimate to both the auditory and visual sources, the location of the sound is less biased than if only one location was asked from the subject (see Alais and Burr, Chap. 2). These findings support the late processing model, suggesting that independent estimates are made for each modality and a task-modulated decision-making stage integrates and weighs evidence across sensory modalities. This contrasts with the scenario where the auditory and visual sources are bound in early processing, resulting in a single location associated with the unified multisensory object, independent of whether the observers are providing an auditory or a visual location estimate. Furthermore, behavioral modeling using casual inference suggests that these sensory estimates are maintained separately (Körding et al. 2007). Finally, reward expectation (Bruns et al. 2014) and emotional valence (Maiworm et al. 2012) can also modulate the ventriloquist effect, suggesting that, at least in part, top-down factors could modulate decision making, consistent with the late integration model.

Evidence from a recent functional magnetic resonance imaging study shows a neural hierarchy of multisensory processes in a ventriloquist paradigm (Rohe and Noppeney 2015). At the bottom of this neural hierarchy, location is represented as if the auditory and visual signals are generated by independent sources. These are processed in the auditory and visual areas. At the next stages of this neural hierarchy, in higher parietal cortices, location is first estimated by integrating the auditory and visual signals by assuming that these signals originate from a common source, weighted by their bottom-up sensory reliabilities. Then the uncertainty about whether the auditory and visual signals are generated by common or independent sources is finally taken into account. This suggests that multisensory interactions are pervasive but governed by different computational principles across the cortical hierarchy. Future studies should further separate out whether the sensory cortex modulation in the ventriloquist illusion is primarily due to amodal attention modulation from the higher cortical areas or specific multisensory binding effects that can exist independent of attention.

4.3.2 *Sound-Induced Flash Illusion*

In the sound-induced flash illusion, brief auditory and visual stimuli are presented rapidly in succession, and the number of auditory stimuli can influence the reported number of visual stimuli. The nature of these illusory flashes is not totally clear; subjectively, observers often report that the illusionary flashes are different from the real flashes. Indeed, if the experimenter offers a third "not-one, not-two" option, many subjects choose that instead (van Erp et al. 2013). Nonetheless, using signal detection theory, it has been demonstrated that the illusory flashes affect sensitivity (and not only bias) to the number of flashes perceived, suggesting that the illusion is due in part to a change in the multisensory sensory representation (McCormick and Mamassian 2008; Kumpik et al. 2014). However, the caveat discussed in Sect. 4.2.2 must be applied here; if the number of sounds systematically shifts the decision criteria toward the number of perceived sounds, what appears to be a sensitivity change could, in fact, be a systematic switching of the decision criterion (e.g., listeners shifting decision criteria for the number of flashes perceived at a trial-by-trial level based on the number of sounds perceived). In contrast to the majority of sound-induced flash illusion experiments that do not fulfill the aforementioned strong test of multisensory binding, a few studies do test perception of another stimulus dimension in the context of the illusion. Mishra et al. (2013) asked observers to report not only the number of flashes but also their color, which is an orthogonal feature dimension. Another study tested contrast perception, again an orthogonal dimension, in addition to the number of events and found that the illusion is likely explained by both an early perceptual change and a late criterion shift in the decision-making process (McCormick and Mamassian 2008).

Human neurophysiological studies provide further support for the early integration model playing a key role in the sound-induced flash illusion (Mishra et al. 2007). Specifically, a difference in event-related potentials from electroencephalographical recordings derived to isolate neural activity associated with the illusory flash revealed an early modulation of activities in the visual cortex after the second sound. Furthermore, the amplitude of this different waveform is larger in the groups of subjects who saw the illusory flash more frequently, pointing to consistent individual differences that underlie this multisensory integration. Similar to the behavioral observation, the overall pattern of cortical activity associated with the induced illusory flash differed markedly from the pattern evoked by a real second flash. There is evidence that this illusion is generated by a complex interaction between the primary sensory cortices and the multimodal superior temporal areas (see Beauchamp, Chap. 8, for a review). Perhaps future animal studies may shed more light on the neural basis of multisensory integration or binding underlying this illusion, although to do so would require that investigations be made in the context of a behavioral paradigm to ensure that there was a single-trial readout of whether the illusion was perceived on that trial.

4.3.3 McGurk Effect

The McGurk effect is often striking; watching a video of a mouth movement that does not match the auditory syllable presented can lead to a percept that is neither of the veridical unisensory percepts but is instead a third one. In their original paper (McGurk and MacDonald 1976), the investigators reported a "fused" percept arising out of a pair of incongruent auditory and visual speech syllables. This illusion has been widely used to understand different aspects of audiovisual processing, in part because the illusion can be measured by a few repetitions of a simple language stimulus. Not generally discussed, however, is the inhomogeneity of this effect across individuals as well as the efficacy of the illusion across different stimuli (Magnotti and Beauchamp 2015). In fact, although 98% of adult subjects in the original study responded to an intermediate/da/ percept when an auditory /ba/ and a visual/ga/ stimuli were presented, only 81% responded to an intermediate/ta/ percept when the unvoiced counterparts were presented (i.e., an auditory /pa/ and a visual/ka/). Across individuals, some participants almost always perceive the McGurk effect, whereas others rarely do (Mallick et al. 2015). Are these discrepancies caused by differences in multisensory binding across individuals or differences in how they integrate sensory information?

Studying the individual differences across subjects in susceptibility to the McGurk illusion can be more revealing about the nature of the underlying multiple processes than interpreting data at the group level. Meta-analyses across different studies using the same paradigm but with different stimulus parameters are equally important. One potential source of variability across experiments using the McGurk illusion is that the contribution of the unisensory components is not explicitly measured (Tiippana 2014). In fact, McGurk and MacDonald commented in their original paper (1976) that by their own observations, lip movements for /ga/ are frequently misread as /da/ in the absence of auditory input, although they did not measure speech reading performance in that study. Similarly, the variations in the specific acoustic samples used in different experiments have also not been examined with respect to their phoneme categorization. Furthermore, there are variations in the nature of the response; listeners respond differently depending on whether they are presented with two alternative forced choice, making a third "other" choice or an open descriptive response. Nevertheless, perceptually assessing the phonemic feature is still not orthogonal to the feature that links the auditory and visual stimuli.

Instead of asking subjects about the phonemic percept of congruent versus incongruent auditory and visual stimuli, one could ask the subjects to judge the pair's temporal synchrony. In this way, a temporal window of integration can be measured for both the congruent and incongruent multisensory stimuli (cf. Sect. 4.2.1.2). Importantly, temporal synchrony is an orthogonal feature of the phonemic judgment that links the auditory and visual stimuli and would satisfy the strong test of multisensory binding as outlined in Sect. 4.2.2. Furthermore, it has been shown that the temporal window of integration correlates well with the amount of McGurk illusion perceived across subjects as well as with other illusions such as the sound-induced flash illusion as described in Sect. 4.3.2 (Stevenson et al. 2012). In one study, the temporal window was measured to be much narrower for incongruent pairs compared with congruent stimuli. In other words, subjects were more sensitive to asynchronies in incongruent audiovisual syllables than in congruent ones. This suggests that the McGurk-incongruent stimuli are not integrated as strongly as the congruent stimuli when the subjects were asked only to attend to the simultaneity of the stimuli and not the content of the utterances (van Wassenhove et al. 2007). This finding is also suggestive of binding at least of congruent McGurk stimuli but leaves questions about binding of incongruent stimuli when illusions are not perceived. Higher level contextual effects (Nahorna et al. 2012, 2015) as well as visual attention (Tiippana et al. 2004) can also influence the strength of the McGurk effect, casting further doubt on binding as the sole explanation for this illusion.

Converging neurophysiological and computational modeling evidence suggests that audiovisual speech processing is best modeled as a two-stage process (Peelle and Sommers 2015; Magnotti and Beauchamp 2017). Visual input could alter the processing of auditory information through early integration instantiated by a cross-modal reset (or perturbation) of low-frequency neural oscillations in the auditory cortex (Mercier et al. 2015; also see Keil and Senkowski, Chap. 10, for an in-depth discussion). However, the cues related to speech gestures are better modeled as a late integration process, with the posterior superior temporal sulcus likely playing a role in weighting individual auditory and visual inputs (Nath and Beauchamp 2012; also see Beauchamp, Chap. 8). In summary, an illusory trial whereby an intermediate percept was reported using McGurk stimuli does not necessarily show that the auditory and visual information were bound (even though this is often referred to as the "fused" percept, implicitly assuming binding). Rather, the report of the third percept is evidence that auditory and visual information have been integrated and influenced the decision making in the syllable classification task. Paradoxically, the nonillusionary (or the nonfused) trials could have elicited a bound percept, especially if the auditory and visual stimuli were presented with relatively low asynchrony, even though the integrated audiovisual information did not result in a third syllable categorization, possibly due to the relative strength of the unisensory evidence. In other words, the presence of the McGurk effect is evidence for multisensory integration (but maybe not binding). Furthermore, the absence of the McGurk effect is not necessarily evidence for two distinct objects. Future studies should aim to explicitly separate how the early and late integration models could affect McGurk perception.

4.4 Competing Objects in the Audiovisual Scene

Even though most naturalistic environments comprise numerous auditory and visual objects vying for our limited attentional resources, stimulus competition is not often examined in the laboratory setting. One exception is the "pip and pop" paradigm as discussed in Sect. 4.2.1.2. In that case, an auditory stimulus helps resolve visual competition. However, there are relatively few multisensory experiments that address how visual information resolves competition between auditory stimuli.

4.4.1 Prediction from Unisensory Object-Based Attention Theory

Introducing stimulus competition allows experimenters to draw on the object-based attention literature (Shinn-Cunningham et al. 2017). Doing so leads to predictions about the processing advantages offered by binding auditory and visual stimuli into a single audiovisual object. Stimulus competition also provides a more naturalistic and taxing environment, making the perceptual benefit of multisensory binding easier to detect.

Based on the theory developed in object-based attention literature (cf. Sect. 4.2.1), the expectation would be that if the auditory and visual stimuli were bound as a multisensory object, features in both sensory modalities belonging to the same object would be enhanced. Conversely, if the auditory and visual stimuli came from different objects, even though they can be integrated to make a judgment, there should be a measurable behavioral cost associated with dividing attention across two objects.

Chamber music provides a situation in which we can test how visual input shapes the ability of listeners to parse a complex auditory scene. A string quartet comprises four instruments: first violin, second violin, viola, and cello. The bowing action of each player yields some temporal information about each part of the music. If the observer were to look at the movement of the first violinist while listening to the cello part, a behavioral cost would be expected due to attention spreading across two objects (here, players). Conversely, if the observer were to both watch the movements of and listen to the cellist, a behavioral benefit would be expected. The bow would provide information about the timing of the cellist's notes; therefore, to test binding, an experimenter could ask the observer to pick when the music is modulated to another key, something not discernible from the bowing alone. It is temporal coherence that binds the auditory and visual representations of the cellist, and judging temporal aspects of the music is susceptible to bias from the visual modality as vision directly informs the timing of the notes. However, listening for a specific change in key tests an orthogonal dimension to the temporal feature underlying binding because the player's bowing motion provides no information about pitch. In this hypothetical experiment, if the observer were better at parsing the specific cello notes played when looking at the cellist, then it meets the strong test of binding and suggests that the observer was attending to a bound multisensory object.

4.4.2 Effect of Spatial Cues

In a previous sound-induced flash illusion experiment, it was concluded that the probability of an illusory percept was not influenced by the degree of spatial separation between the auditory and visual stimuli (Innes-Brown and Crewther 2009). A similar study suggested that the visual sensitivity, not the audiovisual spatial proximity, was the determining factor of the illusory percept (Kumpik et al. 2014).

However, one recent study that used competing stimuli showed that the sound-induced flash illusion can be modulated by spatial congruence between the auditory and visual stimuli (Bizley et al. 2012). Subjects were asked to report the number of flashes and beeps perceived as opposed to an orthogonal feature like color. Nevertheless, the results show that stimulus competition can provide different outcomes; in this case, the influence of spatial congruence for multisensory processing in a complex auditory scene.

4.4.3 Effect of Temporal Coherence

By using artificial stimuli with naturalistic dynamics, it is possible to test the potential benefits of temporal coherence between auditory and visual stimuli when performing a task based on an orthogonal feature. One such study used competing stimuli to examine whether a visual stimulus being temporally coherent with the target auditory stream would show better behavioral performance compared with when the visual stream was temporally coherent with the masker auditory stream (Maddox et al. 2015). Subjects were asked to report brief pitch or timbre deviants in one of two ongoing independently amplitude-modulated sound streams; the timing of the brief pitch or timbre deviants were independent of the amplitude modulation imposed on each of the sound streams. They were also instructed to attend a radius-modulated disk that changed coherently with the amplitude of either the target stream or the masker stream and were also asked to report occasional color changes of the visual disk. Performance was better when the visual disk was temporally coherent with the target auditory stream compared with when it was coherent with the masker stream. Importantly, because the modulations of the visual stimulus were orthogonal to the pitch or timbre deviants and offered no information to their timing (cf. Fig. 4.2), Maddox et al. (2015) suggested that the behavioral benefit observed was through binding of the temporally coherent audiovisual streams forming an audiovisual object whose properties were subsequently enhanced (thus satisfying the "strong test" for binding). In other words, when the auditory target stream and the visual stream were bound into a single multisensory object, performance was improved because the observers no longer had to divide attention across two sensory objects (Fig. 4.3). Although not tested, they also hypothesized that future experiments should be able to show an equivalent auditory enhancement of visual perception, hinted at already by the pip and pop illusion as discussed in Sect. 4.2.1.2.

4.5 Summary

The temporal evolution of the auditory and visual information available in real-life cocktail party environments is inevitably complex. Experimental paradigms used in laboratories should strive to expand beyond the canonical multisensory studies to

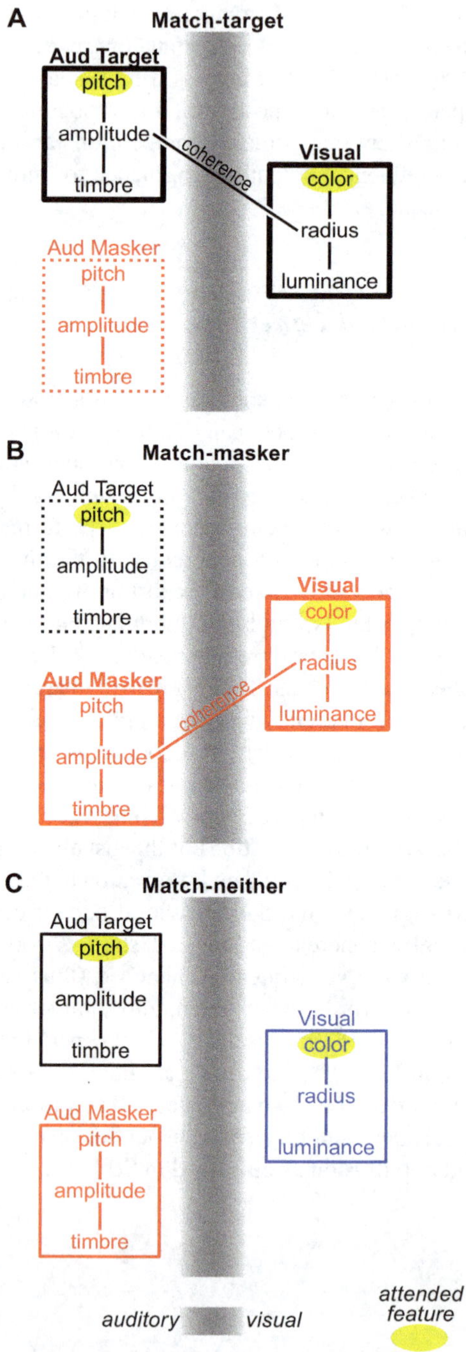

Fig. 4.3 Conceptual model of binding leading to cross-modal object formation in a task whereby subjects were asked to attend to a pitch change in an auditory (Aud; *left*) target stream while ignoring an auditory masker stream and a color perturbation in the visual stimulus (*right*); attended

address this complexity. As presented in Sect. 4.3, the collective findings from these well-studied illusions still leave much to be learned regarding the nature by which multisensory stimuli are grouped and processed and to what extent binding is a prerequisite for such phenomena.

One particular issue is the use of brief stimuli which are not representative of naturalistic signals. Although this is sometimes seen as a necessary sacrifice for experimental control, this was not always the case. In one of the classic ventriloquism illusion experiments (Jackson 1953), it was shown that the bias in the perceived location of the sound of a steam whistle accompanied by the sight of steam was larger than the bias of a bell sound accompanied by a light pulse. This finding has been interpreted over the years to mean that arbitrary combinations of auditory and visual stimuli with no strong assumption of unity (see Sect. 4.2.1.1) lead to less effective multisensory integration. Alternatively, the temporally rich and naturally occurring combination of the sight and sound of a kettle may promote object formation due to their temporal coherence. The original experimental setup of realistic and temporally rich stimuli—steam coming out of a singing kettle—might be too elaborate to replicate in most psychophysics laboratories.

"Three brass tubes each containing a 50-watt soldering iron element were inserted into the rear of each whistle. Water was led from a header tank through a battery of taps, controllable by the experimenter, to a fine jet which was pressed into a tufnol sleeve round the central of the three heater units in each whistle. Thus, when the heater units had been allowed to attain their working temperature, momentary release of one tap caused a visible cloud of steam to rise from the corresponding whistle" (Jackson 1953, p. 57).

However, modern experimenters need not resort to such drastic measures. Digitally presenting coherent stimuli across time that can be parametrically manipulated is a relatively recent capability, especially if synchronicity of the auditory and visual stimuli are to be guaranteed. These technical challenges should no longer limit experimental possibilities, meaning researchers can now explore temporally rich stimuli and move away from the canonical paradigms involving only brief stimuli. This will also present an opportunity for experimenters to assess the dynamic evolution of multisensory integration over time driven by accumulation of sensory evidence.

Fig. 4.3 (continued) features highlighted in *yellow ellipses*. *Boxes:* connected sets of features in each sensory stream. Cross-modal temporal coherence, if present, is shown as a *line* connecting the coherent features. (**A**) Amplitude of the auditory target stream is coherent with the visual size (match-target condition); (**B**) Amplitude of the auditory masker stream is coherent with the visual size (match-masker condition); (**C**) No visual features are coherent with any features in the auditory streams. Cross-modal binding of the coherent auditory and visual streams enhances each stream's features, resulting in a benefit in the match-target condition (**A**), a disadvantage in the match-masker condition (**B**), and no effect in the match-neither condition (**C**). Enhancement/suppression resulting from object formation is reflected in the strength of the box drawn around each stream's feature (i.e., *solid lines*, enhancement, *dashed lines*, suppression). Reproduced from Maddox et al. (2015)

Despite Cherry's (1953) listing of visual cues as a potential solution to the cocktail party problem more than half a century ago, only recent audiovisual paradigms have started addressing how visual information can help listeners segregate sounds in complex auditory scenes. With new experimental paradigms and more specific frameworks for delineating audiovisual integration and binding, the field is poised to gain substantial insights into how humans communicate in noisy everyday environments.

Compliance with Ethics Requirements Adrian K. C. Lee declares that he has no conflict of interest.

Ross K. Maddox declares that he has no conflict of interest.

Jennifer K. Bizley declares that she has no conflict of interest.

References

Alais, D., Blake, R., & Lee, S. H. (1998). Visual features that vary together over time group together over space. *Nature Neuroscience, 1*(2), 160–164.

Alais, D., & Burr, D. (2004). The ventriloquist effect results from near-optimal bimodal integration. *Current Biology, 14*(3), 257–262.

Atilgan, H., Town, S. M., Wood, K. C., Jones, G. P., Maddox, R. K., Lee, A. K. C., & Bizley, J. K. (2018). Integration of visual information in auditory cortex promotes auditory scene analysis through multisensory binding. *Neuron, 97*, 640–655.

Behrmann, M., Zemel, R. S., & Mozer, M. C. (1998). Object-based attention and occlusion: Evidence from normal participants and a computational model. *Journal of Experimental Psychology: Human Perception and Performance, 24*(4), 1011–1036.

Bizley, J. K., & Cohen, Y. E. (2013). The what, where and how of auditory-object perception. *Nature Reviews Neuroscience, 14*(10), 693–707.

Bizley, J. K., Nodal, F. R., Bajo, V. M., Nelken, I., & King, A. J. (2007). Physiological and anatomical evidence for multisensory interactions in auditory cortex. *Cerebral Cortex, 17*(9), 2172–2189.

Bizley, J. K., Shinn-Cunningham, B. G., & Lee, A. K. C. (2012). Nothing is irrelevant in a noisy world: Sensory illusions reveal obligatory within-and across-modality integration. *Journal of Neuroscience, 32*(39), 13402–13410.

Bizley, J. K., Jones, G. P., & Town, S. M. (2016a). Where are multisensory signals combined for perceptual decision-making? *Current Opinion in Neurobiology, 40*, 31–37.

Bizley, J. K., Maddox, R. K., & Lee, A. K. C. (2016b). Defining auditory-visual objects: Behavioral tests and physiological mechanisms. *Trends in Neuroscience, 39*(2), 74–85.

Blake, R., & Lee, S.-H. (2005). The role of temporal structure in human vision. *Behavioral and Cognitve Neuroscience Reviews, 4*(1), 21–42.

Blaser, E., Pylyshyn, Z. W., & Holcombe, A. O. (2000). Tracking an object through feature space. *Nature, 408*(6809), 196–199.

Bruns, P., Maiworm, M., & Röder, B. (2014). Reward expectation influences audiovisual spatial integration. *Attention Perception & Psychophysics, 76*(6), 1815–1827.

Chandrasekaran, C., Trubanova, A., Stillittano, S., Caplier, A., & Ghazanfar, A. A. (2009). The natural statistics of audiovisual speech. *PLoS Computational Biology, 5*(7), e1000436.

Chen, Y. C., & Spence, C. (2013). The time-course of the cross-modal semantic modulation of visual picture processing by naturalistic sounds and spoken words. *Multisensory Research, 26*, 371–386.

Chen, Y.-C., & Spence, C. (2017). Assessing the role of the 'unity assumption' on multisensory integration: A review. *Frontiers in Psychology, 8*, 445.
Cherry, E. (1953). Some experiments on the recognition of speech, with one and with two ears. *The Journal of the Acoustical Society of America, 25*(5), 975–979.
Colin, C., Radeau, M., Deltenre, P., & Morais, J. (2001). Rules of intersensory integration in spatial scene analysis and speechreading. *Psychologica Belgica, 41*, 131–144.
Colonius, H., & Diederich, A. (2004). Multisensory interaction in saccadic reaction time: A time-window-of-integration model. *Journal of Cognitive Neuroscience, 16*(6), 1000–1009.
Colonius, H., & Diederich, A. (2010). The optimal time window of visual-auditory integration: A reaction time analysis. *Frontiers in Integrative Neuroscience, 4*, 11.
Culling, J. F., & Stone, M. A. (2017). Energetic masking and masking release. In J. C. Middlebrooks, J. Z. Simon, A. N. Popper, & R. R. Fay (Eds.), *The auditory system at the cocktail party* (pp. 41–74). New York: Springer International.
Cusack, R., & Roberts, B. (2000). Effects of differences in timbre on sequential grouping. *Perception & Psychophysics, 62*(5), 1112–1120.
Desimone, R., & Duncan, J. (1995). Neural mechanisms of selective visual attention. *Annual Review of Neuroscience, 18*(1), 193–222.
Doyle, M. C., & Snowden, R. J. (2001). Identification of visual stimuli is improved by accompanying auditory stimuli: The role of eye movements and sound location. *Perception, 30*(7), 795–810.
Durlach, N. I., & Braida, L. D. (1969). Intensity perception. I. Preliminary theory of intensity resolution. *The Journal of the Acoustical Society of America, 46*(2), 372–383.
Falchier, A., Schroeder, C. E., Hackett, T. A., Lakatos, P., Nascimento-Silva, S., Ulbert, I., Karmos, G., & Smiley, J. F. (2010). Projection from visual areas V2 and prostriata to caudal auditory cortex in the monkey. *Cerebral Cortex, 20*(7), 1529–1538.
Feldman, J. (2003). What is a visual object? *Trends in Cognitive Sciences, 7*(6), 252–256.
Green, D., & Swets, J. (1966). *Signal detection theory and psychophysics*. New York: Wiley.
Harrington, L. K., & Peck, C. K. (1998). Spatial disparity affects visual-auditory interactions in human sensorimotor processing. *Experimental Brain Research, 122*(2), 247–252.
Howard, I., & Templeton, W. (1966). *Human spatial orientation*. New York: Wiley.
Innes-Brown, H., & Crewther, D. (2009). The impact of spatial incongruence on an auditory-visual illusion. *PLoS One, 4*(7), e6450.
Jackson, C. V. (1953). Visual factors in auditory localization. *Quarterly Journal of Experimental Psychology, 5*(2), 52–65.
Knudsen, E. I. (2007). Fundamental components of attention. *Annual Review of Neuroscience, 30*, 57–78.
Körding, K. P., Beierholm, U., Ma, W. J., Quartz, S., Tenenbaum, J. B., & Shams, L. (2007). Causal inference in multisensory perception. *PLoS One, 2*(9), e943.
Kumpik, D. P., Roberts, H. E., King, A. J., & Bizley, J. K. (2014). Visual sensitivity is a stronger determinant of illusory processes than auditory cue parameters in the sound-induced flash illusion. *Journal of Vision, 14*(7), 12.
Lakatos, P., Chen, C.-M., Connell, M. N., Mills, A., & Schroeder, C. E. (2007). Neuronal oscillations and multisensory interaction in primary auditory cortex. *Neuron, 53*(2), 279–292.
Laurienti, P. J., Kraft, R. A., Maldjian, J. A., Burdette, J. H., & Wallace, M. T. (2004). Semantic congruence is a critical factor in multisensory behavioral performance. *Experimental Brain Research, 158*(4), 405–414.
Lee, A. K. C. (2017). Imaging the listening brain. *Acoustics Today, 13*(3), 35–42.
Lovelace, C. T., Stein, B. E., & Wallace, M. T. (2003). An irrelevant light enhances auditory detection in humans: A psychophysical analysis of multisensory integration in stimulus detection. *Cognitive Brain Research, 17*(2), 447–453.
Maddox, R. K., & Shinn-Cunningham, B. G. (2012). Influence of task-relevant and task-irrelevant feature continuity on selective auditory attention. *Journal of the Association for Research in Otolaryngology, 13*(1), 119–129.

Maddox, R. K., Atilgan, H., Bizley, J. K., & Lee, A. K. C. (2015). Auditory selective attention is enhanced by a task-irrelevant temporally coherent visual stimulus in human listeners. *eLife, 4*, e04995.
Magnotti, J. F., & Beauchamp, M. S. (2015). The noisy encoding of disparity model of the McGurk effect. *Psychonomic Bulletin & Review, 22*, 701–709.
Magnotti, J. F., & Beauchamp, M. S. (2017). A causal inference model explains perception of the McGurk effect and other incongruent audiovisual speech. *PLoS Computional Biology, 13*(2), e1005229.
Maiworm, M., Bellantoni, M., Spence, C., & Röder, B. (2012). When emotional valence modulates audiovisual integration. *Attention, Perception & Psychophysics, 74*(6), 1302–1311.
Mallick, D. B., Magnotti, J. F., & Beauchamp, M. S. (2015). Variability and stability in the McGurk effect: Contributions of participants, stimuli, time, and response type. *Psychonomic Bulletin & Review, 22*(5), 1299–1307.
McCormick, D., & Mamassian, P. (2008). What does the illusory-flash look like? *Vision Research, 48*(1), 63–69.
McGurk, H., & MacDonald, J. (1976). Hearing lips and seeing voices. *Nature, 264*(5588), 746–748.
Mercier, M. R., Molholm, S., Fiebelkorn, I. C., Butler, J. S., Schwartz, T. H., & Foxe, J. J. (2015). Neuro-oscillatory phase alignment drives speeded multisensory response times: An electrocorticographic investigation. *The Journal of Neuroscience, 35*(22), 8546–8557.
Meredith, M. A., Nemitz, J. W., & Stein, B. E. (1987). Determinants of multisensory integration in superior colliculus neurons. I. Temporal factors. *The Journal of Neuroscience, 7*(10), 3215–3229.
Micheyl, C., & Oxenham, A. J. (2010). Objective and subjective psychophysical measures of auditory stream integration and segregation. *Journal of the Association for Research in Otolaryngology, 11*(4), 709–724.
Middlebrooks, J. C., Simon, J. Z., Popper, A. N., & Fay, R. R. (2017). *The auditory system at the cocktail party*. New York: Springer International.
Mishra, J., Martinez, A., Sejnowski, T. J., & Hillyard, S. A. (2007). Early cross-modal interactions in auditory and visual cortex underlie a sound-induced visual illusion. *The Journal of Neuroscience, 27*(15), 4120–4131.
Mishra, J., Martinez, A., & Hillyard, S. A. (2013). Audition influences color processing in the sound-induced visual flash illusion. *Vision Research, 93*, 74–79.
Nahorna, O., Berthommier, F., & Schwartz, J.-L. (2012). Binding and unbinding the auditory and visual streams in the McGurk effect. *The Journal of the Acoustical Society of America, 132*(2), 1061–1077.
Nahorna, O., Berthommier, F., & Schwartz, J.-L. (2015). Audio-visual speech scene analysis: Characterization of the dynamics of unbinding and rebinding the McGurk effect. *The Journal of the Acoustical Society of America, 137*(1), 362–377.
Nath, A. R., & Beauchamp, M. S. (2012). A neural basis for interindividual differences in the McGurk effect, a multisensory speech illusion. *NeuroImage, 59*(1), 781–787.
O'Craven, K. M., Downing, P. E., & Kanwisher, N. (1999). fMRI evidence for objects as the units of attentional selection. *Nature, 401*(6753), 584–587.
Odgaard, E. C., Arieh, Y., & Marks, L. E. (2004). Brighter noise: Sensory enhancement of perceived loudness by concurrent visual stimulation. *Cognitive, Affective, & Behavorial Neuroscience, 4*(2), 127–132.
Peelle, J. E., & Sommers, M. S. (2015). Prediction and constraint in audiovisual speech perception. *Cortex, 68*, 169–181.
Raposo, D., Sheppard, J. P., Schrater, P. R., & Churchland, A. K. (2012). Multisensory decision-making in rats and humans. *The Journal of Neuroscience, 32*(11), 3726–3735.
Rohe, T., & Noppeney, U. (2015). Cortical hierarchies perform Bayesian causal inference in multisensory perception. *PLoS Biology, 13*(2), e1002073.
Shamma, S. A., Elhilali, M., & Micheyl, C. (2011). Temporal coherence and attention in auditory scene analysis. *Trends in Neuroscience, 34*(3), 114–123.

Shams, L., Kamitani, Y., & Shimojo, S. (2000). Illusions. What you see is what you hear. *Nature, 408*(6814), 788.

Shinn-Cunningham, B. G., Best, V., & Lee, A. K. C. (2017). Auditory object formation and selection. In J. C. Middlebrooks, J. Z. Simon, A. N. Popper, & R. R. Fay (Eds.), *The auditory system at the cocktail party* (pp. 7–40). New York: Springer International.

Spence, C. (2013). Just how important is spatial coincidence to multisensory integration? Evaluating the spatial rule. *Annals of the New York Academy of Sciences, 1296*(1), 31–49.

Spence, C., & Driver, J. (2004). *Crossmodal space and crossmodal attention*. Oxford: Oxford University Press.

Spence, C., & McDonald, J. (2004). The crossmodal consequences of the exogenous spatial orienting of attention. In G. A. Calvert, C. Spence, & B. E. Stein (Eds.), *The handbook of multisensory processing* (pp. 3–25). Cambridge, MA: MIT Press.

Stein, B. E., & Stanford, T. R. (2008). Multisensory integration: Current issues from the perspective of the single neuron. *Nature Reviews Neuroscience, 9*(4), 255–266.

Stevenson, R. A., Zemtsov, R. K., & Wallace, M. T. (2012). Individual differences in the multisensory temporal binding window predict susceptibility to audiovisual illusions. *Journal of Experimental Psychology: Human Perception and Performance, 38*(6), 1517–1529.

Thelen, A., Talsma, D., & Murray, M. M. (2015). Single-trial multisensory memories affect later auditory and visual object discrimination. *Cognition, 138*, 148–160.

Tiippana, K. (2014). What is the McGurk effect? *Frontiers in Psychology, 5*, 725.

Tiippana, K., Andersen, T. S., & Sams, M. (2004). Visual attention modulates audiovisual speech perception. *European Journal of Cognitive Psychology, 16*(3), 457–472.

Treisman, A. (1998). Feature binding, attention and object perception. *Philosophical Transactions of the Royal Society B: Biological Sciences, 353*(1373), 1295–1306.

Van der Burg, E., Olivers, C. N. L., Bronkhorst, A. W., & Theeuwes, J. (2008). Pip and pop: Nonspatial auditory signals improve spatial visual search. *Journal of Experimental Psychology: Human Perception and Performance, 34*(5), 1053–1065.

van Erp, J. B. F., Philippi, T. G., & Werkhoven, P. (2013). Observers can reliably identify illusory flashes in the illusory flash paradigm. *Experimental Brain Research, 226*(1), 73–79.

van Wassenhove, V., Grant, K. W., & Poeppel, D. (2007). Temporal window of integration in auditory-visual speech perception. *Neuropsychologia, 45*(3), 598–607.

Wallace, M. T., & Stevenson, R. A. (2014). The construct of the multisensory temporal binding window and its dysregulation in developmental disabilities. *Neuropsychologia, 64*, 105–123.

Welch, R. B., & Warren, D. H. (1980). Immediate perceptual response to intersensory discrepancy. *Psychological Bulletin, 80*(3), 638–667.

Chapter 5
Hearing in a "Moving" Visual World: Coordinate Transformations Along the Auditory Pathway

Shawn M. Willett, Jennifer M. Groh, and Ross K. Maddox

Abstract This chapter reviews the literature on how auditory signals are transformed into a coordinate system that facilitates interactions with the visual system. Sound location is deduced from cues that depend on the position of the sound with respect to the head, but visual location is deduced from the pattern of light illuminating the retina, yielding an eye-centered code. Connecting sights and sounds originating from the same position in the physical world requires the brain to incorporate information about the position of the eyes with respect to the head. Eye position has been found to interact with auditory signals at all levels of the auditory pathway that have been tested but usually yields a code that is in a hybrid reference frame: neither head nor eye centered. Computing a coordinate transformation, in principle, may be easy, which could suggest that the looseness of the computational constraints may permit hybrid coding. A review of the behavioral literature addressing the effects of eye gaze on auditory spatial perception and a discussion of its consistency with physiological observations concludes the chapter.

S. M. Willett (✉)
Department of Neurobiology, Center for Cognitive Neuroscience, Duke University, Durham, NC, USA
e-mail: shawn.willett@duke.edu

J. M. Groh
Department of Neurobiology, Center for Cognitive Neuroscience, Duke University, Durham, NC, USA

Department of Psychology and Neuroscience, Center for Cognitive Neuroscience, Duke University, Durham, NC, USA
e-mail: jmgroh@duke.edu

R. K. Maddox
Department of Biomedical Engineering, University of Rochester, Rochester, NY, USA

Department of Neuroscience, University of Rochester, Rochester, NY, USA

Del Monte Institute for Neuroscience, University of Rochester, Rochester, NY, USA

Center for Visual Science, University of Rochester, Rochester, NY, USA
e-mail: ross.maddox@rochester.edu

A. K. C. Lee et al. (eds.), *Multisensory Processes*, Springer Handbook of Auditory Research 68, https://doi.org/10.1007/978-3-030-10461-0_5

Keywords Eye centered · Frontal eye field · Head centered · Hybrid · Inferior colliculus · Intraparietal cortex · Multimodal · Multisensory integration · Reference frame · Sound localization · Superior colliculus

5.1 Introduction

No sensory system is an island. The auditory system works in concert with other sensory systems to help organisms understand the events occurring in their environments. The process of integrating sensory information from different senses usually proceeds so seamlessly that animals are not aware of it, and it only becomes obvious in cases where the brain is swayed by one sense to overlook the evidence in another sense. Two classic audiovisual examples involve ventriloquism, in which sounds are erroneously perceived as coming from the mouths of puppets, and the McGurk effect (McGurk and MacDonald 1976), in which the sound /bɑ/ is dubbed to a video of a person mouthing /ga/, leading to a nonveridical perception of /dɑ/ (see Lee, Maddox, and Bizley, Chap. 4, for an in-depth discussion of these multisensory illusions).

Illusions such as these reflect a deep intertwining of sensory pathways, with communication occurring between the pathways at multiple levels and taking multiple forms. In the case of interactions between hearing and vision specifically, eye movements play a critical role. In humans and monkeys, the eyes move about three times per second and cover about an 80° range of space. Every time the eyes move, the visual input stream is disrupted and shifted to a new location on the retina. In contrast, the input of the auditory system depends on the locations of sounds with respect to the head and ears. Eye movements in relation to the head, then, prevent a simple static connection between the visual and auditory domains. Rather, one or both sensory systems must adjust its processing based on these eye movements to be able to communicate with the other system.

This chapter reviews what is known about where and how this happens in the brain (Sects. 5.3, 5.4, and 5.5) and its consequences for auditory perception (Sect. 5.6) and attention (Sect. 5.7).

5.2 The Why and How of Linking Visual and Auditory Signals in Space

Combining visual and auditory information can be useful to help resolve ambiguities in sensory input. In the McGurk effect, for example, some phonemes are acoustically similar, such as /bɑ/ versus /gɑ/ or /fɑ/ versus /sɑ/, but the lip movements associated with generating those sounds look very different. Thus, watching someone's lips move while listening to their speech can greatly facilitate comprehension. However, it is critical that the visually observed lip movements used to resolve auditory ambiguities belong to the person who is actually speaking. At a cocktail party with many talkers, determining which person's lips to associate with which person's voice is necessary to derive any benefit from lip reading.

In principle, this can be accomplished by matching stimuli having a common spatial origin, but the visual and auditory systems use very different methods of determining spatial location. The optics of the eye creates an image of the visual scene on the retina. For sound, the brain must deduce location by comparing sound loudness and timing differences across the two ears as well as from direction-dependent spectral variations. These different methods mean that the original information available to the brain specifies locations in different reference frames. The retina provides the brain with information about the eye-centered location of visual stimuli. The cues on which sound localization are based provide information about the location of sounds with respect to the head and ears.

However, it is critical to note that although the *cues* are head centered, *it does not follow that the brain representations are*. In fact, as described in Sects. 5.3 and 5.4, there is no known auditory brain representation that appears to encode sound location in a strictly head-centered frame of reference. Rather, eye movements and the resulting changes in eye position with respect to the head and ears modulate auditory processing at multiple stages throughout the auditory pathway and in multiple ways.

5.3 Auditory Reference Frames in the Superior Colliculus

Interactions between eye movements and auditory processing were first found in the intermediate and deep layers of the superior colliculus (SC) of monkeys (Jay and Sparks 1984; Lee and Groh 2012) and cats (Populin and Yin 1998; Populin et al. 2004). Attention focused on the SC because it was known to play a role in guiding saccadic eye movements (Robinson 1972; Sparks 1975), which can be made to visual, auditory (Zahn et al. 1979; Zambarbieri et al. 1982), and tactile (Groh and Sparks 1996) targets. It was also known that the SC exhibited responses to auditory stimuli in anesthetized animals such as hamsters (Chalupa and Rhoades 1977), mice (Drager and Hubel 1975), and cats (Meredith and Stein 1986a, b). Furthermore, stimulation studies (Robinson 1972) and recording studies involving visual stimuli (Mays and Sparks 1980) suggested that the SC likely used an eye-centered reference frame specifying the direction and amplitude of the eye movement necessary to look at the saccade goal. Jay and Sparks (1987a, b) therefore postulated that the SC must convert auditory information, originally determined from head-centered cues, to an eye-centered reference frame to accurately move the eyes to auditory targets.

Answering this question required evaluating responses to sounds as a function of both their position with respect to the head and their position with respect to the eyes, i.e., with the eyes in several different positions with respect to the head (Fig. 5.1A). The shift in initial eye position is key because it forces the eye-centered and head-centered reference frames out of alignment. If both the eyes and head are

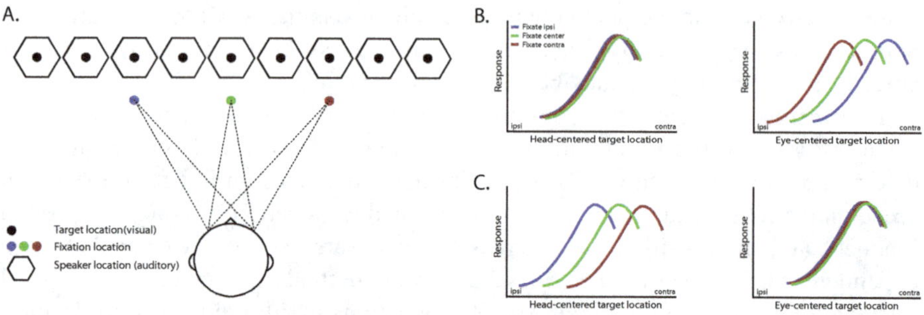

Fig. 5.1 Schematics of behavioral paradigm and hypothetical neurons with perfect eye or head-centered encoding. (**A**) Monkeys typically fixated on one of three locations and then made a saccade to either a visual or auditory target. Perfectly head-centered cells (**B**) and perfectly eye-centered cells (**C**) exhibit different behavior depending on how the target location is defined. *Left,* receptive fields in head-centered space; *right*, receptive fields in eye-centered space. A head-centered response pattern exhibits well-aligned response curves across the different fixation patterns when the target location is defined with respect to the head (**B**, *left*), whereas the responses of an eye-centered response pattern align better when the target location is defined with respect to the eyes (**C**, *right*). *ipsi*, Ipsilateral; *contra*, contralateral. (**B**, **C**) Modified from Maier and Groh (2009)

oriented forward (or aligned in any direction), then the eye-centered and head-centered reference frames are in register, meaning no coordinate transformation is needed to accurately orient the eyes to a head-centered cue.

Jay and Sparks (1984, 1987a) were the first to implement this clever experimental manipulation of shifting initial eye position. They recorded the activity of single neurons while head-restrained monkeys made saccades to visual (LED) or auditory (bandpass-filtered noise) targets from different fixation positions (−24°, 0°, or 24° in horizontal azimuth). They then mapped the receptive field of each neuron as a function of initial fixation location. If a neuron encoded auditory stimuli in a head-centered reference frame, then its responses should be governed by sound location with respect to the head regardless of eye position. A schematic of a perfectly head-centered cell is shown in Fig. 5.1B. A head-centered response pattern would have superimposed receptive fields if the responses are plotted in a head-centered space, but receptive fields would be shifted by the amount of the initial fixation if the responses are plotted in an eye-centered space. In contrast, in a perfectly eye-centered response pattern, receptive fields would be shifted by initial fixation if the responses are plotted in a head-centered space but superimposed if plotted in an eye-centered space. A schematic of a perfectly eye-centered cell is shown in Fig. 5.1C.

Jay and Sparks (1984, 1987a, b) actually found something between these two canonical cases. Specifically, they found that initial eye position affected the majority of auditory responses in the SC but did not appear to produce perfectly eye-centered response patterns. The response of an example cell modulated by eye position is shown in Fig. 5.2A. Each column displays the activity of the same neuron in three different trial conditions. While the target remained at 20° with respect to the head across trials, the monkey fixated at three different locations (−24°, 0° or 24°), meaning that the target was at the same place in reference to the head but in three

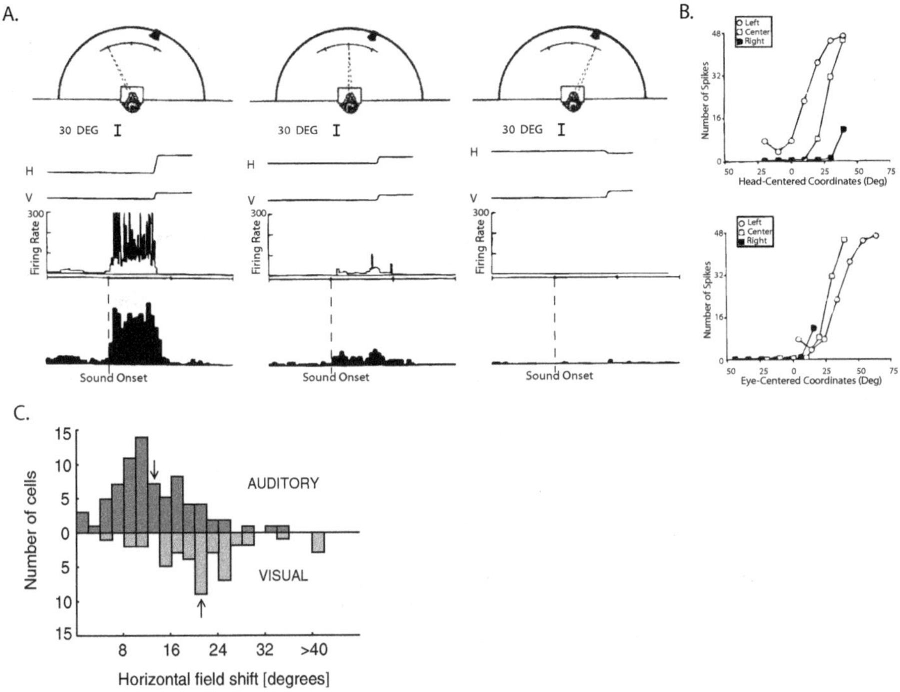

Fig. 5.2 Auditory responses in the superior colliculus (SC) as a function of eye position. (**A**) Response of an SC cell to a sound at a fixed head-centered position (20° right with respect to the head; icon is speaker) while the monkey fixated on three different positions (−24°, 0°, or 24° at *left*, *center*, and *right*, respectively). *Top*, sound and fixation positions are schematized; *middle*, horizontal and vertical eye traces of saccades to this target as well as the instantaneous firing rate of the neuron as a function of time on an individual trial; *bottom*, a peristimulus time histogram for all the trials of that particular condition. This neuron fires much more strongly to this particular head-centered sound when initially fixating at −24° compared with 0° or 24°, consistent with an eye-centered frame of reference. (**B**) Summary of the auditory-response functions of the neuron as measured at different fixation positions when plotted in head-centered coordinates (*top*) or eye-centered coordinates (*bottom*). The response functions are more shifted when plotted in head-centered space but more nearly superimposed in eye-centered space, indicating this neuron encodes sound location in an eye-centered reference frame. (**C**) Population distributions for response function shifts. Average (*arrows*) of the auditory shift is 12.9° while the visual shift is 21.7°. (**A, B**) Modified from Jay and Sparks (1984); (**C**) taken from Maier and Groh (2009)

different locations in reference to the eyes. When the receptive fields for this cell are plotted in head-centered coordinates, the responses are shifted. In contrast, when the receptive fields for this cell are plotted in eye-centered coordinates, the responses are more closely superimposed (Fig. 5.2B). These results indicate this particular cell's response depended not only on the position of the auditory target with respect to the head but also on the position of the eyes in the orbit. Since the fixation locations were 24° apart, one would predict that if auditory receptive fields of SC neurons are encoded in an eye-entered reference frame, the receptive fields would shift 24°, which this particular example neuron appears to do. However, the population results reveal auditory responses were only shifted on average by 12.9°, whereas the visual receptive fields were shifted on average by 21.7°, closer to the expected amount for an eye-centered reference frame (Fig. 5.2C). The auditory receptive field shift is only about one-half of what is expected and indicates that auditory sensory responses in the SC are neither head nor eye centered but rather are encoded in a hybrid reference frame. How, then, are primates able to accurately move their eyes toward auditory stimuli (Metzger et al. 2004)?

It took a study nearly three decades later to start unraveling this mystery. Lee and Groh (2012) advanced understanding of the coordinate transform by teasing apart the time course of activity in the SC (Fig. 5.3A). It had long been known that SC

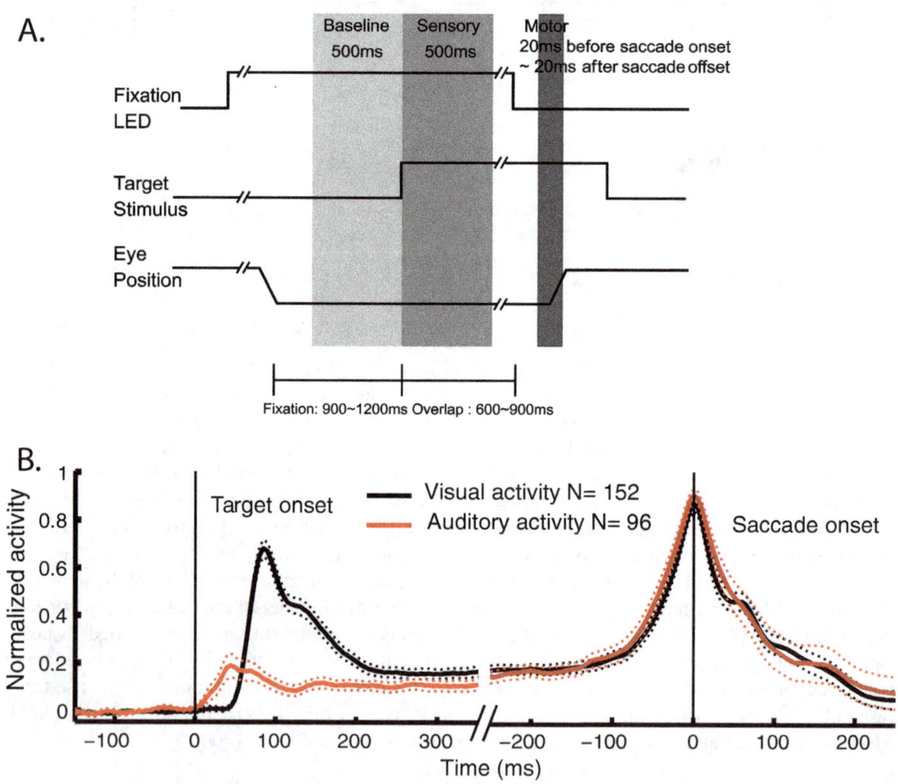

Fig. 5.3 Analysis of sensory-related versus motor-related SC activity. (**A**) Schematic of the time course of the experiment. Monkeys fixated on a fixation LED after which a target (visual or auditory) was presented. After a delay, the fixation light was extinguished, cuing the monkey to saccade to the target. Neural activity was evaluated before stimulus onset (baseline period, 500 ms), after target onset (sensory period, 500 ms), and around the time of the saccade (motor period, 20 ms before saccade onset to 20 ms after saccade offset). (**B**) Population peristimulus time histogram showing the different phases of the SC population response. The activity is normalized to the largest response of the cell and then averaged across the population and aligned to the stimulus onset (*left* 0) and the saccade onset (*right* 0). The SC population clearly displays a response to sensory and saccade onsets that are referred to as the sensory phase and motor phase, respectively. Adapted from Lee and Groh (2012)

neurons exhibit both "sensory" and "motor" activity, defined as activity time locked to the stimulus and the movement, respectively (Fig. 5.3B; Wurtz and Goldberg 1972; Sparks 1978). Indeed, Jay and Sparks (1987a) demonstrated that the motor burst occurs for auditory saccades, but they had not quantitatively analyzed the coordinate frame of these signals. Using essentially the same procedures as Jay and Sparks but analyzing the motor period as well as the sensory period, Lee and Groh (2012) found that the auditory reference frame evolved over time. In the sensory period, the auditory reference frame was encoded in a hybrid reference frame, as previously reported by Jay and Sparks. However, in the motor period, auditory-evoked signals appear to correspond to a target location in an eye-centered reference frame. The population results are shown in Fig. 5.4, which plots index values reflecting

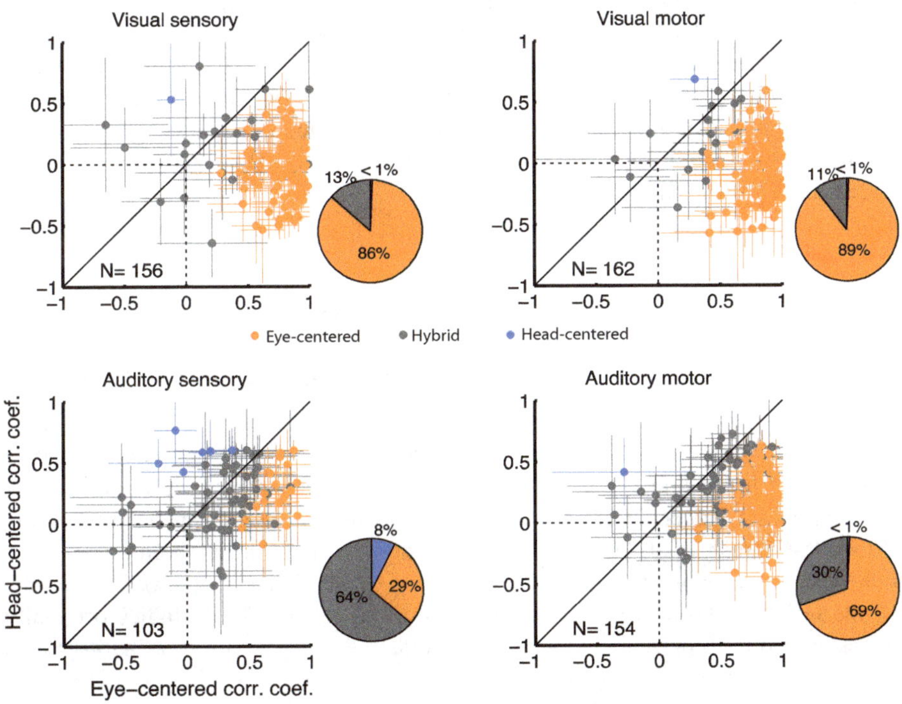

Fig. 5.4 Reference frames of SC cells during the sensory and motor periods to auditory and visual targets. The eye-centered correlation coefficient (corr. coef.) is a measure of how well response functions align in an eye-centered coordinate frame, and the head-centered correlation coefficient is a measure of how well response functions align in a head-centered coordinate frame: −1 indicates perfect anticorrelation, 0 indicates no correlation, and 1 indicates perfect correlation with respect to the designated reference frame. *Orange units* are classified as eye centered because the 95% confidence intervals on the head- and eye-centered correlation coefficients (*crosshairs*) exclude the head-centered reference frame. *Blue units* are classified as head centered due to the exclusion of the eye-centered reference frame. Gray units are classified as hybrid because neither reference frame can be excluded. Overall, visual signals in the SC are strongly eye centered, whereas auditory signals transition from mainly hybrid during the sensory period to mainly eye centered during the motor period. The eye-centered auditory activity during the motor period of the SC is the only place in the brain where a reasonably pure reference frame for auditory signals has been identified. Modified from Lee and Groh (2012)

how well head-centered versus eye-centered reference frames describe the activity for each neuron in the population. Neurons exhibiting predominantly eye-centered signals are plotted in orange and cluster below the unity line, whereas neurons exhibiting predominantly head-centered signals are plotted in blue and cluster above it. For visual signals, both sensory and motor periods are dominated by eye-centered signals. In stark contrast, for auditory signals, the sensory period is predominantly hybrid, but the motor period is dominated by eye-centered response patterns. This shift from hybrid encoding in the sensory period to more eye-centered encoding in the motor period of auditory stimuli likely allows for accurate saccades to auditory targets regardless of initial eye position (Metzger et al. 2004).

The intermediate and deep layers of the SC comprise a comparatively "late" sensory structure, situated well on the oculomotor side of the sensorimotor continuum in the brain. Because the auditory reference frame already appears to be hybrid in the SC, where does the process of adjusting the auditory reference frame begin? The SC receives inputs from four structures with auditory activity: parietal cortex, frontal eye fields, auditory cortex, and the inferior colliculus (Sparks and Hartwich-Young 1989). Sections 5.4 and 5.5 outline what is known about the auditory reference frame in these structures.

5.4 Reference Frames Throughout the Brain

5.4.1 Reference Frames in the Parietal and Frontal Cortices

The parietal cortex is known to exhibit activity related to both auditory (Stricanne et al. 1996; Linden et al. 1999) and visual cues as well as to eye and limb movements (Andersen and Buneo 2002) and is thought to play a role in spatial processing (Mullette-Gillman et al. 2005, 2009). Early studies from Andersen and colleagues indicated that changes in eye position affected visual signals in the parietal cortex (Andersen and Mountcastle 1983; Andersen et al. 1985). These studies originally characterized the representation as eye centered, with eye position contributing to the gain of the response; however, the study design involved confounds that rendered the reference frame ambiguous (Mullette-Gillman et al. 2009). A more recent quantitative analysis indicated that at least in the banks of the intraparietal sulcus, this visual representation was essentially a hybrid between eye- and head-centered coordinates (Mullette-Gillman et al. 2005, 2009). This finding was exciting from the standpoint of visual-auditory integration because it suggested some "movement" of the visual reference frame to meet auditory signals in a common middle ground. Indeed, the auditory signals, although weaker and less prevalent than the visual signals, also showed eye position effects, and the net result was a hybrid reference frame similar to the visual reference frame. Unlike in the SC, this reference frame was stable across time and did not become eye centered at the time of the saccade (Mullette-Gillman et al. 2009).

Much like the SC, the frontal eye fields (FEFs) are integral to generating eye movements (Robinson and Fuchs 1969; Schiller et al. 1979) to visual cues

(Mohler et al. 1973; Schiller et al. 1980). However, until the mid-1990s, it remained unknown if the FEFs contributed to the generation of saccades to auditory stimuli. Russo and Bruce (1994) investigated the response of FEF neurons to auditory and visually evoked saccades from multiple initial fixation positions. Although Russo and Bruce found the responses of FEF neurons vary with changes in eye position for both modalities, they did not quantitatively investigate which frame of reference these neurons used to encode space. More recently, studies have indicated that auditory signals in FEFs are largely encoded in a hybrid reference frame in both sensory and motor periods (Caruso, Pages, Sommer, and Groh, unpublished observations). Although this might seem inconsistent with a native visual eye-centered reference frame, the available evidence indicates that in the FEFs, the visual code is only about 60% eye centered (Sajad et al. 2015; Caruso et al. 2017). These data suggest that visual and auditory signals in the FEFs are similar to each other but do not employ a completely pure eye- or head-centered coordinate frame. The coding of auditory cues in eye-centered coordinates thus appears to be uniquely reflected in the motor response of SC neurons.

5.4.2 Reference Frames in the Auditory Pathway: Inferior Colliculus and Auditory Cortex

The studies described so far concern auditory signals in association cortex or the oculomotor system. These areas could encode auditory stimuli in hybrid coordinates due to hybrid input from the auditory system or they could receive a head-centered input and transform it to a more hybrid reference frame. In what reference frame, then, do auditory areas encode auditory stimuli?

It is intuitive to assume that neurons in the auditory system would use a head-centered reference frame to encode the target location because the cues originally needed to compute auditory target location in space are head centered, relying on differences in the level and timing of the sound between the two ears. However, current evidence, so far, does not bear this theory out. Several studies investigating core auditory cortex (Werner-Reiss et al. 2003; Fu et al. 2004), belt auditory cortex (Maier and Groh 2010), and inferior colliculus (Groh et al. 2001; Zwiers et al. 2004) identified numerous examples of neurons sensitive to the combination of sound and eye position (Porter et al. 2006, 2007). In aggregate, the reference frame of signals in both structures is generally hybrid, similar to the SC (sensory phase), parietal cortex, and FEF. These data are shown in Fig. 5.5 using a displacement index. Values of 0 correspond to head centered, values of 1 indicate eye centered, and values of 0.5 indicate an intermediate or hybrid reference frame in which head- and eye-centered information is equally weighted. Both the auditory cortex (Fig. 5.5, *yellow bars*) and the inferior colliculus (Fig. 5.5, *orange bars*) have a mean distribution centered around a displacement index score of 0.5, showing both regions encode auditory targets with a hybrid reference frame, similar to those used in the parietal cortex (Fig. 5.5, *pink bars*) and during the sensory phase of the SC response (Fig. 5.5, *blue bars*).

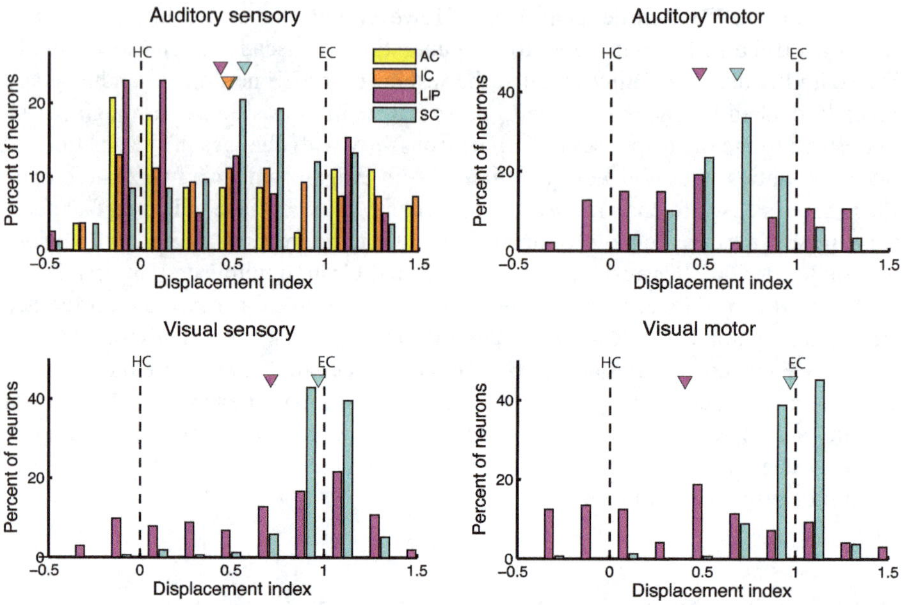

Fig. 5.5 Comparison of the reference frames for auditory and visual stimuli in the auditory cortex (AC), the inferior colliculus (IC), lateral/medial intraparietal cortex (LIP), and the SC. A displacement index value of 0 corresponds to a head-centered (HE) reference frame (*vertical dashed line*); a value of 0.5 indicates a hybrid reference frame; and a value of 1 indicates an eye-centered (EC) reference frame (*vertical dashed line*). *Inverted triangles*, mean displacement index value for each distribution. Again, note the auditory motor period for the SC is the most eye-centered auditory representation measured. Modified from Lee and Groh (2012)

This finding is surprising because it shows that the auditory system, largely thought to be independent of the visual and oculomotor systems, is, in fact, quite dependent on eye position, with auditory signals encoded in a hybrid reference frame throughout the brain, except for the eye-centered encoding by the SCs of auditory evoked saccades (Fig. 5.5). If there are any regions possessing a solely head-centered reference frame, they would need to be before the inferior colliculus in the auditory pathway. The reference frames of these areas, principally the lateral lemniscus, the superior olivary complex, and the cochlear nucleus, have yet to be probed and require further investigation.

5.5 Why Hybrid? Some Insights from Models of Coordinate Transformations

An enduring mystery is why a hybrid reference frame, the most commonly observed scenario, may be useful to the brain. Insights into this question can be gleaned from considering the computational steps involved in transforming signals from one coordinate system to another.

The first models for coordinate transformations of auditory signals were developed by Groh and Sparks (1992). Their vector subtraction model capitalized on the comparatively straightforward mathematics of computing a coordinate transformation. If the brain possesses a signal representing sound location with respect to the head (e.g., suppose there is a sound 24° to the right with respect to the head) and another signal representing eye position with respect to the head (e.g., the eyes might be 10° to the right with respect to the head), then subtraction of the eye position signal from the sound with respect to the head signal (24° − 10°) will yield a signal of sound with respect to the eyes (the sound is 14° to the right with respect to the eyes). This core computation forms the crux of the model and is accomplished through subtracting eye position information via an inhibitory synapse.

At the time this model was created, little was known about how sound location was encoded in the primate brain. As noted in Sects. 5.3 and 5.4, subsequent work has failed to identify any brain regions that encode sound location with respect to the head in studies that explicitly manipulate eye position, so the potential inputs to this coordinate transformation remain unknown. However, another aspect of auditory coding does support this model: the use of rate coding, in which the activities of auditory neurons are monotonically related to the horizontal component of sound location. This coding format has now been identified in the inferior colliculus of primates (Groh et al. 2003) as well as in other mammals (McAlpine and Grothe 2003), the auditory cortex of monkeys (Woods et al. 2006; Werner-Reiss and Groh 2008), cats (Middlebrooks et al. 1994, 1998), and the primate SC (Lee and Groh 2014). Given that eye position signals are also rate coded (Fuchs and Luschei 1970; Luschei and Fuchs 1972), this suggests that, indeed, the core computation of subtraction can be easily accomplished using known signal types. Relatedly, other computational modeling studies argued that a hybrid code can serve as a basis from which any coordinate transformation may be computed (Pouget and Sejnowski 1997; Deneve et al. 2001). But because the same can be said of inputs in pure coordinates (Groh and Sparks 1992), these models may better be interpreted as how the nervous system computes reference frames as opposed to why it implements any particular reference frame.

The ease of this computation may provide an explanation for why the hybrid format is used. Specifically, the computation may be so easy that it is underconstrained. Neural populations are not under strong selective pressure to produce a purely eye-centered code until the point at which a specific behavioral response requires it, namely, the eye-centered coding of a saccadic motor command in the SC (Lee and Groh 2012).

5.6 Behavioral Investigations of the Effect of Gaze on Auditory Reference Frame

The presence of hybrid signals has also led to considerable interest in whether there are behavioral signatures of this computation. The rationale is as follows: if signals that are in a hybrid code are read out under the erroneous assumption that they are actually either purely head centered or purely eye centered, then they should produce errors in sound localization that depend on the position of the sound with respect to the eyes. These errors might vary in magnitude but should be intermediate between the two reference frames. That is, an eccentric fixation position of 20° might be associated with errors in sound localization of 10°. Accurate sound localization would only occur if the eyes were in some privileged position that brought the two reference frames into alignment.

It is readily apparent that this is not the case. Broadly speaking, one's ability to localize sounds is not obviously impaired when the eyes are directed to some position away from straight ahead (if it were, locating a ringing cell phone would prove quite difficult). In monkeys, for which hybrid coding is well observed physiologically, the accuracy of saccades to auditory targets is not adversely affected by starting from different initial fixation positions (Metzger et al. 2004). To be sure, initial fixation position does affect the final eye position for a given target, but this effect is comparable to the undershooting observed for saccades to visual targets in which saccades typically travel about 90% of the distance to a target. Indeed, many studies with human subjects have reported the effects of eye position on various types of sound localization tasks. However, the magnitude of these effects is modest under natural viewing conditions in which the eyes move frequently and may only become large when subjects maintain fixation eccentrically for minutes at a time, as was done in some of these studies. This section and Sect. 5.7 review those studies and then discuss whether they are consistent with the reported neurophysiology.

In binaural lateralization studies with short fixations and frequent eye movements, the effect of gaze on auditory localization appears to depend on the specifics of the paradigm, but the majority of studies find small shifts (less than 10% of the gaze magnitude) that are actually in the opposite direction of gaze. Lewald and Ehrenstein (1996) asked subjects to adjust interaural level difference over earphones while they maintained an eccentric fixation, finding that auditory lateralization shifted away from gaze by 1–3 dB. In a series of follow-up studies, Lewald (1997, 1998) found again that the localization of sounds shifted away from eye position as long as there was an absolute visual reference to compare against the location of the sound. Complicating matters, Lewald (1997) notes that with eccentric eye position, both the perception of a central sound and a central visual stimulus shift away from gaze. That is, if eyes are fixated in the right hemifield, central stimuli appear to shift into the left hemifield. Importantly, if the shift of the visual reference exceeds the shift of the probe sound (the sound to be localized), it could cause the subject's response to shift toward eye position, accounting for mixed results. Lewald and Getzmann (2006) found that horizontal (as well as vertical) auditory localization, again, shifted in the opposite direction as gaze, and Lewald and Ehrenstein (2001) found that the shift was also away from gaze in rear space. In other words, horizontal localization shifts are reflected about the coronal plane as opposed to rotated 180° in azimuth. This result makes sense because horizontal binaural cues undergo the same transformation (reflection rather than rotation). It is thus safe to say that, depending on the specifics of the paradigm, the work of Lewald and colleagues generally finds a modest shift (about 2–4°) in auditory localization in the direction opposite eccentric eye gazes of 45°.

A number of other studies have investigated the effects of gaze on fixation with what seems to be an important experimental difference: lateral fixations were maintained over long periods, multiple seconds to minutes, rather than redirecting gaze on each trial. These studies have more consistent results and typically find a larger effect, around 40% of the gaze magnitude, but in the same direction as gaze (in contrast to the studies discussed above). Weerts and Thurlow (1971) found that when subjects expected the probe sound to come from a visible loudspeaker at ±20° azimuth, localization shifted by 4–8° in the direction of gaze. Through further manipulations, they determined that lateral gaze with no expectation of stimulus origin resulted in smaller shifts, on the order of 2°, but that expectation on its own resulted in no shift at all, demonstrating that gaze direction and subject expectation yielded the biggest overall localization shift. A number of follow-up studies confirmed these results of localization shifts toward eye position (Bohlander 1984). Razavi et al. (2007) showed that those shifts increased with fixation duration, approaching large steady-state shifts after several minutes around 8° when fixation was 20° to the side. Notably, they tested many sound locations and found the shift to be largely consistent across auditory space. Looking at both horizontal and vertical localization, Cui et al. (2010b) found a shift in the same direction as gaze, with a similar asymptotic time course to other studies from that group (Razavi et al. 2007; Cui et al. 2010a). This is in accord with a previous study that tested only vertical localization and vertical gaze offsets (Getzmann 2002).

In short, the difference in the sequencing of fixations from trial to trial appears to be what is driving the differing results between sets of studies. Studies that employ naturalistic (i.e., short) fixations show only modest gaze-driven localization shifts in the opposite direction of gaze (Lewald 1997; Lewald and Getzmann 2006). This is consistent with daily human experience; there is little trouble registering the locations of what is heard with what is seen. Studies that employ nonnaturalistic fixations (i.e., long, constant fixations, often for several minutes at a time) show larger localization shifts in the same direction as gaze (e.g., Razavi et al. 2007). These larger shifts were around 40% of the magnitude of eccentric gaze, consistent with the partial shift of a hybrid reference frame.

The mechanism that explains these results is not known but could be the result of decay in the accuracy of the relevant signals across time. In particular, sensing of eye position relies at least in part on corollary discharge or the copy of the motor command that was issued to bring the eyes to that location in space (Guthrie et al. 1983; Sommer and Wurtz 2008). Memory for such corollary discharge signals may decay in seconds to minutes, producing shifts in the perceived location of sounds with respect to the (missensed) eyes. This idea is similar to the proprioceptive drift of occluded limb position, in which 15–120 seconds after an occluded limb movement, the limb drifts back toward the body (Wann and Ibrahim 1992). Such a model, an initial hybrid shift time locked to an eye movement that decays with continued fixation, allows the disparate behavioral localization results to be reconciled with physiological observations.

5.7 Effects of Gaze Direction on Auditory Spatial Acuity

Most of the behavioral studies of auditory-oculomotor reference frames have been based on looking for gaze-driven biases of absolute sound localization. As seen in Sect. 5.6, studies with natural fixation lengths find small biases, a fact that is in-line with the general ease of registering visual and auditory space in daily life. A related but distinct test of spatial perception is the ability to discriminate subtle differences between the locations of two sounds (or small changes to binaural lateralization cues). Human auditory spatial acuity is best in front of the head and worsens with increasing distance from the median plane (Hafter and Maio 1975; Middlebrooks and Onsan 2012). This is partly a physical acoustical effect (Mills 1958); however, auditory spatial discrimination is poorer for lateral sounds even when controlling for the acoustics, suggesting that the neural resolution of horizontal space coding is better for central locations than for lateral ones (Maddox et al. 2014).

Discrimination paradigms test relative rather than absolute localization, so why would they be useful for studying reference frames? If the neural representation of acoustical space is modified by eye gaze, then it is reasonable to expect that such a modification resulting from directed gaze may improve auditory spatial acuity. This improvement could occur by moving the experimental probe sounds into a part of the nonlinear receptive field with a steeper or shallower slope (e.g., Fig. 5.2B), improving or impairing discrimination performance, respectively.

Maddox et al. (2014) tested that notion by directing eye fixation with an eccentric dot and engaging subjects in a binaural cue discrimination task. They found that in blocks where gaze was held centered, discrimination performance was as expected for both interaural level difference (ILD) and interaural timing difference (ITD) cues and much better for central sounds than for lateral ones (Fig. 5.6). However, when subjects directed their gaze toward the probe sounds, there was an improvement in discrimination of the lateral binaural cues of about 10% (there was also a smaller improvement for centered auditory probes). Such a result, in isolation, could be attributed to auditory attention; visual cues indicating the location of a masked speech stream do improve intelligibility (Best et al. 2007). Surprisingly, though, they found no such benefit in blocks where auditory spatial attention was directed with an acoustic primer, suggesting that eccentric fixation was a crucial component of the observation and that neither spatial attention nor expectation was

Fig. 5.6 (continued) lateralizations were centered about the primer. Subject performance is shown for all conditions of interaural level difference (**C**) and interaural time difference (**D**) stimuli. Center performance was better than side performance. For interaural level difference, performance was better in visual (Vis) directional trials than in visual uninformative trials at both the center and side positions. For interaural time difference, directional visual trials showed improved discrimination when the stimulus was located on the side. Auditory (Aud) primers offered no benefit. Values are means ± SE (across the 15 intrasubject means) and are based on raw percent correct scores. One-tailed paired t-test significance: $^{*}P < 0.00625$, $^{**}P < 0.00125$ (Bonferroni-corrected values of 0.05 and 0.01, respectively). Arcsine-transformed values were used for t-tests and effect sizes. Adapted from Maddox et al. (2014)

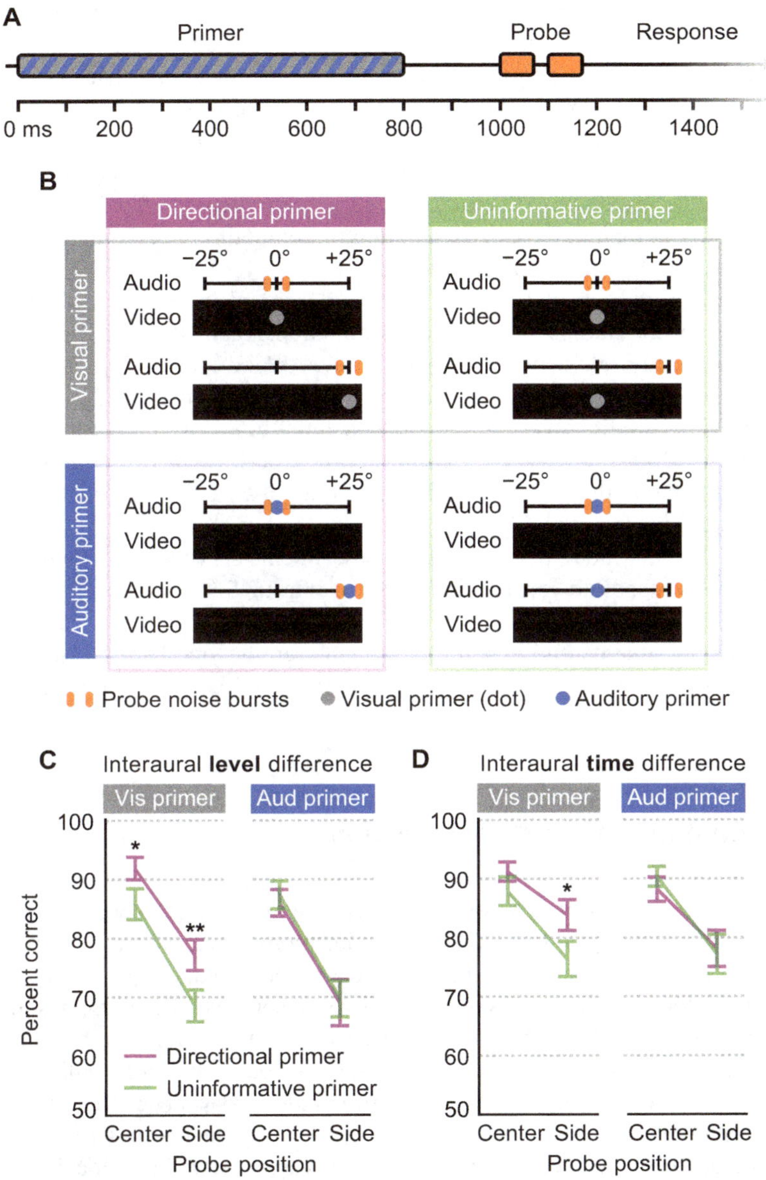

Fig. 5.6 Behavioral paradigm and results showing a gaze-mediated improvement in auditory spatial acuity. (**A**) Time course of a trial. In visual trials, the dot brightened on fixation and darkened after 800 ms; in auditory trials, the primer was a noise burst. The probe noise bursts lasted 70 ms each, with 30 ms between each. The subject responded by button press any time after the stimulus. Primers provided the same timing information whether visual or auditory and directional or uninformative. (**B**) Experimental blocks (one per quadrant). Each quadrant shows an example of a center trial above a side trial. The positions of the visual and auditory primers, where present, are the *gray* and *blue dots*, respectively. In auditory trials, subjects were presented with a black screen and not instructed where to direct their eyes. The probe noise bursts (*orange bars*) of different

the main driver. The authors hypothesize that an enhancement to spatial acuity in the direction of gaze could lead to enhanced spatial release from masking (Marrone et al. 2008) when attempting to selectively attend to one sound that is physically close to another distracting sound, in a sense because the two sound sources become better separated in perceptual space. This would represent an advantage that was distinct from the correct alignment of auditory and visual reference frames, one which would be in-line with the notion that the major benefit afforded by binaural spatial hearing in many species is separation of competing sounds rather than precise localization (Grothe and Pecka 2014).

However, as with absolute localization as described in Sect. 5.6, there is disagreement between studies for spatial discrimination. Again, the duration of fixation is a possible factor, albeit with opposite results. Wood and Bizley (2015) tested discrimination over a broad range of reference azimuths between ±75°. Their subjects maintained fixation over the course of an entire block of trials (lasting minutes at a time) at −30, 0, or 30°. They confirmed that performance was best near the median plane, but in this case, they found no effect of gaze, in stark contrast to previous results (Maddox et al. 2014).

In short, these relative discrimination experiments serve as a convenient complement to the absolute localization experiments in Sect. 5.6; here there is an effect for short fixations that seems to disappear for longer ones. This suggests that relative judgments and absolute judgments are accomplished using at least partly different mechanisms and are differentially affected by the duration of fixation.

5.8 Summary and Future Directions

The auditory and visual systems work together to help animals understand the events happening in the environment. In species with mobile eyes, such as humans and monkeys, such movements must be factored in when comparing the locations of sounds to the locations of images of potential visual sources. The neurophysiological processes involved in this computation appear to span a wide range of brain regions. Although auditory location *cues* depend on the position of the sound with respect to the head, no purely head-centered brain representation has yet been identified.

The perceptual implications of the neurophysiological findings remain somewhat unclear. On the whole, humans and monkeys are clearly able to localize sounds accurately despite movements of the eyes. However, perceptual errors that depend on eye position do occur and can vary in direction and size depending on whether the fixation duration is short (<seconds) or long (>minutes) as well as whether the task involves absolute versus relative judgments.

A number of questions remain to be answered. Is there a purely head-centered reference frame for auditory stimuli anywhere in the brain? Where in the neural hierarchy does this occur? Where and how do eye position signals first reach the auditory pathway? Is the commonly observed hybrid reference frame a "bug" or a not-yet-understood feature? How does the process of integrating visual and auditory spaces accommodate the many simultaneous visual and auditory events that occur in natural scenes? Do the same neural mechanisms underlie eye movements, visual attention, and auditory spatial attention?

These questions can be addressed through continued integration of physiology, behavior, and modeling in this computationally rich system. That the brain somehow manages to weave together information derived from two distinct physical phenomena using completely different sensors in dynamically misaligned physical reference frames is a truly remarkable feat that goes unnoticed in daily life.

Compliance with Ethics Requirements Shawn M. Willett declares that he has no conflict of interest.

Jennifer M. Groh declares that she has no conflict of interest.

Ross K. Maddox declares that he has no conflict of interest.

References

Andersen, R. A., & Buneo, C. A. (2002). Intentional maps in posterior parietal cortex. *Annual Review of Neuroscience, 25*(1), 189–220.

Andersen, R. A., & Mountcastle, V. B. (1983). The influence of the angle of gaze upon the excitability of the light-sensitive neurons of the posterior parietal cortex. *The Journal of Neuroscience, 3*(3), 532–548.

Andersen, R. A., Essick, G. K., & Siegel, R. M. (1985). Encoding of spatial location by posterior parietal neurons. *Science, 230*(4724), 456–458.

Best, V., Ozmeral, E. J., & Shinn-Cunningham, B. G. (2007). Visually-guided attention enhances target identification in a complex auditory scene. *Journal of the Association for Research in Otolaryngology, 8*(2), 294–304.

Bohlander, R. W. (1984). Eye position and visual attention influence perceived auditory direction. *Perceptual and Motor Skills, 59*(2), 483–510.

Caruso, V., Pages, D. S., Sommer, M., & Groh, J. M. (2017). Beyond the labeled line: Variation in visual reference frames from intraparietal cortex to frontal eye fields and the superior colliculus. *Journal of Neurophysiology, 119*(4), 1411–1421.

Chalupa, L. M., & Rhoades, R. W. (1977). Responses of visual, somatosensory, and auditory neurones in the golden hamster's superior colliculus. *The Journal of Physiology, 270*(3), 595–626.

Cui, Q. N., O'Neill, W. E., & Paige, G. D. (2010a). Advancing age alters the influence of eye position on sound localization. *Experimental Brain Research, 206*(4), 371–379.

Cui, Q. N., Razavi, B., O'Neill, W. E., & Paige, G. D. (2010b). Perception of auditory, visual, and egocentric spatial alignment adapts differently to changes in eye position. *Journal of Neurophysiology, 103*(2), 1020–1035.

Deneve, S., Latham, P. E., & Pouget, A. (2001). Efficient computation and cue integration with noisy population codes. *Nature Neuroscience, 4*(8), 826–831.

Drager, U. C., & Hubel, D. H. (1975). Responses to visual stimulation and relationship between visual, auditory, and somatosensory inputs in mouse superior colliculus. *Journal of Neurophysiology, 38*(3), 690–713.

Fu, K.-M. G., Shah, A. S., O'Connell, M. N., McGinnis, T., Eckholdt, H., Lakatos, P., Smiley, J., & Schroeder, C. E. (2004). Timing and laminar profile of eye-position effects on auditory responses in primate auditory cortex. *Journal of Neurophysiology, 92*(6), 3522–3531.

Fuchs, A. F., & Luschei, E. S. (1970). Firing patterns of abducens neurons of alert monkeys in relationship to horizontal eye movement. *Journal of Neurophysiology, 33*(3), 382–392.

Getzmann, S. (2002). The effect of eye position and background noise on vertical sound localization. *Hearing Research, 169*(1–2), 130–139.

Groh, J. M., & Sparks, D. L. (1992). Two models for transforming auditory signals from head-centered to eye-centered coordinates. *Biological Cybernetics, 67*(4), 291–302.

Groh, J. M., & Sparks, D. L. (1996). Saccades to somatosensory targets. I. Behavioral characteristics. *Journal of Neurophysiology, 75*(1), 412–427.

Groh, J. M., Trause, A. S., Underhill, A. M., Clark, K. R., & Inati, S. (2001). Eye position influences auditory responses in primate inferior colliculus. *Neuron, 29*(2), 509–518.

Groh, J. M., Kelly, K. A., & Underhill, A. M. (2003). A monotonic code for sound azimuth in primate inferior colliculus. *Journal of Cognitive Neuroscience, 15*(8), 1217–1231.

Grothe, B., & Pecka, M. (2014). The natural history of sound localization in mammals—A story of neuronal inhibition. *Frontiers in Neural Circuits, 8*(116), 1–19.

Guthrie, B. L., Porter, J. D., & Sparks, D. L. (1983). Corollary discharge provides accurate eye position information to the oculomotor system. *Science, 221*(4616), 1193–1195.

Hafter, E. R., & Maio, J. D. (1975). Difference thresholds for interaural delay. *The Journal of the Acoustical Society of America, 57*(1), 181–187.

Jay, M. F., & Sparks, D. L. (1984). Auditory receptive fields in primate superior colliculus shift with changes in eye position. *Nature, 309*(5966), 345–347.

Jay, M. F., & Sparks, D. L. (1987a). Sensorimotor integration in the primate superior colliculus. I. Motor convergence. *Journal of Neurophysiology, 57*(1), 22–34.

Jay, M. F., & Sparks, D. L. (1987b). Sensorimotor integration in the primate superior colliculus. II. Coordinates of auditory signals. *Journal of Neurophysiology, 57*(1), 35–55.

Lee, J., & Groh, J. M. (2012). Auditory signals evolve from hybrid- to eye-centered coordinates in the primate superior colliculus. *Journal of Neurophysiology, 108*(1), 227–242.

Lee, J., & Groh, J. M. (2014). Different stimuli, different spatial codes: A visual map and an auditory rate code for oculomotor space in the primate superior colliculus. *PLoS One, 9*(1), e85017.

Lewald, J. (1997). Eye-position effects in directional hearing. *Behavioural Brain Research, 87*(1), 35–48.

Lewald, J. (1998). The effect of gaze eccentricity on perceived sound direction and its relation to visual localization. *Hearing Research, 115*(1–2), 206–216.

Lewald, J., & Ehrenstein, W. H. (1996). The effect of eye position on auditory lateralization. *Experimental Brain Research, 108*(3), 473–485.

Lewald, J., & Ehrenstein, W. H. (2001). Effect of gaze direction on sound localization in rear space. *Neuroscience Research, 39*(2), 253–257.

Lewald, J., & Getzmann, S. (2006). Horizontal and vertical effects of eye-position on sound localization. *Hearing Research, 213*(1–2), 99–106.

Linden, J. F., Grunewald, A., & Andersen, R. A. (1999). Responses to auditory stimuli in macaque lateral intraparietal area II. Behavioral modulation. *Journal of Neurophysiology, 82*(1), 343–358.

Luschei, E. S., & Fuchs, A. F. (1972). Activity of brain stem neurons during eye movements of alert monkeys. *Journal of Neurophysiology, 35*(4), 445–461.

Maddox, R. K., Pospisil, D. A., Stecker, G. C., & Lee, A. K. C. (2014). Directing eye gaze enhances auditory spatial cue discrimination. *Current Biology, 24*(7), 748–752.

Maier, J. X., & Groh, J. M. (2009). Multisensory guidance of orienting behavior. *Hearing Research, 258*(1–2), 106–112.

Maier, J. X., & Groh, J. M. (2010). Comparison of gain-like properties of eye position signals in inferior colliculus versus auditory cortex of primates. *Frontiers in Integrative Neuroscience, 4*, 121.

Marrone, N., Mason, C. R., & Kidd, G. (2008). Tuning in the spatial dimension: Evidence from a masked speech identification task. *The Journal of the Acoustical Society of America, 124*(2), 1146–1158.

Mays, L. E., & Sparks, D. L. (1980). Dissociation of visual and saccade-related responses in superior colliculus neurons. *Journal of Neurophysiology, 43*(1), 207–232.

McAlpine, D., & Grothe, B. (2003). Sound localization and delay lines—Do mammals fit the model? *Trends in Neurosciences, 26*(7), 347–350.

McGurk, H., & MacDonald, J. (1976). Hearing lips and seeing voices. *Nature, 264*(5588), 746–748.

Meredith, A. M., & Stein, B. E. (1986a). Spatial factors determine the activity of multisensory neurons in cat superior colliculus. *Brain Research, 365*(2), 350–354.

Meredith, M. A., & Stein, B. E. (1986b). Visual, auditory, and somatosensory convergence on cells in superior colliculus results in multisensory integration. *Journal of Neurophysiology, 56*(3), 640–662.

Metzger, R. R., Mullette-Gillman, O. D. A., Underhill, A. M., Cohen, Y. E., & Groh, J. M. (2004). Auditory saccades from different eye positions in the monkey: Implications for coordinate transformations. *Journal of Neurophysiology, 92*(4), 2622–2627.

Middlebrooks, J. C., & Onsan, Z. A. (2012). Stream segregation with high spatial acuity. *The Journal of the Acoustical Society of America, 132*(6), 3896–3911.

Middlebrooks, J. C., Clock, A. E., Xu, L., & Green, D. M. (1994). A panoramic code for sound location by cortical neurons. *Science, 264*(5160), 842–843.

Middlebrooks, J. C., Xu, L., Eddins, A. C., & Green, D. M. (1998). Codes for sound-source location in nontonotopic auditory cortex. *Journal of Neurophysiology, 80*(2), 863–881.

Mills, A. W. (1958). On the minimum audible angle. *The Journal of the Acoustical Society of America, 30*(4), 237–246.

Mohler, C. W., Goldberg, M. E., & Wurtz, R. H. (1973). Visual receptive fields of frontal eye field neurons. *Brain Research, 61*, 385–389.

Mullette-Gillman, O. D. A., Cohen, Y. E., & Groh, J. M. (2005). Eye-centered, head-centered, and complex coding of visual and auditory targets in the intraparietal sulcus. *Journal of Neurophysiology, 94*(4), 2331–2352.

Mullette-Gillman, O. D. A., Cohen, Y. E., & Groh, J. M. (2009). Motor-related signals in the intraparietal cortex encode locations in a hybrid, rather than eye-centered reference frame. *Cerebral Cortex, 19*(8), 1761–1775.

Populin, L. C., & Yin, T. C. T. (1998). Sensitivity of auditory cells in the superior colliculus to eye position in the behaving cat. In A. R. Palmer, A. Q. Summerfield, & R. Meddis (Eds.), *Psychophysical and physiological advances in hearing* (pp. 441–448). London: Whurr.

Populin, L. C., Tollin, D. J., & Yin, T. C. T. (2004). Effect of eye position on saccades and neuronal responses to acoustic stimuli in the superior colliculus of the behaving cat. *Journal of Neurophysiology, 92*(4), 2151–2167.

Porter, K. K., Metzger, R. R., & Groh, J. M. (2006). Representation of eye position in primate inferior colliculus. *Journal of Neurophysiology, 95*(3), 1826–1842.

Porter, K. K., Metzger, R. R., & Groh, J. M. (2007). Visual- and saccade-related signals in the primate inferior colliculus. *Proceedings of the National Academy of Sciences of the United States of America, 104*(45), 17855–17860.

Pouget, A., & Sejnowski, T. J. (1997). Spatial transformations in the parietal cortex using basis functions. *Journal of Cognitive Neuroscience, 9*(2), 222–237.

Razavi, B., O'Neill, W. E., & Paige, G. D. (2007). Auditory Spatial perception dynamically realigns with changing eye position. *The Journal of Neuroscience, 27*(38), 10249–10258.

Robinson, D. A. (1972). Eye movements evoked by collicular stimulation in the alert monkey. *Vision Research, 12*(11), 1795–1808.

Robinson, D. A., & Fuchs, A. F. (1969). Eye movements evoked by stimulation of frontal eye fields. *Journal of Neurophysiology, 32*(5), 637–648.

Russo, G. S., & Bruce, C. J. (1994). Frontal eye field activity preceding aurally guided saccades. *Journal of Neurophysiology, 71*(3), 1250–1253.

Sajad, A., Sadeh, M., Keith, G. P., Yan, X., Wang, H., & Crawford, J. D. (2015). Visual-motor transformations within frontal eye fields during head-unrestrained gaze shifts in the monkey. *Cerebral Cortex, 25*(10), 3932–3952.

Schiller, P. H., True, S. D., & Conway, J. L. (1979). Effects of frontal eye field and superior colliculus ablations on eye movements. *Science, 206*(4418), 590–592.

Schiller, P. H., True, S. D., & Conway, J. L. (1980). Deficits in eye movements following frontal eye-field and superior colliculus ablations. *Journal of Neurophysiology, 44*(6), 1175–1189.

Sommer, M. A., & Wurtz, R. H. (2008). Brain circuits for the internal monitoring of movements. *Annual Review of Neuroscience, 31*, 317–338.

Sparks, D. L. (1975). Response properties of eye movement-related neurons in the monkey superior colliculus. *Brain Research, 90*(1), 147–152.

Sparks, D. L. (1978). Functional properties of neurons in the monkey superior colliculus: Coupling of neuronal activity and saccade onset. *Brain Research, 156*(1), 1–16.

Sparks, D. L., & Hartwich-Young, R. (1989). The deep layers of the superior colliculus. *Reviews of Oculomotor Research, 3*, 213–255.

Stricanne, B., Andersen, R. A., & Mazzoni, P. (1996). Eye-centered, head-centered, and intermediate coding of remembered sound locations in area LIP. *Journal of Neurophysiology, 76*(3), 2071–2076.

Wann, J. P., & Ibrahim, S. F. (1992). Does limb proprioception drift? *Experimental Brain Research, 91*(1), 162–166.

Weerts, T. C., & Thurlow, W. R. (1971). The effects of eye position and expectation on sound localization. *Perception & Psychophysics, 9*(1), 35–39.

Werner-Reiss, U., & Groh, J. M. (2008). A rate code for sound azimuth in monkey auditory cortex: Implications for human neuroimaging studies. *The Journal of Neuroscience, 28*(14), 3747–3758.

Werner-Reiss, U., Kelly, K. A., Trause, A. S., Underhill, A. M., & Groh, J. M. (2003). Eye position affects activity in primary auditory cortex of primates. *Current Biology, 13*(7), 554–562.

Wood, K. C., & Bizley, J. K. (2015). Relative sound localisation abilities in human listeners. *The Journal of the Acoustical Society of America, 138*(2), 674–686.

Woods, T. M., Lopez, S. E., Long, J. H., Rahman, J. E., & Recanzone, G. H. (2006). Effects of stimulus azimuth and intensity on the single-neuron activity in the auditory cortex of the alert macaque monkey. *Journal of Neurophysiology, 96*(6), 3323–3337.

Wurtz, R. H., & Goldberg, M. E. (1972). Activity of superior colliculus in behaving monkey. III. Cells discharging before eye movements. *Journal of Neurophysiology, 35*(4), 575–586.

Zahn, J. R., Abel, L. A., Dell'Osso, L. F., & Daroff, R. B. (1979). The audioocular response: Intersensory delay. *Sensory Processes, 3*(1), 60.

Zambarbieri, D., Schmid, R., Magenes, G., & Prablanc, C. (1982). Saccadic responses evoked by presentation of visual and auditory targets. *Experimental Brain Research, 47*(3), 417–427.

Zwiers, M. P., Versnel, H., & Van Opstal, A. J. (2004). Involvement of monkey inferior colliculus in spatial hearing. *The Journal of Neuroscience, 24*(17), 4145–4156.

Chapter 6
Multisensory Processing in the Auditory Cortex

Andrew J. King, Amy Hammond-Kenny, and Fernando R. Nodal

Abstract The capacity of the brain to combine and integrate information provided by the different sensory systems has a profound impact on perception and behavior. This is especially the case for audition, with many studies demonstrating that the ability of listeners to detect, discriminate, or localize sounds can be altered in the presence of other sensory cues. For example, the availability of congruent visual stimuli can make it easier to localize sounds or to understand speech, benefits that are most apparent when auditory signals are weak or degraded by the presence of background noise. Multisensory convergence has been demonstrated at most levels of the auditory pathway, from the cochlear nucleus to the auditory cortex. This is particularly the case in extralemniscal nuclei from the midbrain upward but has also been observed in the tonotopically organized lemniscal or core projections. In addition to inheriting multisensory signals from subcortical levels, the auditory cortex receives visual and somatosensory inputs from other cortical areas. Although non-auditory stimuli can evoke spiking activity in auditory cortex, they typically modulate auditory responses. These interactions appear to provide contextual cues that signal the presence of an upcoming sound, but they can also increase the information conveyed by cortical neurons about the location or identity of sounds and may even recalibrate cortical responses when the information provided by different sensory modalities is conflicting. Identifying the neural circuitry responsible for the behavioral consequences of multisensory integration remains an area of intense investigation.

Keywords Attention · Auditory pathway · Cross-modal plasticity · Dorsal cochlear nucleus · Eye movement · Inferior colliculus · Somatosensory · Sound localization · Spectral cue · Speech · Superior colliculus · Thalamus · Ventriloquism illusion · Visual · Vocalization

---

A. J. King (✉) · A. Hammond-Kenny · F. R. Nodal
Department of Physiology, Anatomy and Genetics, University of Oxford, Oxford, UK
e-mail: andrew.king@dpag.ox.ac.uk; amy.hammond-kenny@merton.ox.ac.uk; fernando.nodal@dpag.ox.ac.uk

A. K. C. Lee et al. (eds.), *Multisensory Processes*, Springer Handbook of Auditory Research 68, https://doi.org/10.1007/978-3-030-10461-0_6

6.1 Introduction

Creating a unified sensory percept requires the integration of information from different sensory modalities. This process is traditionally viewed as occurring in two distinct phases in the brain. First, unisensory signals are processed by dedicated neural pathways, which are assumed to be largely independent and hierarchically organized. Second, once modality-specific computations have been performed, sensory information is combined and integrated in certain higher order association areas that implement different aspects of multisensory perception. In the cortex, classical multisensory areas have been described in the frontal, parietal, and temporal lobes, where their functions are thought to range from linking multiple sensory signals with the execution of particular motor actions to the merging of communication signals provided by the eyes and ears (reviewed by Cappe et al. 2012).

That sensory pathways are organized in this fashion stems from the different forms of energy (light, sound) that need to be detected. This necessitates the use of specialized transduction mechanisms for converting each form of energy into neural activity and imposes constraints on the associated neural circuits in order to overcome the differences between each sensory modality, such as the lack of spatial information at the cochlea or the differing temporal dynamics of visual and auditory processing. Furthermore, some of our perceptions, for example, the color of a flower or the pitch of someone's voice, do not have obvious equivalents in other sensory modalities. Nevertheless, it is often the case that we can identify or locate an object, such as a familiar person speaking, by using more than one of our senses. Although this cross-modal redundancy is extremely useful for perceptual stability should one set of cues disappear, such as when that person stops speaking or walks outside our field of view, sensory processing most commonly occurs in a multisensory context and the simultaneous availability of information across different modalities can have profound effects on perception and behavior.

A good example of this is provided by speech perception. If we want to understand the basis for this vital ability, it is necessary to consider not only how the brain responds to auditory information but also the motor aspects of speech production and, consequently, the associated visual articulation cues. Orofacial movements during speech production provide temporally correlated cues (Fig. 6.1; Chandrasekaran et al. 2009) that, when combined with acoustic signals, improve the detection and comprehension of speech, particularly if those signals are degraded by the presence of background sounds (Sumby and Pollack 1954; also see Grant and Bernstein, Chap. 3). The tendency to merge auditory-visual speech cues is further illustrated by the well-known McGurk illusion (McGurk and MacDonald 1976): pairing a voice articulating one syllable with a face articulating a different syllable can result in the perception of a novel token that represents a fusion of those syllables.

This work clearly indicates the capacity of the brain to integrate the informational content of auditory-visual speech. If the signals available in each modality are first processed independently and only subsequently combined at a specialized integration stage, one might expect the neural basis for the influence of vision on

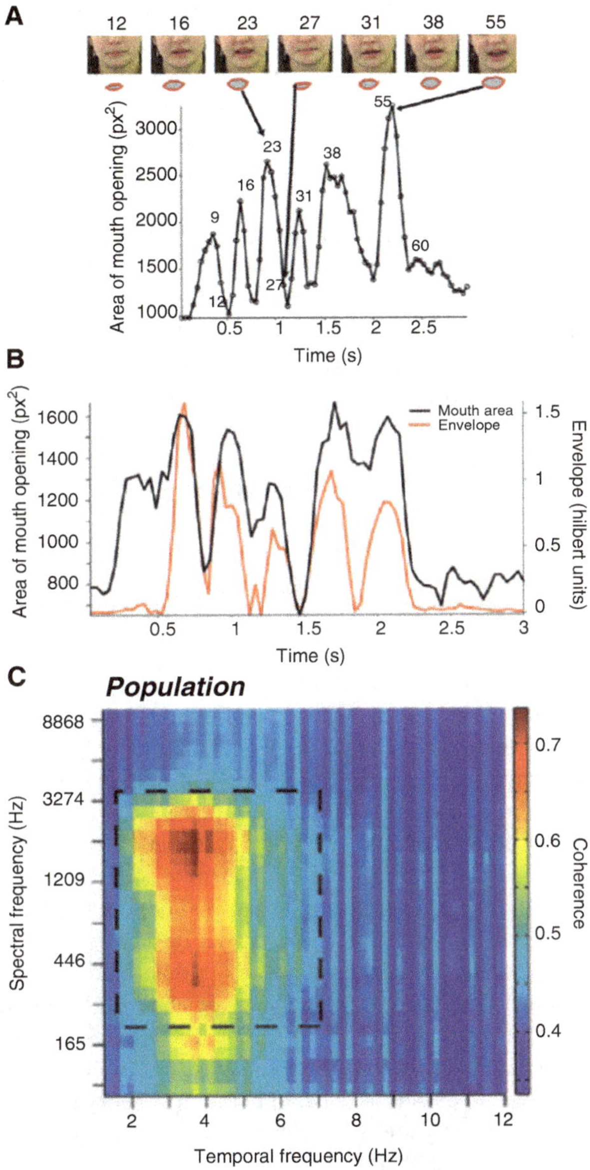

Fig. 6.1 Visual and auditory statistics of human speech. (**A**) *Top*, example facial gestures at different frames from a video of a speaker uttering a sentence, with the *red ellipses* below each frame representing the area of the mouth opening. *Bottom:* graph shows the estimated area for each mouth contour in pixels squared as a function of time in seconds. *Numbers* refer to corresponding frames in the video. *Arrows* point to specific frames in the time series depicting different size of mouth opening. (**B**) Variation in the area of the mouth opening (*black*) and the broadband auditory envelope (*orange*) for a single sentence from a single subject as a function of time in seconds. (**C**) Heat map illustrating the robust coherence between the mouth opening and auditory signal as a function of both spectral frequency band and temporal modulation frequency for 20 subjects. *Dashed–line rectangle*, region of maximal coherence between the visual and auditory signals. Adapted from Chandrasekaran et al. (2009), with permission

auditory speech intelligibility to reside in higher order speech-related areas such as the superior temporal sulcus (STS; see Beauchamp, Chap. 8). Although this is undoubtedly the case (Sekiyama et al. 2003; McGettigan et al. 2012), there is growing evidence that auditory and visual speech signals also interact as early as the primary auditory cortex (Schroeder et al. 2008; Okada et al. 2013). Furthermore, both cortical and subcortical auditory brain regions have been implicated in the various cross-modal effects that have been described for other dimensions of auditory perception. Indeed, it is a general property of sensory systems that the availability of congruent multisensory cues can result in faster responses as well as improvements in the ability to detect, discriminate, or localize stimuli (Murray and Wallace 2012). It is therefore important to consider where and how those interactions take place as well as the nature of the information provided by the "nondominant" modality if we are to understand the impact of vision and other sensory modalities on auditory processing and perception.

This chapter considers these issues in the context of the auditory pathway as a whole but with a focus on visual and somatosensory influences on the auditory cortex and the implications of these effects for its primary role in hearing. Although similar questions can be asked about the functional significance of multisensory influences on processing in the visual or somatosensory cortex, the auditory cortex has been at the vanguard of research in this area. Consequently, these studies have the potential not only to improve our understanding of the computations performed by auditory cortical neurons but also to reveal general principles of how multisensory interactions influence perception and behavior.

6.2 Multisensory Versus Auditory Brain Areas

Conceptually, it is difficult to classify a given brain area as unisensory if stimuli belonging to different sensory modalities can influence the activity of the neurons found there. However, multisensory influences take different forms, ranging from a change in action potential firing in response to more than one type of sensory stimulus to cross-modal modulation of the spiking responses to one modality even if the other modality cues are by themselves ineffective in driving the neurons (Fig. 6.2). In the case of the auditory cortex, there is considerable evidence for modulatory effects of nonauditory inputs on responses to sound. These interactions have been found to be particularly prevalent in functional imaging experiments, which also show that visual cues alone can activate certain parts of the auditory cortex in humans (Calvert et al. 1997; Pekkola et al. 2005) and nonhuman primates (Kayser et al. 2007). Similar results have been obtained using electrophysiological measurements, with local field potential recordings demonstrating widespread effects of visual or somatosensory stimuli on sound-evoked responses in both primary and secondary areas of the auditory cortex (Ghazanfar et al. 2005; Kayser et al. 2008).

Multisensory convergence in the auditory cortex appears to be more limited, however, when the spiking responses of individual neurons or small clusters of neurons are considered. This may be because cortical local field potentials primarily reflect summed synaptic currents and their accompanying return currents and therefore capture the input activity of the neurons (Einevoll et al. 2013). Nevertheless, multisensory influences on the spiking behavior of auditory cortical neurons again range from a change in firing rate when otherwise ineffective stimuli are paired with a sound to responses evoked directly by visual or somatosensory

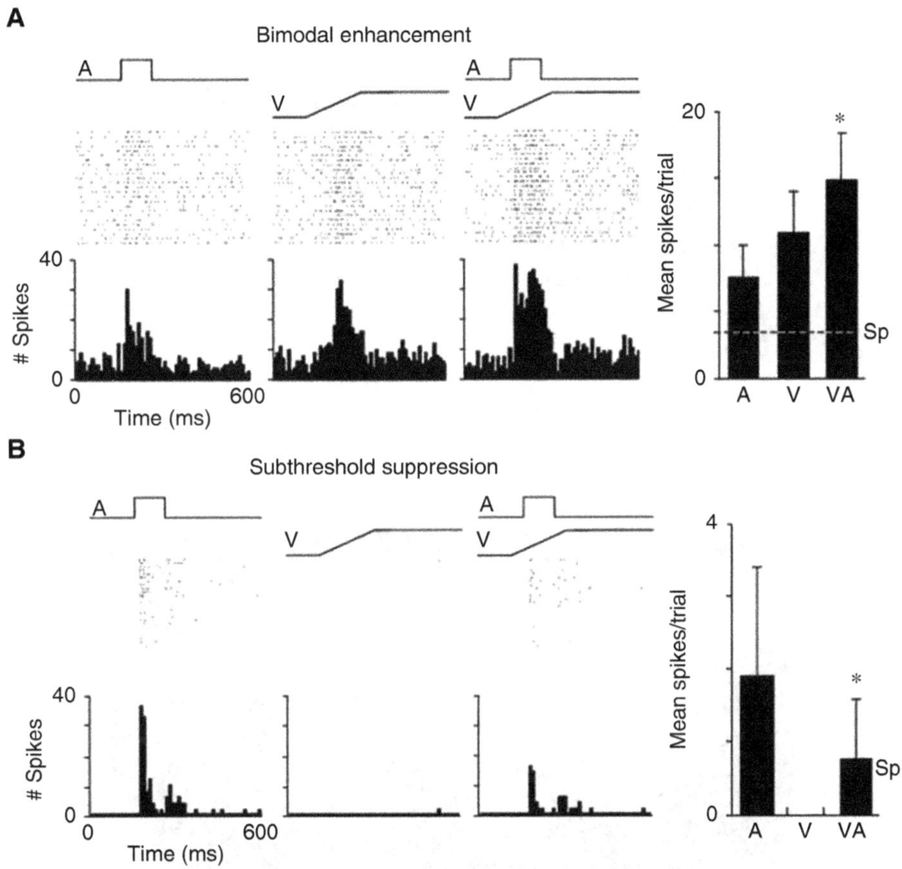

Fig. 6.2 Multisensory responses of neurons recorded in the auditory field of the anterior ectosylvian sulcus (FAES) of a cat to auditory, visual, and combined auditory-visual stimulation. (**A**) Example of a neuron that gave a suprathreshold spiking response to both auditory (*square wave A; top left*) and visual (*ramp V; top center*) stimuli presented alone and that generated a significantly enhanced response when the same stimuli were combined (*square wave and ramp together AV; top right*). (**B**) different FAES neuron that was activated by the auditory (*top left*) but not the visual stimulus (*top center*); in this case, presenting the two stimuli together led to a suppression of the auditory response (*top right*). In both (**A**) and (**B**), responses are shown in the form of raster plots (where each dot represents a spike with the response to multiple stimulus presentations arranged vertically; *center*), the corresponding peristimulus time histograms (*bottom*), and bar charts of the mean ± SD evoked activity for each stimulus type (*right*). $^*P < 0.05$, paired *t*-test. *Sp*, spontaneous activity level. Adapted from Meredith and Allman (2009), with permission

stimuli (Fu et al. 2003; Bizley et al. 2007). This apparent continuum of multisensory properties could reflect differences in the way sensory inputs converge on neurons either in the cortex itself (Fig. 6.3; Clemo et al. 2012) or at an earlier level in the processing hierarchy.

It is unclear what functions spiking responses to nonauditory stimuli in auditory cortex might serve, unless they convey signals that can be processed as if they were

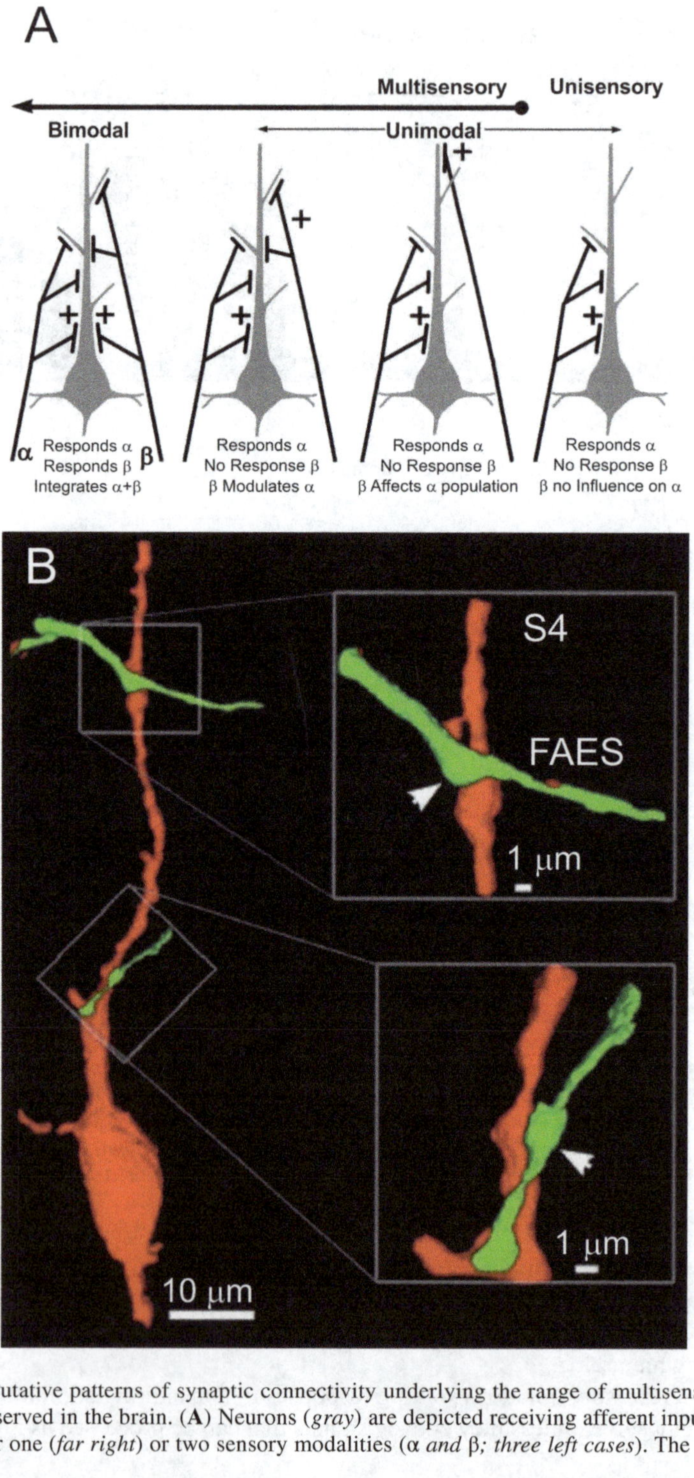

Fig. 6.3 Putative patterns of synaptic connectivity underlying the range of multisensory interactions observed in the brain. (**A**) Neurons (*gray*) are depicted receiving afferent inputs (*black*) from either one (*far right*) or two sensory modalities (α *and* β*; three left cases*). The simplified

auditory in origin. Indeed, it is possible that they are simply a consequence of synaptic inputs rising above threshold. In ferrets (*Mustela putorius*), the incidence of these nonauditory spiking responses increases between primary and high-level auditory cortical areas (Bizley et al. 2007), which likely reflects the greater density of projections to the latter from extralemniscal thalamic nuclei (Winer and Lee 2007) and from other sensory cortices (Bizley et al. 2007). Consequently, the relative proportion of neurons that receive subthreshold, modulatory inputs versus suprathreshold inputs that are capable of driving spiking activity is likely to be indicative of a progression from areas with a unisensory primary function to those more involved in merging independent inputs from different sensory modalities.

Another aspect to consider is the expected neural output of multisensory integration and to what extent it might vary in different parts of the brain. Electrophysiological recordings from multisensory neurons in the superior colliculus (SC) have led to the identification of several key principles by which different sensory inputs interact to govern the spiking activity of these neurons (King and Palmer 1985; Wallace et al. 1998). The SC is characterized by the presence of topographically aligned visual, auditory, and somatosensory maps. In such an organizational structure, the different modality signals arising from a particular location, and therefore potentially from the same source, can be represented by the same neurons. The strongest enhancement of the unisensory responses of SC neurons has been shown to occur when the component stimuli are weakly effective in eliciting a response and when those stimuli occur at approximately the same time and originate from the same region of space. By contrast, pairing strongly effective unisensory stimuli typically produces little or no enhancement as do multisensory signals that are widely separated in time or space. Indeed, this can result in a reduction of the firing rate elicited by unisensory stimulation. That these principles operate clearly makes sense because the relative timing and location of sensory signals are important factors in determining whether they belong to the same object and should therefore be bound together or to different objects.

Similar principles of multisensory integration have been observed in cortical neurons (Stein and Wallace 1996) and for various behavioral tasks, including the sensory-guided orienting responses with which the SC is likely to be involved (Stein et al. 1988; Bell et al. 2005). However, attempts to apply them to population and more indirect measures of neural activity, such as functional magnetic resonance imaging (fMRI), have turned out to be less straightforward (Stevenson et al. 2014). Moreover, it is an oversimplification to assume that improved perceptual abilities necessarily result from increased neural activity. Although functions in which the

Fig. 6.3 (continued) convergence patterns vary among the multisensory neurons so that although modality α evokes a spiking response in each case, modality β can result in a continuum of effects from producing a spiking response to modulating the response to α at the level of either the neuron or the neuronal population. (**B**) *Left*, confocal images of a neuron in the cat higher level somatosensory cortical area S4 (*red*) contacted by axons that originated in auditory FAES (*green*). *Right*, each axodendritic point of contact is enlarged to show the putative bouton swelling (*arrow*). Adapted from Clemo et al. (2012), with permission auditory cortex plays a pivotal role, including speech perception and sound localization, can be enhanced by the availability of other sensory cues, cortical neurons frequently exhibit cross-modal suppression. Thus, the response to the primary auditory stimulus is often reduced when combined with visual or somatosensory stimuli (Bizley et al. 2007; Meredith and Allman 2009). Furthermore, some studies have stressed the contribution of changes in the timing rather than the magnitude of the responses in the presence of multisensory stimuli (Chandrasekaran et al. 2013).

Other studies in which single neuron recordings were made from the auditory cortex have also highlighted the importance of quantifying multisensory interactions in ways that go beyond simple changes in the number of action potentials evoked. Application of information theoretic analyses to the spike discharge patterns recorded from neurons in the auditory cortex has revealed that visual cues can enhance the reliability of neural responses and hence the amount of information transmitted even if the overall firing rate either does not change or is suppressed (Bizley et al. 2007; Kayser et al. 2010). This finding is consistent with earlier work demonstrating that the location and identity of sounds can be encoded by the temporal firing pattern of auditory cortical neurons (Furukawa and Middlebrooks 2002; Nelken et al. 2005). Moreover, as first highlighted by the principle of inverse effectiveness in the superior colliculus (Meredith and Stein 1986), it is important to take response magnitude into account when characterizing the effects of multisensory stimulation on neuronal firing patterns. Indeed, Kayser et al. (2010) showed that multisensory stimulation can have opposite effects on the magnitude and reliability of cortical responses according to how strong the responses are to sound alone, with these opposing modes of multisensory integration potentially having different functions. Thus, enhanced responses to weakly effective stimuli are likely to facilitate the detection of near-threshold events, whereas suppressed but more reliable responses may be particularly relevant for sound discrimination at stimulus levels that are more typical of everyday listening.

Population measures of auditory cortical activity in human (Luo et al. 2010; Thorne et al. 2011) and nonhuman primates (Lakatos et al. 2007; Kayser et al. 2008) also indicate that nonauditory inputs can modulate the phase of low-frequency oscillatory activity in the auditory cortex. This is thought to alter the excitability of the cortex, increasing the amplitude of responses evoked by temporally correlated auditory inputs and thereby providing another way in which visual or other sensory inputs can modulate neuronal responses to accompanying sounds without necessarily evoking spiking activity. Indeed, it has even been proposed that phase resetting may represent one of the "canonical" operating principles used by the brain to integrate different types of information (van Atteveldt et al. 2014; see Keil and Senkowski, Chap. 10).

Together, these findings stress the importance of investigating multisensory interactions at multiple levels, from the activity of individual neurons to more population-based signals, including local field potentials (LFPs), EEG, MEG, and fMRI, and of employing the appropriate metrics in each case to quantify the magnitude and nature of integration. This is particularly important for making sense of how nonauditory inputs influence the auditory cortex without altering its fundamental role in hearing.

6.3 Nonauditory Inputs at Different Levels of the Auditory Pathway

In considering the potential role of multisensory interactions in the auditory cortex, it is essential to examine the origin of nonauditory inputs as well as their entry point into the auditory pathway. This can provide insight into the type of stimulus-related information those inputs convey and the extent to which the signals provided have already been processed and integrated by the time they reach the cortex.

At the subcortical level, nonauditory inputs have been identified in most of the main relay stations of the ascending auditory pathway. The complexity of this network, which includes more levels of subcortical processing than in other sensory modalities, makes it challenging to determine the role of these inputs. Furthermore, we currently have a poor understanding of the extent to which the multisensory interactions in the auditory cortex are inherited from the thalamus and therefore reflect bottom-up processing or arise from the convergence of inputs from other cortical areas.

Before discussing where nonauditory influences are found along the auditory pathway, it is first important to consider briefly the route by which acoustic information passes from the cochlea to the cortex (Fig. 6.4). Auditory nerve fibers transmit information from the cochlea to the cochlear nucleus (CN) in the brainstem, which is the first relay for the ascending auditory pathway. The CN comprises three subdivisions (anteroventral, posteroventral, and dorsal), within which are found several different neuron types that differ in their anatomical location, morphology, cellular physiology, synaptic inputs, and spectrotemporal response properties. The output from the CN takes the form of multiple, parallel ascending pathways with different targets. One of these is the superior olivary complex, where sensitivity to binaural localization cues emerges. The various ascending tracts then innervate the nuclei of the lateral lemniscus and all converge in the inferior colliculus (IC) in the midbrain, which therefore provides a relay for the outputs from each of the brainstem auditory centers. The IC comprises a central nucleus, which is surrounded by a dorsal cortex (DCIC), a lateral (or external) cortex (LCIC) and a rostral cortex, which can be distinguished by their connections and response properties. The IC in turn delivers much of the auditory input to the SC, which, as we have seen, is a major site for the integration of multisensory spatial information, and also projects to the medial geniculate body (MGB) in the thalamus, which serves as the gateway to the auditory cortex.

Classically, the ascending auditory pathway is thought to comprise a core or lemniscal projection, which is characterized by a precise tonotopic organization at each level from the CN to the primary auditory cortical fields. In addition, the extralemniscal projection includes parts of the IC, the MGB, and a belt of the auditory cortex surrounding the core tonotopic fields (Fig. 6.4). The defining features of neurons in extralemniscal areas are that they tend to show broader frequency tuning than those in the lemniscal projection and their tonotopic organization is less well defined or even nonexistent. Furthermore, they often receive inputs from other sensory modalities.

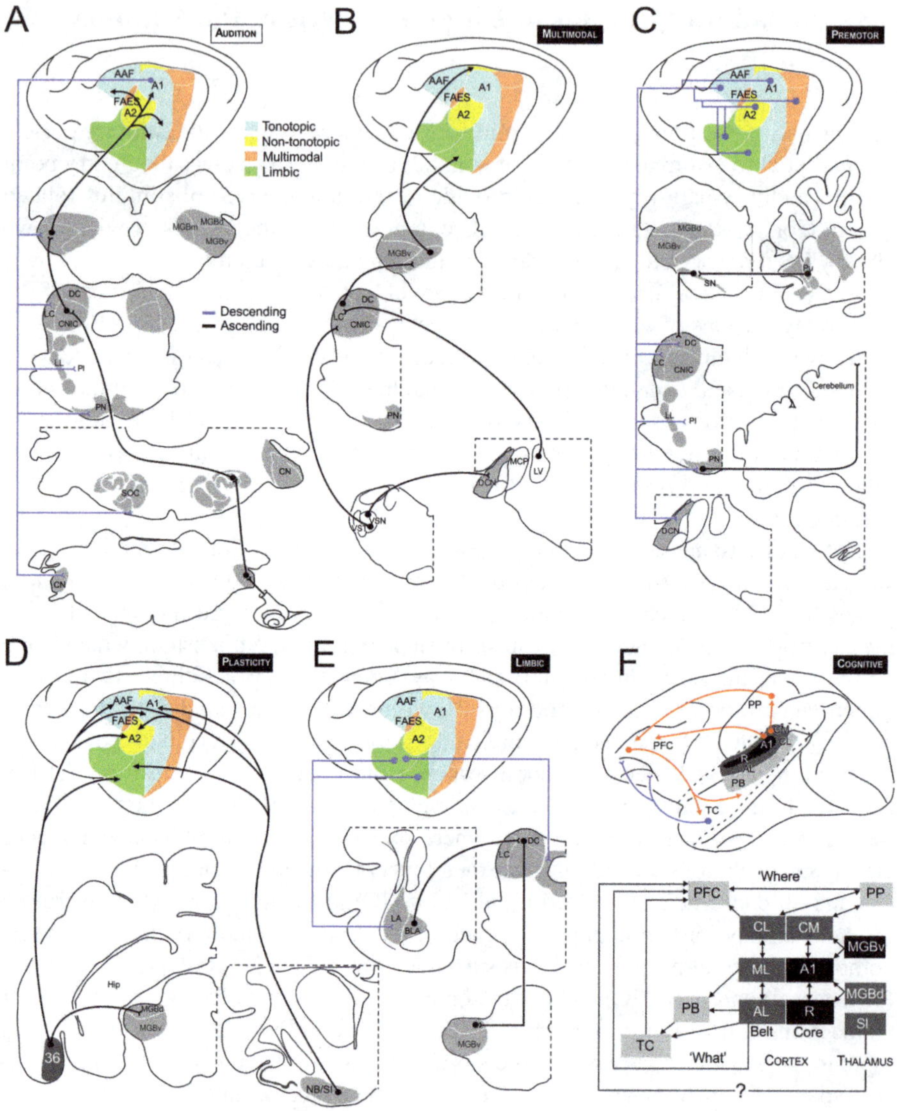

Fig. 6.4 Ascending and descending pathways of the cat auditory cortex with some of the main ascending (*black*) and descending (*blue*) connections shown according to their putative functional role and/or nature of the information transmitted. (**A**) Principal connections of the tonotopic lemniscal pathway. (**B**) Ascending connections in the extralemniscal pathway, highlighting auditory brain areas that receive projections from other sensory systems. (**C**) Descending cortical projections to premotor brain areas that participate in vocalization production and other motor functions. (**D**) Ascending connections associated with plasticity in the auditory cortex because of their cholinergic nature. (**E**) Descending cortical connections to the limbic system that are thought to contribute to emotional responses. (**F**) Putative cognitive streams involved in sound identification and localization (What and Where, respectively) described in macaques on the basis of the connectivity between the auditory cortex and prefrontal cortex. *A1* primary auditory cortex; *A2*, secondary

Although the functional significance of multisensory convergence within the subcortical auditory pathway is often unclear, there are instances where information from other sensory modalities makes an important contribution to the "unisensory" role of the neurons in question. Perhaps the best example is to be found in the dorsal CN (DCN). The complex excitatory and inhibitory interactions displayed by type IV neurons in the DCN of the cat (*Felis catus*) allow these neurons to signal the presence of spectral notches that are generated by the directional filtering properties of the external ear (Yu and Young 2000). Together with the finding that lesions of the pathway by which DCN projection neurons reach the IC result in impaired head orienting responses to broadband sounds (May 2000), this points to a likely role for this nucleus in sound localization. But cats are able to move their ears, shifting the locations at which spectral notches occur relative to the head. Consequently, information about the ongoing position of the pinnae is required to maintain accurate sound localization. This appears to be provided by muscle proprioceptors located in and around the pinna of the external ear, with the DCN combining monaural acoustical cues to sound source direction with somatosensory inputs about the orientation of the pinna (Kanold and Young 2001). More recent work in rats (*Rattus norvegicus*) indicates that multisensory computations in the DCN may also help distinguish moving sound sources from the apparent movement produced by motion of the head, suggesting that the integration of auditory and vestibular inputs helps to create a surprisingly sophisticated representation of spatial information at this early stage of auditory processing (Wigderson et al. 2016).

There is also evidence to suggest that somatosensory projections to the DCN are involved not only in sound localization but also in suppressing the effects of self-generated noises on the central auditory system, such as those produced by vocalizing and masticating (Shore and Zhou 2006). In support of this adaptive filter function, paired stimulation of the auditory and trigeminal nerves shows that neurons in the DCN are capable of multisensory integration and, more importantly, that the majority of multisensory interactions elicited are suppressive (Shore 2005; Koehler and Shore 2013). Interestingly, a related effect has also been described in tinnitus sufferers, whereby some individuals are able to modulate the loudness of their tinnitus by activating the trigeminal system using orofacial stimulation (Pinchoff et al. 1998). It has been suggested that the tinnitus percept arises, at least in part, from increased spontaneous activity in the DCN (Kaltenbach 2007).

Fig. 6.4 (continued) auditory cortex; *AAF*, anterior auditory field; *AL*, anterolateral area of the belt auditory cortex; *BLA*, basolateral nucleus of the amygdala; *CM*, caudomedial area of the belt auditory cortex; *CN*, cochlear nucleus; *CNIC*, central nucleus of the inferior colliculus (IC); *CL*, caudolateral area of the belt auditory cortex; *DC*, dorsal cortex of the IC; *DCN*, dorsal CN; *Hip*, hippocampus; *MGBv*, *-d*, and *-m*, medial geniculate body (ventral, dorsal, and medial divisions, respectively); *LA*, lateral nucleus of the amygdala; *LC*, lateral cortex of the IC; *LL*, lateral lemniscus; *NB/SI*, nucleus basalis/substantia innominata; *PB*, parabelt auditory cortex; *PFC*, prefrontal cortex; *PP*, posterior parietal cortex; *Pl*, paralemniscal area; *PN*, pontine nuclei; *Pu*, putamen; *R*, auditory cortical area; *SI*, suprageniculate nucleus, lateral part; *SN*, substantia nigra; *SOC*, superior olivary complex; *TC*, temporal cortex; *36*, cortical area 36. Adapted from Winer and Lee (2007), with permission

Therefore, it is conceivable that in addition to suppressing neural responses to self-generated sounds, somatosensory inputs to the DCN may reduce abnormal activity associated with phantom sounds, highlighting the therapeutic potential of harnessing somatosensory inputs to the DCN to alleviate tinnitus.

Beyond the DCN, responses to somatosensory stimulation have also been described in the auditory midbrain and, particularly, in the LCIC. This activity again likely reflects the influence of inputs from multiple sources, which include the dorsal column nuclei, the spinal trigeminal (Sp5) nucleus, and the somatosensory cortex (Shore and Zhou 2006; Lesicko et al. 2016). But whereas somatosensory inputs to the DCN originate principally from the pinnae, in accordance with their presumed role in the processing of spectral localization cues, those to the IC suggest a diffuse input from the entire body (Aitkin et al. 1981). Thus, somatosensory responses in the IC are not just inherited from the DCN and may serve to suppress the effects of self-generated noises regardless of their spatial origin.

The first responses to visual stimulation in the auditory pathway appear to be found in the midbrain, and recordings in behaving monkeys (*Macaca mulatta*) have reported that the prevalence of visual influences on IC neurons may be surprisingly high (Porter et al. 2007). This is supported by the presence of sparse inputs from the retina to the DCIC (Morin and Studholme 2014) and from the visual cortex to various subdivisions of the IC (Cooper and Young 1976; Gao et al. 2015). However, the primary source of visual input to the auditory midbrain, and potentially therefore to other parts of the auditory pathway, appears be the SC. Indeed, in ferrets, the nucleus of the brachium of the IC (Doubell et al. 2000) and the LCIC (Stitt et al. 2015) have reciprocal connections with the SC. This provides a source of retinotopically organized input into different regions of the IC, which may play a role in coordinating and updating the alignment of maps of visual and auditory space in the SC (Doubell et al. 2000; Stitt et al. 2015). Potentially related to this is the finding that auditory responses in the monkey IC are modulated by changes in gaze direction (Groh et al. 2001; Zwiers et al. 2004). If accurate gaze shifts are to be made to auditory targets, it is essential that eye position signals are incorporated in the brain's representation of auditory space (see also Willett, Groh, and Maddox, Chap. 5). This is well-known to be the case in the SC (Jay and Sparks 1984; Hartline et al. 1995), and these findings indicate that this process most likely begins in the IC. Beyond a role in spatial processing, nonauditory inputs to the IC could contribute to other aspects of multisensory behavior. A single case study of a human patient with a unilateral IC lesion reported a weaker McGurk effect for audiovisual speech stimuli in the contralesional hemifield (Champoux et al. 2006), although it is unclear whether this reflects multisensory processing in the IC itself.

The thalamus is the final subcortical level in the auditory pathway at which multisensory processing occurs. In addition to inheriting nonauditory inputs via ascending projections from earlier stages in the pathway, the medial division of the MGB (MGBm) is innervated by the spinal cord, whereas the dorsal nucleus of the MGB (MGBd) and the suprageniculate nucleus receive inputs from the SC (Jones and Burton 1974; Katoh and Benedek 1995). An added complication when discussing multisensory processing in the thalamus is that we need to consider not only those subdivisions comprising the auditory thalamus itself (e.g., the lemniscal ventral nucleus of the MGB (MGBv) and the extralemniscal MGBm and MGBd), but also those subdivisions regarded as higher order or multisensory, such as the pulvinar, which project to and receive inputs from auditory as well as other cortical areas (de la Mothe et al. 2006a; Scott et al. 2017). A detailed description of these projections is beyond the scope of this chapter (see Cappe et al. 2012 for a review). However, their existence is important to note, given that they provide a potential route for transferring information between different cortical areas, including those belonging to different sensory modalities (Rouiller and Welker 2000; Sherman 2016). Moreover, cortico-thalamo-cortical circuits can also involve the primary sensory thalamus. Thus, visual and whisker signals are combined in the ventral posterior medial region of the thalamus in mice (*Mus musculus*) (Allen et al. 2017), whereas activation of the primary somatosensory cortex in this species can alter the activity of neurons in the MGB (Lohse et al. 2017).

6.4 Origins of Visual and Somatosensory Inputs to the Auditory Cortex

The studies discussed so far show that multisensory information is incorporated at most stages along the ascending auditory pathway, with nonauditory inputs primarily, but not exclusively, targeting extralemniscal regions. Therefore, at the cortical level, it seems reasonable to expect that nonauditory influences will be most apparent in the cortical belt areas because of their extralemniscal inputs, and this has been confirmed by anatomical and physiological experiments in a range of species (e.g., Bizley et al. 2007). In addition to its subcortical origin, however, multisensory convergence in the auditory cortex has been shown to result from inputs from other sensory as well as higher level association cortical areas.

Anatomical tracing studies have identified direct corticocortical connections between different sensory areas in several species. As with subcortical levels of the auditory pathway, inputs from visual and somatosensory cortical areas are distributed primarily to noncore parts of the auditory cortex, such as the caudomedial belt areas in marmosets (*Callithrix jacchus*; de la Mothe et al. 2006b) and macaque monkeys (Falchier et al. 2010) or the fields on the anterior and posterior ectosylvian gyri in ferrets (Bizley et al. 2007). This is consistent with the greater incidence of multisensory neurons in those regions and with the often nonlemniscal nature of their auditory response properties. Nevertheless, the activity of neurons in the core auditory cortex can be modulated and sometimes even driven by nonauditory inputs. In the ferret, for example, around 20% of neurons in the core areas, the primary auditory cortex and the anterior auditory field, were shown to be sensitive to visual (Bizley et al. 2007) or tactile (Meredith and Allman 2015) stimulation. Although sparse projections from primary or secondary sensory areas were observed in these studies, the greatest proportion of retrograde labeling following tracer injections in the core auditory cortex was found in visual area 20 (Bizley et al. 2007) and in the rostral suprasylvian sulcal somatosensory area (Meredith and Allman 2015). This would suggest that core auditory areas in ferrets are mainly innervated by higher order visual and somatosensory cortical areas. Direct connections between A1 and other primary and secondary sensory cortical areas have also been described in rodents (Fig. 6.5; Budinger et al. 2006; Stehberg et al. 2014). Similarly, in marmosets, the core auditory cortex is innervated by the secondary somatosensory cortex and by the STS (Cappe and Barone 2005), whereas other studies in primates suggest that nonauditory influences on A1 most likely originate from the thalamus as well as from multisensory association areas like the STS (Smiley and Falchier 2009).

Most of the anatomical studies have used retrograde tracer injections to reveal the origins of projections to the auditory cortex. Although this approach does not provide a clear picture of the extent and the laminar distribution of the terminal fields in the auditory cortex, it is possible to infer something about the nature of the inputs from the laminar origin of the projection neurons. Thus, feedforward corticocortical projections typically originate in the supragranular layers and terminate in granular layer IV, whereas feedback corticocortical projections are more likely to originate in the infragranular layers and to terminate in the supragranular and infragranular layers. After retrograde tracer injections into A1, labeled cells were found predominantly in the infragranular layers of the projecting cortices (Cappe and Barone 2005; Budinger et al. 2006). This suggests that the core auditory cortex receives mainly feedback projections from other cortical areas and is consistent with physiological evidence in monkeys that somatosensory inputs target the supragranular layers and have a modulatory influence on A1 activity (Lakatos et al. 2007). However, feedforward projections to the auditory cortex cannot be excluded because several studies have also reported retrogradely labeled cells in the supragranular layers of other cortical areas (Cappe and Barone 2005; Budinger et al. 2006).

The relative contributions of thalamocortical and corticocortical projections to multisensory processing in the auditory cortex are poorly understood. However, Budinger et al. (2006) estimated that approximately 60% of nonauditory inputs to gerbil A1 originate subcortically, with the remaining 40% arising from other sensory or multisensory cortical areas. It is therefore clear that a hierarchy of multisensory

Fig. 6.5 (continued) experiments. *Numbers next to the arrows* connecting the cortical areas represent the number of labeled cells found in the supragranular layers minus the number in the infragranular layers divided by the total of labeled cells; positive values indicate putative feedforward projections and negative values indicate putative feedback projections. Although the strongest connections to the primary sensory cortices come from their modality-specific thalamic nuclei, cross-modal inputs arise from other sensory cortices and the (extralemniscal) thalamus. *DLG*, dorsal lateral geniculate nucleus; *LD*, laterodorsal thalamic nucleus; *LP*, lateral posterior thalamic nucleus; *MGBmz*, MGB marginal zone; *Po*, posterior thalamic nuclear group; *S1*, primary somatosensory cortex; *Sg*, suprageniculate nucleus; *V1*, primary visual cortex; *VL*, ventrolateral thalamic nucleus; *VLG*, ventral lateral geniculate nucleus; *VPL*, ventral posterolateral thalamic nucleus; *VPM*, ventral posteromedial thalamic nucleus. Adapted from Henschke et al. (2015), with permission

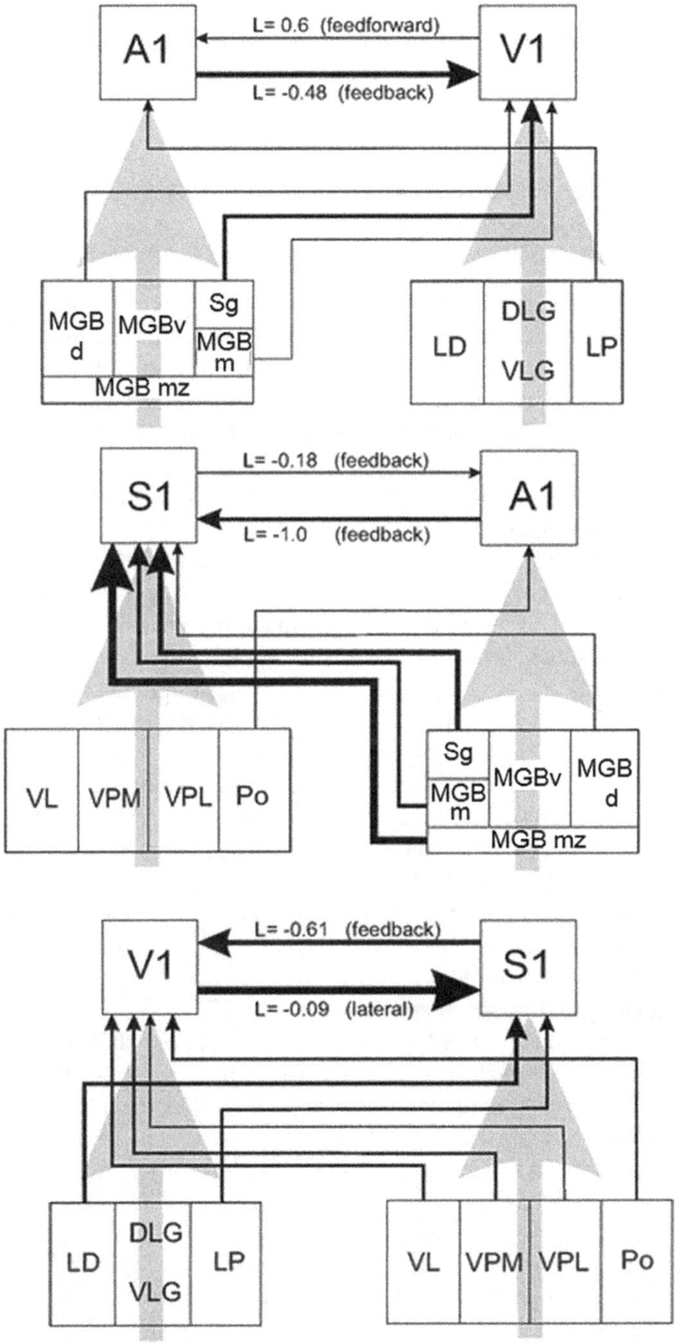

Fig. 6.5 Summary of the direct thalamocortical and corticocortical connections of the primary auditory, visual, and somatosensory cortices in the Mongolian gerbil (*Meriones unguiculatus*). *Thickness of the lines* indicates the strength of the connections as revealed by retrograde tracing

processing exists within the auditory pathway and that the auditory cortex in particular is likely to be involved in various functions that depend on the integration of information across different sensory modalities.

6.5 Functional Significance of Multisensory Interactions in the Auditory Cortex

Because there are so many subcortical and cortical sources of nonauditory inputs in the auditory pathway, it is challenging to pinpoint specific functions for the cross-modal influences that can be observed at the level of the cortex. Indeed, establishing a causal relationship between multisensory interactions at the neural and behavioral levels is particularly difficult because this field of research has yet to benefit to any great degree from the experimental approaches, such as optogenetics, that are now available for interrogating the functions of specific neural circuits (Olcese et al. 2013; Wasserman et al. 2015).

Nevertheless, insights into what those functions might be can be obtained by knowing the sources of input to particular auditory cortical areas and, of course, by measuring how the responses of the neurons change in the presence of stimuli belonging to other sensory modalities. In addition to amplifying the responses of auditory cortical neurons, particularly to relatively weak sounds, visual stimuli have been shown to induce more specific effects on the sensitivity and even the selectivity of these neurons. As discussed in Sect. 6.1, speech perception can be profoundly influenced by the talker's facial gestures, with studies in macaque monkeys demonstrating that neural responses to conspecific vocalizations are enhanced when accompanied by a video clip of an animal vocalizing but not when paired with a disk presented to mimic the opening of the mouth (Ghazanfar et al. 2005; Ghazanfar 2009). Similarly, by pairing complex naturalistic audiovisual stimuli, including videos and the accompanying sounds of conspecific animals, Kayser et al. (2010) found that the information gain in the auditory cortical responses was reduced when the auditory and visual cues were no longer matched in their dynamics or semantic content.

These visual influences have been measured in different auditory cortical areas, including A1. The complexity of the visual information involved in interpreting articulation cues makes it unlikely that the auditory cortex receives this information directly from early visual cortices. Instead, simultaneous recordings in the auditory cortex and STS showed that spiking activity in the auditory cortex was coordinated with the oscillatory dynamics of the STS (Ghazanfar et al. 2008). Thus, in the case of communication signals, the integration of multisensory information in the auditory cortex likely depends, at least in part, on top-down inputs from this area of association cortex and probably also from other cortical areas that have been shown to be entrained by lip movements during speech (Park et al. 2016).

The other major area where the functional significance of cross-modal interactions in the auditory cortex is starting to become clear is sound localization. An intact auditory cortex is required for normal spatial hearing, and inactivation studies in cats suggest that this reflects the contribution of A1 plus certain higher level auditory cortical fields, such as the posterior auditory field (PAF; Malhotra and Lomber 2007). Studies such as these have contributed to the notion that segregated cortical processing streams exist for different auditory functions (Fig. 6.4F). Although the extent to which a division of labor exists across the auditory cortex in the processing of different sound features remains controversial (Schnupp et al. 2011; Rauschecker 2018), these findings raise the possibility that the way nonauditory stimuli influence the processing of spatial and nonspatial sound properties may be area specific.

Studies in ferrets have provided some support for this hypothesis. As expected from the extensive subcortical processing that takes place in the auditory pathway, neurons across different auditory cortical fields encode both spatial and nonspatial sound features. However, neurons located in the auditory fields located on the posterior ectosylvian gyrus in this species are more sensitive to stimulus periodicity and timbre than to spatial location (Bizley et al. 2009). In keeping with a potentially greater role in stimulus identification, this region receives inputs from areas 20a and 20b, which have been implicated in visual form processing (Manger et al. 2004). Conversely, neurons that are most informative about the azimuthal location of auditory, visual or paired auditory-visual stimuli are found on the anterior ectosylvian gyrus (Bizley and King 2008), which is innervated by a region of extrastriate visual cortex that has been implicated in spatial processing (Fig. 6.6; Philipp et al. 2006; Bizley et al. 2007).

The interpretation of these results needs to be treated with some caution because relatively little research has so far been carried out on higher level visual or auditory cortical fields in ferrets, so a detailed understanding of the functions of these areas is not yet available. However, the cross-modal reorganization observed following deafness is consistent with the notion that visual inputs target auditory cortical areas with related functions. Relative to hearing animals, congenitally deaf cats exhibit superior visual localization in the peripheral field and lower movement detection thresholds (Lomber et al. 2010). Cooling PAF, one of the key auditory cortical fields involved in spatial hearing, produced a selective loss of this enhanced visual localization, whereas deactivating the dorsal zone of the auditory cortex raised visual motion detection thresholds to values typical of hearing animals (Fig. 6.7). These findings therefore suggest that cross-modal plasticity occurs in cortical regions that share functions with the nondeprived sensory modality.

In keeping with the effects of matching naturalistic auditory-visual stimuli in nonhuman primates, the presence of spatially congruent visual stimuli produced an overall gain in the spatial information available in the responses of ferret auditory cortical neurons (Fig. 6.8; Bizley and King 2008). However, these effects were found to vary from neuron to neuron, and the largest proportion of neurons that showed an increase in transmitted spatial information when visual and auditory stimuli were presented together was actually found in the posterior suprasylvian field, where sensitivity to sound periodicity and timbre is most pronounced.

Although these studies have shown that information coding in the auditory cortex can be enhanced by the availability of matching visual cues, relatively few

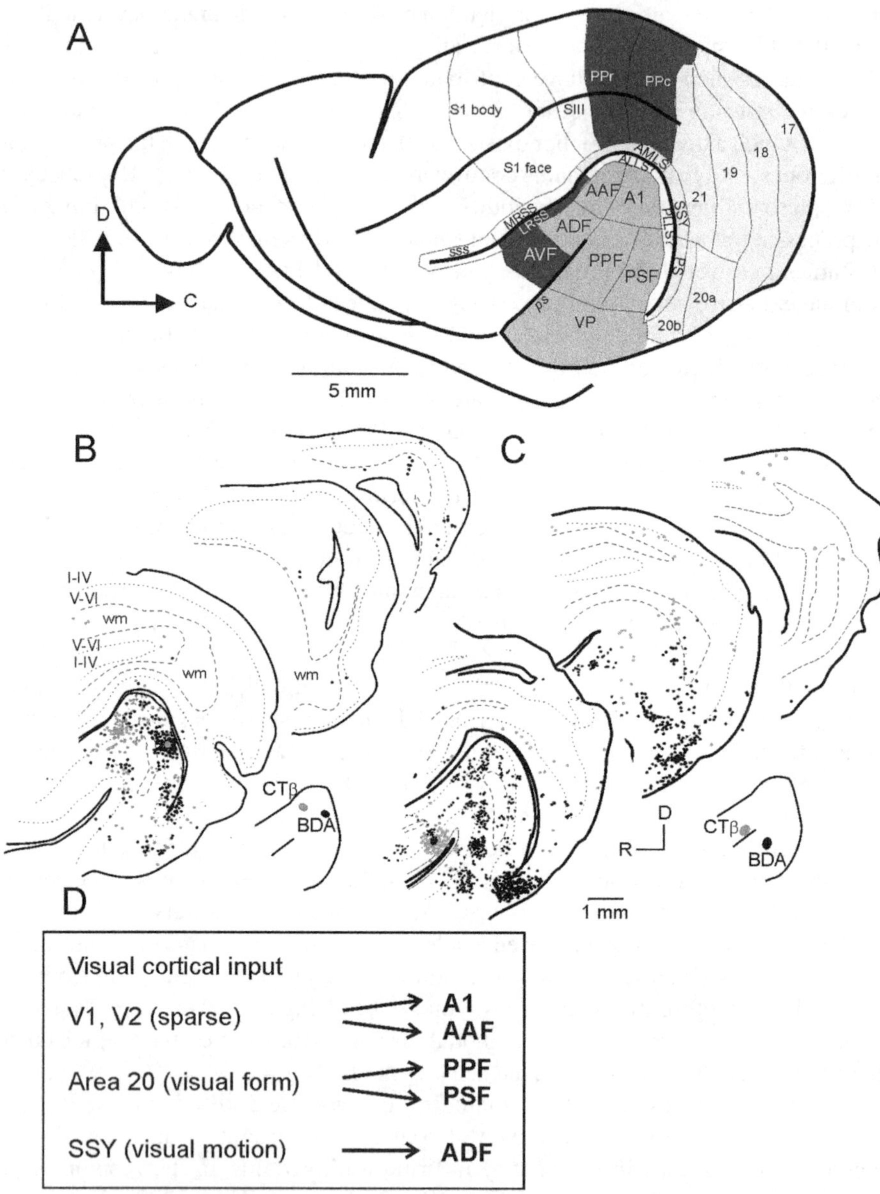

Fig. 6.6 Visual inputs to ferret auditory cortex. (**A**) Visual (areas 17-20, PS, SSY, AMLS), posterior parietal (PPr, PPc), somatosensory (S1, SIII, MRSS), and auditory (A1, AAF, PPF, PSF, and ADF) areas are shown. In addition, LRSS and AVF are multisensory regions, although many of the areas usually classified as modality specific also contain some multisensory neurons. (**B**) Location of neurons in the visual cortex that project to the auditory cortex. Tracer injections made into the core auditory cortex (A1: biotinylated dextran amine, *black*; AAF: cholera toxin subunit β, *gray*) result in retrograde labeling in the early visual areas. *Dotted lines*, limit between cortical layers IV and V; *dashed lines*, delimit the white matter. (**C**) Tracer injections made into belt the auditory cortex.

have measured neuronal activity while the animals carry out multisensory tasks (e.g., Brosch et al. 2005; Chandrasekaran et al. 2013). Consequently, the behavioral relevance of the cross-modal effects observed under anesthesia or in awake, nonbehaving animals remains speculative. Moreover, where auditory cortical recordings have been made in behaving animals, there are indications that task engagement can be accompanied by the emergence of responses to nonauditory stimuli (Brosch et al. 2005; Lakatos et al. 2009) and that the modulatory nature of these stimuli may differ. Thus, visible mouth movements improve the ability of monkeys to detect vocalizations, with this behavioral advantage accompanied by shorter latency responses by auditory cortical neurons rather than changes in the magnitude or variability of their firing rates (Fig. 6.9; Chandrasekaran et al. 2013). This again stresses the importance of considering both rate and temporal codes when investigating the impact of multisensory integration at the level of the auditory cortex.

Measuring cortical activity during behavior has provided other insights into the neural basis for cross-modal influences on auditory perception. Because visual information is normally more accurate and reliable in the spatial domain, it can provide a reference for calibrating the perception of auditory space. This is particularly the case during development when vision plays a key role in aligning the neural maps of space in the SC, as revealed by the changes produced in the auditory spatial receptive fields when the visual inputs are altered (reviewed in King 2009). This cross-modal plasticity compensates for growth-related changes and individual differences in the relative geometry of sense organs. But as illustrated by the ventriloquism illusion and related phenomena (Zwiers et al. 2003), vision can also be used in adulthood to alter the perceived location of sound sources to resolve short-term spatial conflicts between these modalities. The neural basis for the ventriloquism illusion is poorly understood, but event-related potential and fMRI measurements have revealed changes in the activity levels in the auditory cortex on trials in which participants experienced a shift in perceived sound location in the direction of a misaligned visual stimulus (Fig. 6.10; Bonath et al. 2007, 2014).

Fig. 6.6 (continued) *Gray*, retrograde labeling after an injection of CTβ into the anterior fields (on the borders of ADF and AVF); *black*, retrograde labeling resulting from a BDA injection into the posterior fields PPF and PSF. Note the difference in the extent and distribution of labeling after injections into the core and belt areas of auditory cortex. (**D**) Summary of sources of visual cortical input to different regions of auditory cortex, with their likely functions indicated. *ADF*, anterior dorsal field; *ALLS*, anterolateral lateral suprasylvian visual area; *AMLS*, anteromedial lateral suprasylvian visual area; *AVF*, anterior ventral field; *BDA*, biotinylated dextran amine; *C*, caudal; *CTβ*, cholera toxin subunit β; *D*, dorsal; *I-VI*, cortical layers; *LRSS*, lateral bank of the rostral suprasylvian sulcus; *MRSS*, medial bank of the rostral suprasylvian sulcus; *PLLS*, posterolateral lateral suprasylvian area; *PPF*, posterior pseudosylvian field; *PSF*, posterior suprasylvian field; *pss*, pseudosylvian sulcus; *PPc*, caudal posterior parietal cortex; *PPr*, rostral posterior parietal cortex; *PS*, posterior suprasylvian area; *R*, rostral; *S1*, primary somatosensory cortex; *SIII*, tertiary somatosensory cortex; *SSY*, suprasylvian cortex; *VP*, ventroposterior area; *wm*, white matter. Adapted from Bizley et al. (2007), with permission

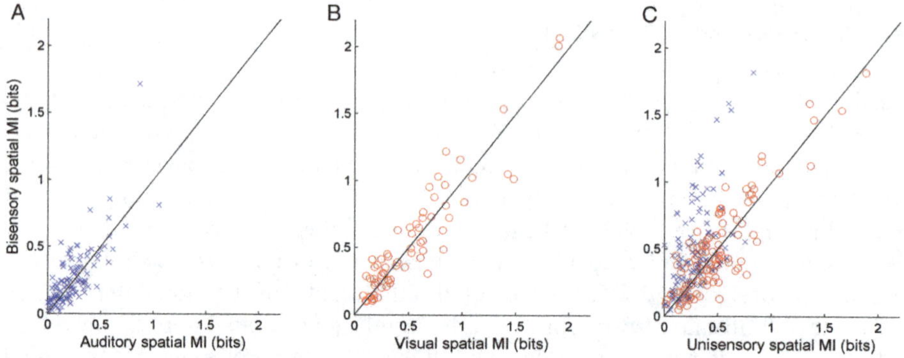

Fig. 6.7 Pairing auditory and visual stimulation produces an overall increase in the spatial information conveyed by ferret auditory cortex neurons that were driven by auditory (**A**), visual (**B**), or both auditory and visual (**C**) stimuli. Each symbol (*blue crosses*, auditory; *red circles*, visual) shows the estimated mutual information (MI) between the stimulus location and the spike trains evoked by unisensory stimulation (*x*-axis) and by combined visual-auditory stimulation (*y*-axis) for a different neuron. Higher values indicate that the responses conveyed more information about the location of the stimuli so points above the line mean that more information was transmitted in response to combined visual-auditory stimulation than in the unisensory condition. Reproduced from Bizley and King (2008), with permission

Task	PAF deactivation	DZ deactivation
Visual localization in the peripheral field	Deficit	No deficit
Movement detection	No deficit	Deficit

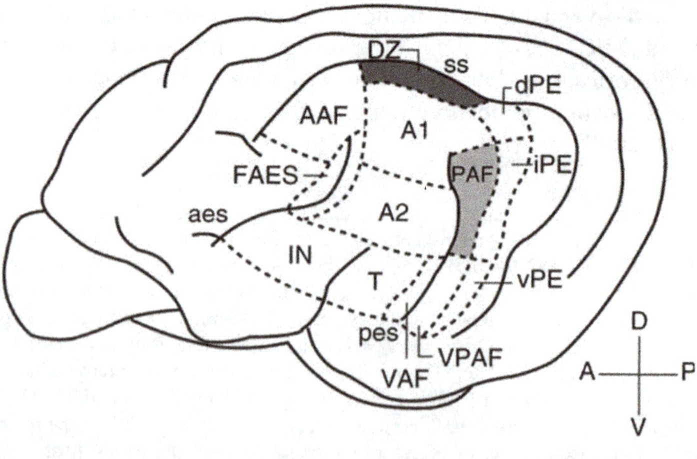

Fig. 6.8 *Top*, double dissociation of visual functions in the auditory cortex of the congenitally deaf cat. Bilateral deactivation of the PAF, but not the DZ, resulted in the loss of enhanced visual localization in the far periphery. On the other hand, bilateral deactivation of the DZ, but not the PAF, resulted in higher movement detection thresholds. *Bottom*, lateral view of the cat cerebrum highlighting the locations of the PAF and DZ. *A*, anterior; *aes*, anterior ectosylvian sulcus; *dPE*, dorsal posterior ectosylvian area; *DZ*, dorsal zone of auditory cortex; *IN*, insular region; *iPE*, intermediate posterior ectosylvian area; *P*, posterior; *PAF*, posterior auditory field; *pes*, posterior ectosylvian sulcus; *ss*, suprasylvian sulcus; *T*, temporal region; *V*, ventral; *VAF*, ventral auditory field; *VPAF*, ventral posterior auditory field; *vPE*, ventral posterior ectosylvian area. Reproduced from Lomber et al. (2010), with permission

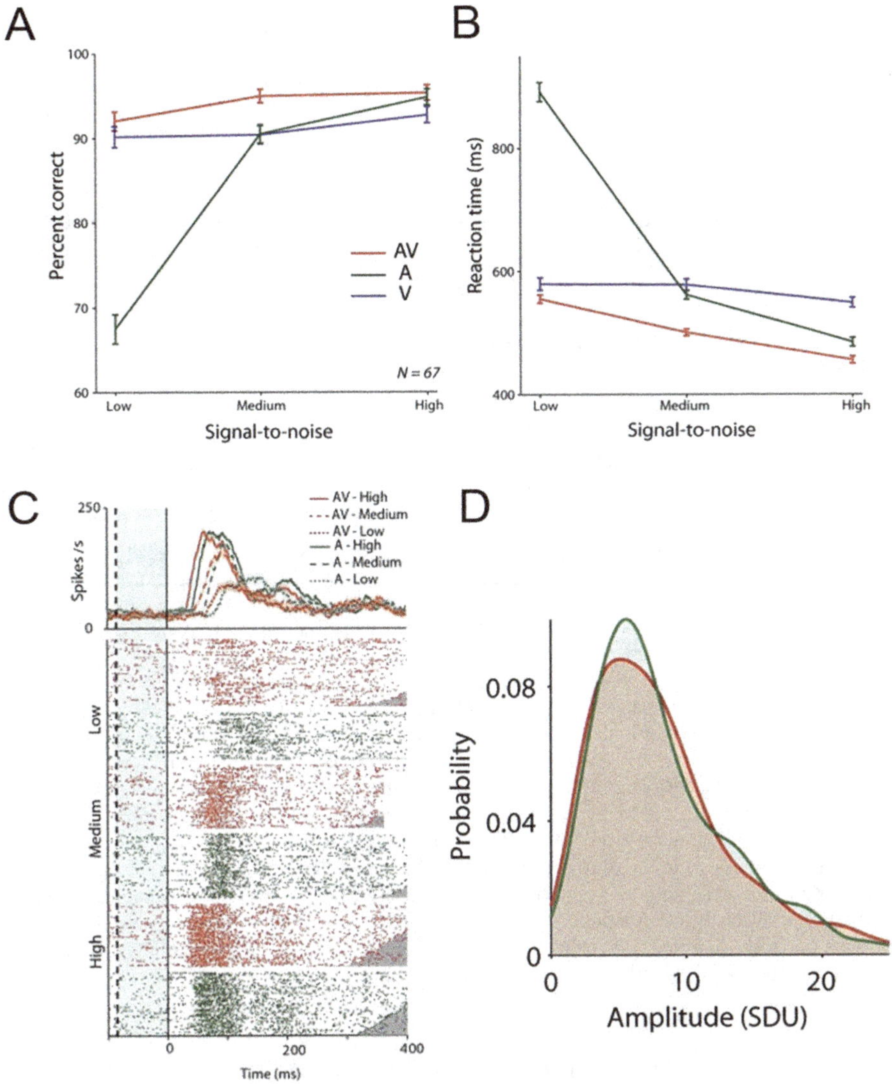

Fig. 6.9 Auditory cortical correlate of the ability of monkeys to detect auditory-visual vocalizations. Accuracy (**A**) and reaction time (**B**) for three different signal-to-noise levels for monkeys trained to detect auditory-visual vocalizations and their component auditory or visual stimulus are shown. Note the superior performance when both modality cues are available. Values are means ± SE. (**C**) Peristimulus time histogram (*top*) and rasters (*bottom*) showing the responses to auditory (A), visual (V), and auditory-visual stimulation (AV) at the three signal-to-noise levels. *Solid line*, onset of the auditory stimulus; *dashed line*, onset of the visual stimulus; *blue shading*, time period when only visual input was present. The auditory cortex responds faster with the addition of mouth motion. (**D**) Probability density of peak magnitudes for the spiking responses in the AV (*red*) and A (*green*) conditions. The *x*-axis depicts the change in normalized response magnitude in standard deviation units (SDU); the *y*-axis depicts the probability of observing that response magnitude. No systematic changes in the magnitude or variability of the firing rate were observed with the addition of mouth motion. Adapted from Chandrasekaran et al. (2013), with permission

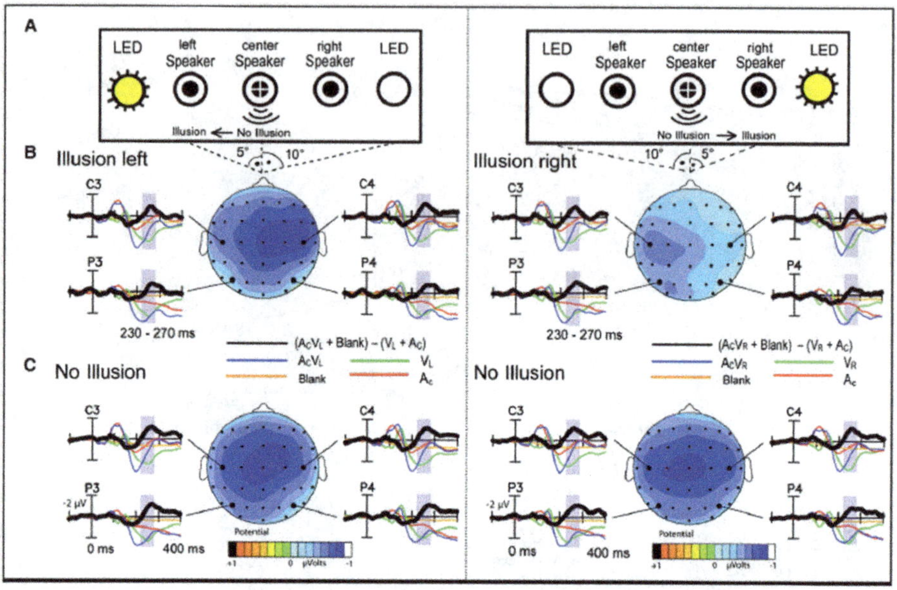

Fig. 6.10 Auditory cortical correlates of the ventriloquism illusion. (**A**) Tones were presented from left, center, or right loudspeakers, either alone or in combination with flashes from a LED on the right or left side. *Left*, stimulus combination of central tone (A_C) plus left flash (V_L); *right*, A_C plus right flash (V_R) combination. (**B**) Grand averaged event-related potential (ERP) waveforms to auditory (*red*), visual (*green*), blank (*orange*), and auditory-visual (*blue*) stimuli together with the multisensory difference waves ([AV + blank] – [A + V]; *thick black*) recorded from central (C3, C4) and parietal (P3, P4) electrodes on trials where the ventriloquist illusion was present (i.e., subjects perceived the sound as coming from the speaker on the same side as the flash). Topographical voltage maps are of the N260 component measured as mean amplitude over 230–270 ms (*shaded areas*) in the multisensory difference waves. Note larger amplitude contralateral to the side of the flash and perceived sound. (**C**) Grand average ERPs and topographical voltage distributions of the N260 component on trials where the ventriloquist illusion was absent (i.e., subjects correctly reported the sound location to be at the center). Note bilaterally symmetrical voltage distributions of N260. Reproduced from Bonath et al. (2007), with permission

The growing evidence that the auditory cortex may provide a substrate for visual influences on spatial hearing raises an important question. In the SC, each of the sensory representations is topographically organized and together they form overlapping maps of space. A shift in the visual map is therefore readily translated into a corresponding adjustment in the representation of auditory space by systematically retuning the neurons to a new set of spatial cue values, as illustrated by recordings from the optic tectum, the homologous structure to the SC, in barn owls (Knudsen 2002). In the mammalian cortex, however, there is no map of auditory space, and it is currently thought that the sound source azimuth is likely to be encoded by a comparison of activity between neurons with heterogeneous spatial sensitivity within each hemisphere (Stecker et al. 2005; Keating et al. 2015). Although it remains unclear how visual inputs, whether they originate subcortically or from other parts of the cortex, might "map" onto this arrangement, the finding that the ventriloquism illusion is associated with a change in the balance of activity between the left and right auditory cortical areas (Bonath et al. 2014) raises testable hypotheses.

Maintaining concordant multisensory representations of space in the brain requires continuous recalibration because the spatial information provided by each modality is, at least initially, encoded using different reference frames (see Willett, Groh, and Maddox, Chap. 5). Thus, visual signals are encoded using eye-centered retinal coordinates, whereas auditory signals are head centered because the location of a sound source is derived from interaural time and level differences in conjunction with the spectral localization cues generated by the head and each external ear. An important strategy used by the brain to cope with this is to incorporate information about current gaze direction into the brain's representation of auditory space. As stated in Sect. 6.3, this process begins in the midbrain and is widespread in the monkey auditory cortex (Werner-Reiss et al. 2003), with at least some of the effects of eye position likely to arise from feedback from the parietal or frontal cortex (Fu et al. 2004). The importance of oculomotor information has also been demonstrated behaviorally by the surprising finding that looking toward a sound while keeping the head still significantly enhances the discrimination of both interaural level and time differences (Maddox et al. 2014).

6.6 Concluding Remarks

It is increasingly clear that focusing exclusively on the responses of neurons to the acoustic properties of sound is insufficient to understand how activity in the central auditory pathway, and the cortex in particular, underpins perception and behavior. Because increasingly naturalistic conditions are being used to study auditory processing, more attention is being paid to the interplay between the senses. It is now known that multisensory interactions are a property of many neurons in the auditory pathway, just as they are for other sensory systems. These interactions most commonly take the form of a modulation of auditory activity, with other sensory inputs providing contextual cues that signal the presence of an upcoming sound, thereby making it easier to hear. Additionally, they may convey more specific information about the location or identity of multisensory objects and events and enhance or recalibrate the tuning properties of the auditory neurons without changing their primary role in hearing.

Although the application of more sophisticated analytical approaches has provided valuable insights into how multisensory signals are encoded by individual auditory neurons, there is currently little understanding of the way in which populations of neurons interact to represent those signals. Moreover, given that multisensory interactions are so widespread in the brain, it remains a daunting task to decipher the specific circuits that underlie a particular behavior. Indeed, it is becoming increasingly clear that multiple circuits exist for mediating the influence of one modality on another, as shown by recent experiments in mice illustrating the different routes by which activity in the auditory cortex can suppress that in the visual cortex (Iurilli et al. 2012; Song et al. 2017). Identification of behaviorally relevant circuits is a necessary step toward an improved understanding of the cellular and synaptic mechanisms underlying multisensory interactions.

The effects of multisensory processing on perception are well documented in humans but understandably less so in other species. As more is learned about the brain regions and cell types that mediate multisensory interactions, it will be necessary to develop new behavioral paradigms to probe their role in merging different sensory stimuli and resolving conflicts between them. This will enable further assessment of the role of attention and task engagement in multisensory processing, which has so far been largely restricted to studies in primates, as well as investigation into the role of sensory experience in shaping the connections and response properties of neurons in the auditory cortex and elsewhere in the brain so that they integrate other sensory inputs that are commonly associated with sounds.

Acknowledgments The authors' research is supported by the Wellcome Trust through Principal Research Fellowship WT108369/Z/2015/Z to Andrew J. King.

Compliance with Ethics Requirements Andrew J. King declares that he has no conflict of interest.

Amy Hammond-Kenny declares that she has no conflict of interest.

Fernando R. Nodal declares that he has no conflict of interest.

References

Aitkin, L. M., Kenyon, C. E., & Philpott, P. (1981). The representation of the auditory and somatosensory systems in the external nucleus of the cat inferior colliculus. *Journal of Comparative Neurology, 196*(1), 25–40.

Allen, A. E., Procyk, C. A., Brown, T. M., & Lucas, R. J. (2017). Convergence of visual and whisker responses in the primary somatosensory thalamus (ventral posterior medial region) of the mouse. *The Journal of Physiology, 595*(3), 865–881.

Bell, A. H., Meredith, M. A., Van Opstal, A. J., & Munoz, D. P. (2005). Crossmodal integration in the primate superior colliculus underlying the preparation and initiation of saccadic eye movements. *Journal of Neurophysiology, 93*(6), 3659–3673.

Bizley, J. K., & King, A. J. (2008). Visual-auditory spatial processing in auditory cortical neurons. *Brain Research, 1242*, 24–36.

Bizley, J. K., Nodal, F. R., Bajo, V. M., Nelken, I., & King, A. J. (2007). Physiological and anatomical evidence for multisensory interactions in auditory cortex. *Cerebral Cortex, 17*(9), 2172–2189.

Bizley, J. K., Walker, K. M., Silverman, B. W., King, A. J., & Schnupp, J. W. (2009). Interdependent encoding of pitch, timbre, and spatial location in auditory cortex. *The Journal of Neuroscience, 29*(7), 2064–2075.

Bonath, B., Noesselt, T., Martinez, A., Mishra, J., Schwiecker, K., Heinze, H. J., & Hillyard, S. A. (2007). Neural basis of the ventriloquist illusion. *Current Biology, 17*(19), 1697–1703.

Bonath, B., Noesselt, T., Krauel, K., Tyll, S., Tempelmann, C., & Hillyard, S. A. (2014). Audio-visual synchrony modulates the ventriloquist illusion and its neural/spatial representation in the auditory cortex. *NeuroImage, 98*, 425–434.

Brosch, M., Selezneva, E., & Scheich, H. (2005). Nonauditory events of a behavioral procedure activate auditory cortex of highly trained monkeys. *The Journal of Neuroscience, 25*(29), 6797–6806.

Budinger, E., Heil, P., Hess, A., & Scheich, H. (2006). Multisensory processing via early cortical stages: Connections of the primary auditory cortical field with other sensory systems. *Neuroscience, 143*(4), 1065–1083.

Calvert, G. A., Bullmore, E. T., Brammer, M. J., Campbell, R., Williams, S. C., McGuire, P. K., Woodruff, P. W., Iversen, S. D., & David, A. S. (1997). Activation of auditory cortex during silent lipreading. *Science, 276*(5312), 593–596.

Cappe, C., & Barone, P. (2005). Heteromodal connections supporting multisensory integration at low levels of cortical processing in the monkey. *European Journal of Neuroscience, 22*(11), 2886–2902.

Cappe, C., Rouiller, E. M., & Barone, P. (2012). Cortical and thalamic pathways for multisensory and sensorimotor interplay. In M. M. Murray & M. T. Wallace (Eds.), *The neural bases of multisensory processes* (pp. 15–30). Boca Raton, FL: CRC Press.

Champoux, F., Tremblay, C., Mercier, C., Lassonde, M., Lepore, F., Gagné, J. P., & Théoret, H. (2006). A role for the inferior colliculus in multisensory speech integration. *NeuroReport, 17*(15), 1607–1610.

Chandrasekaran, C., Trubanova, A., Stillittano, S., Caplier, A., & Ghazanfar, A. A. (2009). The natural statistics of audiovisual speech. *PLoS Computational Biology, 5*(7), e1000436.

Chandrasekaran, C., Lemus, L., & Ghazanfar, A. A. (2013). Dynamic faces speed up the onset of auditory cortical spiking responses during vocal detection. *Proceedings of the National Academy of Sciences of the United States of America, 110*(48), E4668–E4677.

Clemo, H. R., Keniston, L. P., & Meredith, M. A. (2012). Structural basis of multisensory processing convergence. In M. M. Murray & M. T. Wallace (Eds.), *The neural bases of multisensory processes* (pp. 3–14). Boca Raton, FL: CRC Press.

Cooper, M. H., & Young, P. A. (1976). Cortical projections to the inferior colliculus of the cat. *Experimental Neurology, 51*(2), 488–502.

de la Mothe, L. A., Blumell, S., Kajikawa, Y., & Hackett, T. A. (2006a). Thalamic connections of the auditory cortex in marmoset monkeys: Core and medial belt regions. *Journal of Comparative Neurology, 496*(1), 72–96.

de la Mothe, L. A., Blumell, S., Kajikawa, Y., & Hackett, T. A. (2006b). Cortical connections of the auditory cortex in marmoset monkeys: Core and medial belt regions. *Journal of Comparative Neurology, 496*(1), 27–71.

Doubell, T. P., Baron, J., Skaliora, I., & King, A. J. (2000). Topographical projection from the superior colliculus to the nucleus of the brachium of the inferior colliculus in the ferret: Convergence of visual and auditory information. *European Journal of Neuroscience, 12*(12), 4290–4308.

Einevoll, G. T., Kayser, C., Logothetis, N. K., & Panzeri, S. (2013). Modelling and analysis of local field potentials for studying the function of cortical circuits. *Nature Reviews Neuroscience, 14*(11), 770–785.

Falchier, A., Schroeder, C. E., Hackett, T. A., Lakatos, P., Nascimento-Silva, S., Ulbert, I., Karmos, G., & Smiley, J. F. (2010). Projection from visual areas V2 and prostriata to caudal auditory cortex in the monkey. *Cerebral Cortex, 20*(7), 1529–1538.

Fu, K.-M. G., Johnston, T. A., Shah, A. S., Arnold, L., Smiley, J., Hackett, T. A., Garraghty, P. E., & Schroeder, C. E. (2003). Auditory cortical neurons respond to somatosensory stimulation. *The Journal of Neuroscience, 23*(20), 7510–7515.

Fu, K.-M. G., Shah, A. S., O'Connell, M. N., McGinnis, T., Eckholdt, H., Lakatos, P., Smiley, J., & Schroeder, C. E. (2004). Timing and laminar profile of eye-position effects on auditory responses in primate auditory cortex. *Journal of Neurophysiology, 92*(6), 3522–3531.

Furukawa, S., & Middlebrooks, J. C. (2002). Cortical representation of auditory space: Information-bearing features of spike patterns. *Journal of Neurophysiology, 87*(4), 1749–1762.

Gao, P. P., Zhang, J. W., Fan, S. J., Sanes, D. H., & Wu, E. X. (2015). Auditory midbrain processing is differentially modulated by auditory and visual cortices: An auditory fMRI study. *NeuroImage, 123*, 22–32.

Ghazanfar, A. A. (2009). The multisensory roles for auditory cortex in primate vocal communication. *Hearing Research, 258*(1-2), 113–120.

Ghazanfar, A. A., Maier, J. X., Hoffman, K. L., & Logothetis, N. K. (2005). Multisensory integration of dynamic faces and voices in rhesus monkey auditory cortex. *The Journal of Neuroscience, 25*(20), 5004–5012.

Ghazanfar, A. A., Chandrasekaran, C., & Logothetis, N. K. (2008). Interactions between the superior temporal sulcus and auditory cortex mediate dynamic face/voice integration in rhesus monkeys. *The Journal of Neuroscience, 28*(17), 4457–4469.

Groh, J. M., Trause, A. S., Underhill, A. M., Clark, K. R., & Inati, S. (2001). Eye position influences auditory responses in primate inferior colliculus. *Neuron, 29*(2), 509–518.

Hartline, P. H., Vimal, R. L., King, A. J., Kurylo, D. D., & Northmore, D. P. (1995). Effects of eye position on auditory localization and neural representation of space in superior colliculus of cats. *Experimental Brain Research, 104*(3), 402–408.

Henschke, J. U., Noesselt, T., Scheich, H., & Budinger, E. (2015). Possible anatomical pathways for short-latency multisensory integration processes in primary sensory cortices. *Brain Structure and Function, 220*(2), 955–977.

Iurilli, G., Ghezzi, D., Olcese, U., Lassi, G., Nazzaro, C., Tonini, R., Tucci, V., Benfenati, F., & Medini, P. (2012). Sound-driven synaptic inhibition in primary visual cortex. *Neuron, 73*(4), 814–828.

Jay, M. F., & Sparks, D. L. (1984). Auditory receptive fields in primate superior colliculus shift with changes in eye position. *Nature, 309*(5966), 345–347.

Jones, E. G., & Burton, H. (1974). Cytoarchitecture and somatic sensory connectivity of thalamic nuclei other than the ventrobasal complex in the cat. *Journal of Comparative Neurology, 154*(4), 395–432.

Kaltenbach, J. A. (2007). The dorsal cochlear nucleus as a contributor to tinnitus: mechanisms underlying the induction of hyperactivity. *Progress in Brain Research, 166*, 89–106.

Kanold, P. O., & Young, E. D. (2001). Proprioceptive information from the pinna provides somatosensory input to cat dorsal cochlear nucleus. *The Journal of Neuroscience, 21*(19), 7848–7858.

Katoh, Y. Y., & Benedek, G. (1995). Organization of the colliculo-suprageniculate pathway in the cat: A wheat germ agglutinin-horseradish peroxidase study. *Journal of Comparative Neurology, 352*(3), 381–397.

Kayser, C., Petkov, C. I., Augath, M., & Logothetis, N. K. (2007). Functional imaging reveals visual modulation of specific fields in auditory cortex. *The Journal of Neuroscience, 27*(8), 1824–1835.

Kayser, C., Petkov, C. I., & Logothetis, N. K. (2008). Visual modulation of neurons in auditory cortex. *Cerebral Cortex, 18*(7), 1560–1574.

Kayser, C., Logothetis, N. K., & Panzeri, S. (2010). Visual enhancement of the information representation in auditory cortex. *Current Biology, 20*(1), 19–24.

Keating, P., Dahmen, J. C., & King, A. J. (2015). Complementary adaptive processes contribute to the developmental plasticity of spatial hearing. *Nature Neuroscience, 18*(2), 185–187.

King, A. J. (2009). Visual influences on auditory spatial learning. *Philosophical Transactions of the Royal Society of London B: Biological Sciences, 364*(1515), 331–339.

King, A. J., & Palmer, A. R. (1985). Integration of visual and auditory information in bimodal neurones in the guinea-pig superior colliculus. *Experimental Brain Research, 60*(3), 492–500.

Knudsen, E. I. (2002). Instructed learning in the auditory localization pathway of the barn owl. *Nature, 417*(6886), 322–328.

Koehler, S. D., & Shore, S. E. (2013). Stimulus-timing dependent multisensory plasticity in the guinea pig dorsal cochlear nucleus. *PLoS One, 8*(3), e59828.

Lakatos, P., Chen, C. M., O'Connell, M. N., Mills, A., & Schroeder, C. E. (2007). Neuronal oscillations and multisensory interaction in primary auditory cortex. *Neuron, 53*(2), 279–292.

Lakatos, P., O'Connell, M. N., Barczak, A., Mills, A., Javitt, D. C., & Schroeder, C. E. (2009). The leading sense: Supramodal control of neurophysiological context by attention. *Neuron, 64*(3), 419–430.
Lesicko, A. M., Hristova, T. S., Maigler, K. C., & Llano, D. A. (2016). Connectional modularity of top-down and bottom-up multimodal inputs to the lateral cortex of the mouse inferior colliculus. *The Journal of Neuroscience, 36*(43), 11037–11050.
Lohse, M., Bajo, V. M., & King, A. J. (2017). Types and distribution of multisensory interactions in auditory thalamus. *Association for Research in Otolaryngology Abstracts*, 280.
Lomber, S. G., Meredith, M. A., & Kral, A. (2010). Cross-modal plasticity in specific auditory cortices underlies visual compensations in the deaf. *Nature Neuroscience, 13*(11), 1421–1427.
Luo, H., Liu, Z., & Poeppel, D. (2010). Auditory cortex tracks both auditory and visual stimulus dynamics using low-frequency neuronal phase modulation. *PLoS Biology, 8*(8), e1000445.
Maddox, R. K., Pospisil, D. A., Stecker, G. C., & Lee, A. K. (2014). Directing eye gaze enhances auditory spatial cue discrimination. *Current Biology, 24*(7), 748–752.
Malhotra, S., & Lomber, S. G. (2007). Sound localization during homotopic and heterotopic bilateral cooling deactivation of primary and nonprimary auditory cortical areas in the cat. *Journal of Neurophysiology, 97*(1), 26–43.
Manger, P. R., Nakamura, H., Valentiniene, S., & Innocenti, G. M. (2004). Visual areas in the lateral temporal cortex of the ferret (*Mustela putorius*). *Cerebral Cortex, 14*(6), 676–689.
May, B. J. (2000). Role of the dorsal cochlear nucleus in the sound localization behavior of cats. *Hearing Research, 148*(1-2), 74–87.
McGettigan, C., Faulkner, A., Altarelli, I., Obleser, J., Baverstock, H., & Scott, S. K. (2012). Speech comprehension aided by multiple modalities: Behavioural and neural interactions. *Neuropsychologia, 50*(5), 762–776.
McGurk, H., & MacDonald, J. (1976). Hearing lips and seeing voices. *Nature, 264*(5588), 746–748.
Meredith, M. A., & Allman, B. L. (2009). Subthreshold multisensory processing in cat auditory cortex. *NeuroReport, 20*(2), 126–131.
Meredith, M. A., & Allman, B. L. (2015). Single-unit analysis of somatosensory processing in the core auditory cortex of hearing ferrets. *European Journal of Neuroscience, 41*(5), 686–698.
Meredith, M. A., & Stein, B. E. (1986). Visual, auditory, and somatosensory convergence on cells in superior colliculus results in multisensory integration. *Journal of Neurophysiology, 56*(3), 640–662.
Morin, L. P., & Studholme, K. M. (2014). Retinofugal projections in the mouse. *Journal of Comparative Neurology, 522*(16), 3733–3753.
Murray, M. M., & Wallace, M. T. (Eds.). (2012). *The neural bases of multisensory processes*. Boca Raton, FL: CRC Press.
Nelken, I., Chechik, G., Mrsic-Flogel, T. D., King, A. J., & Schnupp, J. W. H. (2005). Encoding stimulus information by spike numbers and mean response time in primary auditory cortex. *Journal of Computational Neuroscience, 19*(2), 199–221.
Okada, K., Venezia, J. H., Matchin, W., Saberi, K., & Hickok, G. (2013). An fMRI study of audiovisual speech perception reveals multisensory interactions in auditory cortex. *PLoS One, 8*(6), e68959.
Olcese, U., Iurilli, G., & Medini, P. (2013). Cellular and synaptic architecture of multisensory integration in the mouse neocortex. *Neuron, 79*(3), 579–593.
Park, H., Kayser, C., Thut, G., & Gross, J. (2016). Lip movements entrain the observers' low-frequency brain oscillations to facilitate speech intelligibility. *eLife, 5*, e14521.
Pekkola, J., Ojanen, V., Autti, T., Jääskeläinen, I. P., Möttönen, R., Tarkiainen, A., & Sams, M. (2005). Primary auditory cortex activation by visual speech: an fMRI study at 3 T. *NeuroReport, 16*(2), 125–128.
Philipp, R., Distler, C., & Hoffmann, K. P. (2006). A motion-sensitive area in ferret extrastriate visual cortex: an analysis in pigmented and albino animals. *Cerebral Cortex, 16*(6), 779–790.
Pinchoff, R. J., Burkard, R. F., Salvi, R. J., Coad, M. L., & Lockwood, A. H. (1998). Modulation of tinnitus by voluntary jaw movements. *American Journal of Otolaryngology, 19*(6), 785–789.

Porter, K. K., Metzger, R. R., & Groh, J. M. (2007). Visual- and saccade-related signals in the primate inferior colliculus. *Proceedings of the National Academy of Sciences of the United States of America, 104*(45), 17855–17860.

Rauschecker, J. P. (2018). Where, when, and how: Are they all sensorimotor? Towards a unified view of the dorsal pathway in vision and audition. *Cortex, 98*, 262–268.

Rouiller, E. M., & Welker, E. (2000). A comparative analysis of the morphology of corticothalamic projections in mammals. *Brain Research Bulletin, 53*(6), 727–741.

Schnupp, J., Nelken, I., & King, A. (2011). *Auditory neuroscience: Making sense of sound.* Cambridge, MA: MIT Press.

Schroeder, C. E., Lakatos, P., Kajikawa, Y., Partan, S., & Puce, A. (2008). Neuronal oscillations and visual amplification of speech. *Trends in Cognitive Sciences, 12*(3), 106–113.

Scott, B. H., Saleem, K. S., Kikuchi, Y., Fukushima, M., Mishkin, M., & Saunders, R. C. (2017). Thalamic connections of the core auditory cortex and rostral supratemporal plane in the macaque monkey. *Journal of Comparative Neurology, 525*(16), 3488–3513.

Sekiyama, K., Kanno, I., Miura, S., & Sugita, Y. (2003). Auditory-visual speech perception examined by fMRI and PET. *Neuroscience Research, 47*(3), 277–287.

Sherman, S. M. (2016). Thalamus plays a central role in ongoing cortical functioning. *Nature Neuroscience, 16*(4), 533–541.

Shore, S. E. (2005). Multisensory integration in the dorsal cochlear nucleus: Unit responses to acoustic and trigeminal ganglion stimulation. *European Journal of Neuroscience, 21*(12), 3334–3348.

Shore, S. E., & Zhou, J. (2006). Somatosensory influence on the cochlear nucleus and beyond. *Hearing Research, 216-217*, 90–99.

Smiley, J. F., & Falchier, A. (2009). Multisensory connections of monkey auditory cerebral cortex. *Hearing Research, 258*(1-2), 37–46.

Song, Y.-H., Kim, J.-H., Jeong, H.-W., Choi, I., Jeong, D., Kim, K., & Lee, S. H. (2017). A neural circuit for auditory dominance over visual perception. *Neuron, 93*(4), 940–954.

Stecker, G. C., Harrington, I. A., & Middlebrooks, J. C. (2005). Location coding by opponent neural populations in the auditory cortex. *PLoS Biology, 3*(3), e78.

Stehberg, J., Dang, P. T., & Frostig, R. D. (2014). Unimodal primary sensory cortices are directly connected by long-range horizontal projections in the rat sensory cortex. *Frontiers in Neuroanatomy, 8*, 93.

Stein, B. E., & Wallace, M. T. (1996). Comparisons of cross-modality integration in midbrain and cortex. *Progress in Brain Research, 112*, 289–299.

Stein, B. E., Huneycutt, W. S., & Meredith, M. A. (1988). Neurons and behavior: The same rules of multisensory integration apply. *Brain Research, 448*(2), 355–358.

Stevenson, R. A., Ghose, D., Fister, J. K., Sarko, D. K., Altieri, N. A., Nidiffer, A. R., Kurela, L. R., Siemann, J. K., James, T. W., & Wallace, M. T. (2014). Identifying and quantifying multisensory integration: A tutorial review. *Brain Topography, 27*(6), 707–730.

Stitt, I., Galindo-Leon, E., Pieper, F., Hollensteiner, K. J., Engler, G., & Engel, A. K. (2015). Auditory and visual interactions between the superior and inferior colliculi in the ferret. *European Journal of Neuroscience, 41*(10), 1311–1320.

Sumby, W. H., & Pollack, I. (1954). Visual contribution to speech intelligibility in noise. *The Journal of Acoustical Society of America, 26*(2), 212–215.

Thorne, J. D., De Vos, M., Viola, F. C., & Debener, S. (2011). Cross-modal phase reset predicts auditory task performance in humans. *The Journal of Neuroscience, 31*(10), 3853–3861.

Van Atteveldt, N., Murray, M. M., Thut, G., & Schroeder, C. E. (2014). Multisensory integration: Flexible use of general operations. *Neuron, 81*(6), 1240–1253.

Wallace, M. T., Meredith, M. A., & Stein, B. E. (1998). Multisensory integration in the superior colliculus of the alert cat. *Journal of Neurophysiology, 80*(2), 1006–1010.

Wasserman, S. M., Aptekar, J. W., Lu, P., Nguyen, J., Wang, A. L., Keles, M. F., Grygoruk, A., Krantz, D. E., Larsen, C., & Frye, M. A. (2015). Olfactory neuromodulation of motion vision circuitry in *Drosophila. Current Biology, 25*(4), 467–472.

Werner-Reiss, U., Kelly, K. A., Trause, A. S., Underhill, A. M., & Groh, J. M. (2003). Eye position affects activity in primary auditory cortex of primates. *Current Biology, 13*(7), 554–562.

Wigderson, E., Nelken, I., & Yarom, Y. (2016). Early multisensory integration of self and source motion in the auditory system. *Proceedings of the National Academy of Sciences of the United States of America, 113*(29), 8308–8313.
Winer, J. A., & Lee, C. C. (2007). The distributed auditory cortex. *Hearing Research, 229*(1-2), 3–13.
Yu, J. J., & Young, E. D. (2000). Linear and nonlinear pathways of spectral information transmission in the cochlear nucleus. *Proceedings of the National Academy of Sciences of the United States of America, 97*(22), 11780–11786.
Zwiers, M. P., Van Opstal, A. J., & Paige, G. D. (2003). Plasticity in human sound localization induced by compressed spatial vision. *Nature Neuroscience, 6*(2), 175–181.
Zwiers, M. P., Versnel, H., & Van Opstal, A. J. (2004). Involvement of monkey inferior colliculus in spatial hearing. *The Journal of Neuroscience, 24*(17), 4145–4156.

Chapter 7
Audiovisual Integration in the Primate Prefrontal Cortex

Bethany Plakke and Lizabeth M. Romanski

Abstract Language and communication rely on the combination of visual and auditory information. The frontal lobes have long been known to support communication processing and receive a wide array of sensory inputs from many brain regions. The ventral frontal lobe, specifically the ventrolateral prefrontal cortex (VLPFC), receives afferents from auditory and visual association cortices. Recordings in nonhuman primates indicate that single neurons in the VLPFC integrate face and vocal stimuli. These multisensory neurons show enhanced and suppressed responses to face and vocalization combinations. Furthermore, studies indicate that ventral prefrontal neurons are affected by the semantic congruence of face-vocalization pairs and by the temporal synchrony of dynamic face-vocalization stimuli. Recordings of VLPFC neurons in macaques performing working memory tasks demonstrate that neurons are context dependent and respond to specific combinations of face and vocal stimuli during memory and decision tasks. Finally, transient inactivation of the prefrontal cortex impairs working memory for face-vocalization stimuli. Thus, results from several studies indicate that the primate prefrontal cortex plays an important role in the processing and integration of face and vocalization information that is essential during communication and social cognition.

Keywords Auditory · Communication · Cross-modal · Face · Frontal lobe · Macaque · Multisensory · Neurophysiology · Vocalization · Ventrolateral prefrontal cortex · Working memory

---

B. Plakke
Department of Psychological Sciences, Kansas State University, Manhattan, KS, USA
e-mail: bplakke@ksu.edu

L. M. Romanski (✉)
Department of Neuroscience, University of Rochester School of Medicine, Rochester, NY, USA
e-mail: Liz_romanski@urmc.rochester.edu

A. K. C. Lee et al. (eds.), *Multisensory Processes*, Springer Handbook of Auditory Research 68, https://doi.org/10.1007/978-3-030-10461-0_7

7.1 Introduction

When one considers the functions of the frontal lobe, one invariably thinks of language. It is well-known that damage to Broca's area in the inferior frontal gyrus, which commonly occurs with a stroke, causes speech difficulties or aphasia. Neuroimaging studies have also revealed consistent, robust activation in the human inferior frontal gyrus during speech comprehension, language processing, and speech production (Price 2012). Many of the processes underlying communication rely on the integration of sensory information. For example, phonological processing while listening to speech involves the integration of auditory perceptual processes and articulation, whereas face-to-face communication typically involves the integration of facial gestures with vocal sounds. Thus, language and, more generally, communication is a multisensory process (Ghazanfar et al. 2013). It comes as no surprise then that the ventral frontal lobe integrates auditory, visual, and motor signals during communication and also during cognitive processes, including working memory, attention, and decision making.

Although neuroimaging studies have advanced understanding the role of the human prefrontal cortex (PFC) during cognitive processes in order to understand the neuronal mechanisms that underlie these processes, neurophysiological recordings have been made in various animal models. Many of the neurophysiological recordings have been focused on determining the role of the PFC in visual-stimulus processing including visual memory, visual attention, and decision making using visual stimuli. Furthermore, neurophysiological studies have primarily focused on the dorsolateral PFC (Fig. 7.1). Therefore, knowledge of visual processing in the PFC far exceeds knowledge concerning the processing of other sensory information in different regions of the PFC. Nonetheless, our environment is a multisensory milieu that requires humans to integrate many pieces of sensory information simultaneously. This realization and the recent focus on natural stimuli has led to an increased interest in exploring multisensory processing. To truly understand cognitive processes such as communication, working memory, and decision making, multisensory paradigms must be utilized. Thus, research into the role of the PFC in the processing of multiple types of sensory stimuli has progressed. Of particular importance is the manner in which the frontal lobe integrates complex auditory and visual information during the process of communication. This chapter reviews recent work to explore this topic.

As a first step toward understanding the role of the PFC in the integration of multisensory information, it is important to appreciate the representation in the PFC of sensory afferents, specifically the regions where afferents from different modalities converge. Early anatomical studies noted that afferents from auditory, visual, and somatosensory cortical regions targeted lateral and orbital PFC (for a review, see Romanski 2012). These studies indicate that the ventrolateral PFC (VLPFC) is a fertile ground in which to examine multisensory integration. Understanding the sources of sensory afferents to the PFC will facilitate the understanding of the nature of multisensory integration in the frontal lobe.

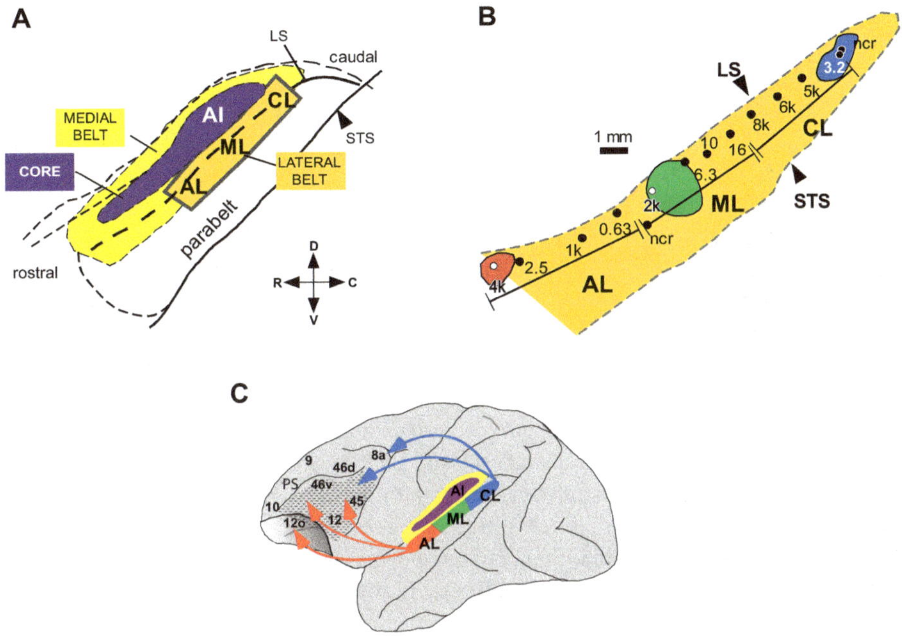

Fig. 7.1 (**A**) Schematic of the macaque auditory cortex with a central core primary region A1 (*purple*), surrounded by a medial (*yellow*) and a lateral belt auditory cortex. The lateral belt is further subdivided into known physiological boundaries including an anterior region (AL), a middle region (ML), and a caudal region (CL). (**B**) After physiological identification of these regions, anatomical tracers were placed into the AL (*red circle*), ML (*green circle*), and CL (*blue region*) in order to trace the connections with the prefrontal cortex (PFC). Numbers are the average frequency response for that penetration. (**C**) Cytoarchitectonic areas of the PFC are shown with the projections from the auditory cortex that innervate the PFC (*arrows*). The dorsolateral PFC includes areas 8a, 46, and 9 and the ventrolateral PFC (*stippled area* below the principal sulcus [PS]) includes areas 12/47 and 45 and the lateral orbital cortex (12o). The projections from the auditory cortex are arranged as a dorsal stream (*blue arrows*) that connects caudal auditory cortex areas CL and ML with the dorsolateral PFC and a ventral stream (*red arrows*) that connects the anterior auditory belt cortex areas AL and ML with the ventral PFC. *LS*, lateral sulcus; *STS*, superior temporal sulcus; *ncr*, location not recordable

7.1.1 Advances in the Understanding of Auditory Cortex

There have been several key scientific advances in the understanding of the organization of the primate auditory cortex that have motivated research on many fronts. Anatomical studies determined that the core region, primary auditory cortex or A1, could be differentiated from the surrounding auditory cortical lateral and medial belt areas using acetylcholinesterase, cytochrome oxidase, and parvalbumin staining density (Jones et al. 1995; Kosaki et al., 1997; Hackett et al. 1998). The lateral belt regions were, in turn, surrounded by a parabelt of auditory cortex with staining characteristics that differentiated it from the belt. Delineation of the auditory cortex core and belt fields by different histochemical staining patterns allowed for precise targeting of anatomical tracers in auditory cortical areas. Moreover, because neurophysiological recordings in the primate auditory cortex were being conducted at this time, the core-belt anatomical localization made it possible to link anatomically different auditory fields with neurophysiological characteristics. In a series of groundbreaking experiments, Rauschecker et al. (1995) devised a way to use band-passed noise bursts with defined center frequencies to evoke responses in nonprimary auditory cortex and distinguish them from those seen in A1. This key development led to the identification of physiologically defined core and belt auditory cortical regions in the primate temporal lobe. Recordings in anesthetized animals demonstrated that the lateral auditory belt could be separated into three tonotopic fields: anterior belt (AL), middle belt (ML), and caudal belt (CL; Fig. 7.1; Rauschecker et al. 1995). Over the past two decades, neurophysiology and neuroimaging studies have determined that these regions differ functionally, with neurons in area AL showing selectivity for sound type and neurons in the caudal auditory belt, area CL showing preferences for sound locations. The establishment of the anatomical, physiological, and functional boundaries of auditory cortical fields in the core, belt, and parabelt inspired a multitude of investigations into the connections and functions of these areas of auditory cortex with temporal, parietal, and prefrontal regions.

7.1.2 Connections of the Auditory Cortex with the Prefrontal Cortex

Although it has long been known that the superior temporal gyrus (STG) projects to PFC (Petrides and Pandya 2002), the results of recent data are helping to define more specific connections between auditory cortical regions and particular prefrontal domains. Historically, there are known fiber paths that connect the temporal lobe with frontal lobe targets, including the uncinate fasciculus and the arcuate fasciculus. These large fiber bands provide the substrate on which more specific afferent input from the auditory cortex could project to the frontal lobe, thereby delivering acoustic information. Decades of research from a number of anatomists have characterized temporal-prefrontal connections and provided evidence that the ventral PFC receives afferents from the middle and rostral temporal lobes (Barbas 1992; Romanski et al. 1999a). However, none of these early anatomical studies clearly demonstrated that the information sent to the PFC was acoustic because the temporal lobe areas studied were only characterized anatomically.

The neurophysiological and neuroanatomical advances described in Sect. 7.1.1 made it possible to pinpoint the primary core and secondary belt regions of the auditory cortex and thus use tract tracing to delineate the differential connectivity of various regions of the auditory cortex. It is now clear that projections from the primary auditory cortex (A1) carry acoustic information to the medial and lateral belt regions (Hackett et al. 2014). In turn, projections from the lateral belt reach the parabelt auditory cortex and the rostral STG. The connections of the lateral belt and parabelt include the rostral temporal lobe, the superior temporal sulcus, and, importantly, the VLPFC (Hackett et al. 1999; Plakke and Romanski 2014). Several anatomical studies investigated the connections of the belt and parabelt regions to create the detailed maps currently available (Hackett et al. 1999; Plakke and Romanski 2014).

By combining neurophysiological recordings in the belt and parabelt with anatomical tract tracing, it is possible to trace acoustic pathways all the way to the PFC. Building on prior research (Rauschecker et al. 1995) that distinguished the boundaries of three distinct fields in the auditory belt (AL, ML, and CL), tracer injections were placed into tonotopically characterized tracks of each region. This allowed for demonstration of precise connectivity of auditory cortical regions with specific locations in the PFC. In particular, the anterior auditory cortex projected to the ventral PFC, whereas the more caudal auditory belt and parabelt projected to the dorsal and caudal PFCs (Romanski et al. 1999b). The rostral and caudal auditory cortical areas have been shown to be involved in object and spatial processing, respectively. Therefore, these connections were seen as "dual streams" that project to prefrontal regions in a topographic manner, similar to the dorsal and ventral streams of the visual system (Fig. 7.1; Romanski et al. 1999b). These dual auditory streams provide spatial auditory information to the dorsal PFC and nonspatial auditory information to the VLPFC. In addition, the VLPFC receives a robust projection from the superior temporal sulcus (STS) that could provide auditory, visual, or multisensory input (see Beauchamp, Chap. 8, for a review). The connections of the PFC with temporal lobe auditory regions have been summarized in previous publications (Fig. 7.2; Plakke and Romanski 2014). Utilizing these auditory maps to guide investigations of prefrontal function, studies have shown that the VLPFC, which includes both the ventral and orbital prefrontal cortices, contains neurons that are responsive to complex sounds (discussed in Sect. 7.2).

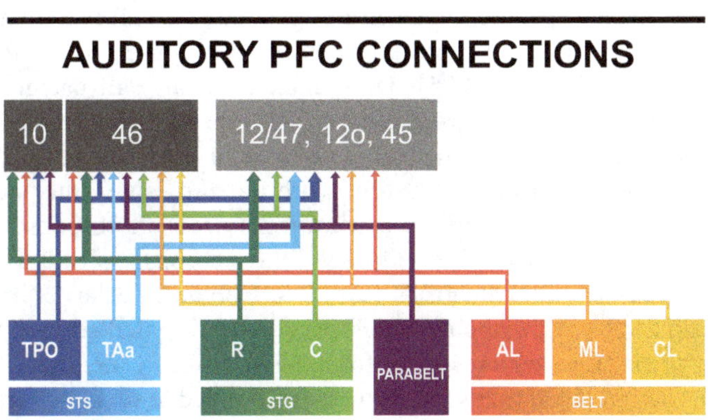

Fig. 7.2 A circuit diagram summary indicating auditory projections from temporal lobe areas (*bottom boxes*) to the PFC (*top boxes*). *Thicker lines* represent more dense anatomical connectivity. *R*, rostral superior temporal gyrus; *C*, caudal superior temporal gyrus; *STG*, superior temporal gyrus; *TPO*, temporo-parieto-occipital area; *TAa*, temporal area. Numerical cytoarchtectonic boundaries from Petrides and Pandya (2002) and Romanski et al. (1999a) are used whereby area 10 refers to the frontal pole, area 46 refers to the area of the PS, and areas 12/47, 12 orbital (12o), and 45 pertain to the cytoarchitectonic regions of the ventrolateral PFC (VLPFC). Reprinted from Plakke and Romanski (2014), with permission

7.2 Responses of Ventral Prefrontal Neurons to Complex Sounds and Vocalizations

Early neurophysiological studies examined widespread regions of the PFC for auditory responsiveness. Auditory-responsive neurons in the frontal lobes of nonhuman primates had weak responses to the stimuli used (Vaadia et al. 1986), were seen sporadically across the cortical surface (Tanila et al. 1992, 1993), or were not tested with complex acoustic stimuli (Bodner et al. 1996) or under controlled conditions. Better testing protocols for auditory responsiveness and a more targeted approach to areas in the PFC were needed.

Armed with the roadmap of "dorsal and ventral streams" of auditory afferents that targeted the primate frontal lobe, recording studies of specific prefrontal regions were possible. Because the rostral auditory cortex projects to the rostral and ventral prefrontal cortices, the VLPFC was targeted as a putative auditory-responsive zone, and auditory-responsive cells were successfully recorded from this area. Initial studies showed that the VLPFC neurons responded mainly to complex sounds (Romanski and Goldman-Rakic 2002). Whereas pure tones did not evoke responses in the majority of prefrontal neurons, monkey vocalizations, human vocalizations, band-passed noise bursts, and other environmental sounds evoked robust responses in VLPFC neurons (Romanski and Goldman-Rakic 2002). The fact that vocal stimuli elicited the strongest prefrontal responses suggested a homology with the human inferior frontal gyrus language regions.

Successive studies focused on the notion that the VLPFC might be preferentially responsive to species-specific vocalizations, and, therefore, studies were aimed at understanding selectivity for vocalization types. In one such study, a large body of species-specific vocalizations grouped by call type and caller that included both common and rarely heard call types was used to examine VLPFC responses (Romanski et al. 2005). Exemplars from all call categories evoked robust responses in VLPFC neurons regardless of the meaning of the call or how familiar the subjects were with the call (i.e., some of the calls are only uttered by infants and juveniles that are not housed with the subjects in this study). Thus, familiarity and call meaning did not constrain the neuronal responses in VLPFC neurons (Plakke et al. 2013b). Individual VLPFC neurons typically responded to 2–3 different vocalization categories (Fig. 7.3; Romanski et al. 2005; Averbeck and Romanski 2006). A hierarchical cluster analysis revealed several pairs of call types that typically elicited a response from a given VLPFC neuron. Examination of these clusters indicated that some neurons responded to tonal or harmonic calls (warbles and coos) while other cells responded to noisy calls (aggressive calls and grunts). This work suggested that neurons in this discrete auditory-responsive zone of the VLPFC were responding on the basis of acoustic similarity. That is, the neurons responded to calls with a similar acoustic structure and not a similar referential meaning.

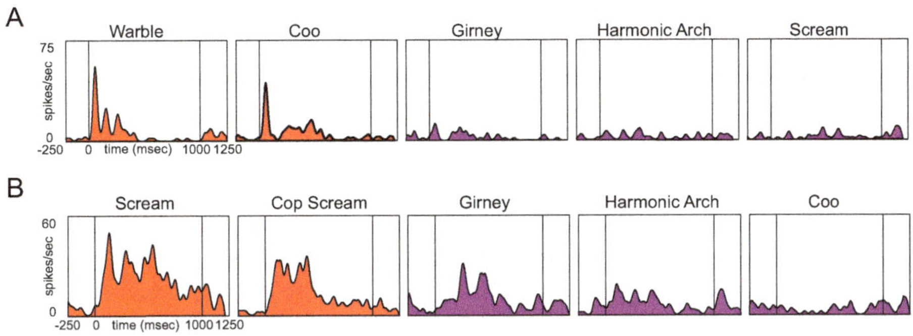

Fig. 7.3 Neuronal responses recorded in VLPFC area 12 to macaque vocalizations. The neuronal responses to 5 types of macaque vocalizations are shown as a spike-density function ([SDF]). A given neuron responded in a similar way to vocalizations that were acoustically similar. (**A**) Cell responded best (*red* SDFs) to the warble and coo stimuli that are acoustically similar. (**B**) Cell responded best (*red* SDFs) to two different types of screams that also have many similar acoustic features. Stimulus onset is at time 0. *Purple,* responses to the other vocalizations

There have been only a few other investigations of VLPFC auditory responses. One study (Gifford et al. 2005) has suggested that prefrontal neurons might respond to vocalizations on the basis of category. Testing in this study involved a habituation-dishabituation paradigm where the "oddball" sound was either within category or outside category. Neuronal responses of PFC cells demonstrated a larger response change when sounds were outside the referential category (Gifford et al. 2005) than when the oddball sound was within the same referential category. However, the sounds used in the testing paradigm also exhibited a change in acoustic features in addition to the change in category. The data have not been confirmed using vocalizations that are similar in meaning but differ in acoustic morphology. Furthermore, given the possibility that the VLPFC is homologous to speech and language regions in the human frontal lobe, it makes sense that neurons would respond on the basis of acoustic structure. In the human brain, Broca's area neurons participate in phonological processing of heard speech where auditory and articulatory signals are combined (Price 2012).

Thus far, only the ventral region of the PFC has been extensively studied regarding auditory responses to complex sound in animal studies. There are many unexplored regions of the frontal lobe with a potential for auditory function. Early studies by Jurgens (2009) and West and Larson (1995) have suggested that the medial PFC plays a role in vocal behavior and could be involved in auditory-vocal interactions. In addition, a recent study (Medalla and Barbas 2014) emphasized connections between the rostral STG and the frontal pole, an area that has received little attention concerning its function in nonhuman primates. Further investigation is certainly warranted and may reveal new domains for higher auditory processing in the frontal lobe. Areas such as the frontal pole and medial PFC receive afferents from the high-level sensory cortex as well as multisensory regions and thus are likely to be multisensory as well.

7.3 Demonstration of Audiovisual Responses in the Primate Frontal Lobe

Establishment of an auditory-responsive domain in the VLPFC was intriguing because it was found to lie adjacent to a face-responsive region in the VLPFC. Neurons in the posterior VLPFC, close to the arcuate sulcus, were found to be responsive to patterns, pictures of objects, and faces (Wilson et al. 1993; O'Scalaidhe et al. 1997). Furthermore, there were patches of the VLPFC where neurons were highly selective for faces (O'Scalaidhe et al. 1997). These "face cells" selectively processed information related to the identity of faces (O'Scalaidhe et al. 1997, 1999) and were adjacent to the auditory-responsive neurons in the VLPFC (Romanski and Goldman-Rakic 2002). Combining these results, it seemed highly likely that multisensory-responsive neurons would be found in the VLPFC.

Multisensory integration has been well documented in neurons of the midbrain (i.e., superior colliculus; see Willet, Groh, and Maddox, Chap. 5), and many of the interactions and principles of multisensory integration in the superior colliculus have been documented in other brain regions. Neurophysiological assessment of face and vocal integration has been less frequent. Two particular studies were pivotal in the analysis of face and vocalization integration. First, David Perrett and colleagues, who had previously described face cells in the inferotemporal cortex, used dynamic video stimuli to demonstrate multisensory-responsive neurons in the STS of awake behaving macaque monkeys (Barraclough et al. 2005). Similarly, Ghazanfar et al. (2005) demonstrated multisensory integration in neurons of the auditory cortex. This was an important discovery because it was previously thought that integration might only occur in higher association cortices. Furthermore, Ghazanfar et al. (2005) utilized carefully constructed and behaviorally characterized species-specific vocalizations and their corresponding facial gestures to test multisensory neurons in the temporal lobe. By using natural communication calls, investigations could assess the circuits involved in the encoding of social communication stimuli.

A number of separate studies have demonstrated the presence in the PFC of visually responsive (O'Scalaidhe et al. 1997), auditory-responsive (Romanski and Goldman-Rakic 2002), and even somatosensory-responsive neurons (Brody et al. 2003). Accordingly, an area such as the frontal lobe, with a multitude of afferents converging, would be an obvious candidate region in which to investigate multisensory integration of these modalities. It is especially important to remember that language is a multisensory phenomenon. During social interactions, integration of vocal information, facial expression, mouth movements, and hand gestures occur. Therefore, VLPFC neurons may combine visual and auditory information from communication relevant face and vocal stimuli in the frontal lobes of nonhuman primates.

7.3.1 Dynamic Face and Vocalization Stimuli Evoke Prefrontal Multisensory Interactions

Given that the percentage of neurons across the lateral surface of the PFC responsive to visual stimuli is much greater than the number responsive to auditory stimuli (55% vs. 18%), it is reasonable to assume that multisensory cells are more likely to be located in the regions where auditory responsive cells reside. Consequently, in searching for multisensory neurons in the PFC, the VLPFC auditory-responsive region was targeted.

In Sugihara et al. (2006), short dynamic movie clips of familiar monkeys vocalizing were presented while single units were recorded from the VLPFC. Neuronal responses to just the soundtrack (auditory), just the video track (visual), or both together as in the usual movie clip (audiovisual stimulus) revealed that approximately half the neurons recorded in the VLPFC were multisensory. They were either bimodal, in that they responded to unimodal auditory and unimodal visual stimuli when presented separately, or they were multisensory because of an enhanced or decreased response to the combined audiovisual stimulus (face and vocalization) compared with the response to the unimodal stimuli (Sugihara et al. 2006).

The presence of both unimodal and multisensory neurons in a single electrode track suggests the likely existence of many more multisensory responsive cells if an unlimited set of stimuli could be used to test them. Although sporadic audiovisual responses could be found over a wide area of the lateral PFC, there was a preponderance of multisensory neurons in the same anterolateral VLPFC region where complex auditory responses had been previously described (Romanski et al. 2005).

7.3.2 Types of Multisensory Interactions

Prefrontal multisensory neurons exhibited a number of interactions to presentations of auditory and visual stimuli. As has been demonstrated in other multisensory areas of the brain, VLPFC multisensory neurons demonstrated multisensory enhancement, that is, an increase during audiovisual stimulation that was significantly greater than the best unimodal response (Fig. 7.4A, B). However, a slightly larger percentage of the population exhibited multisensory suppression where the response to audiovisual stimuli was less than to the most effective unimodal stimulus (Fig. 7.4C, D). A subadditive response is an example of multisensory suppression where the response to the combined stimuli is less than would be expected from the simple linear sum. Enhanced multisensory response also occurred as either linear or nonlinear interactions. In some cases of multisensory enhancement, the combination of the auditory and visual stimuli evoked a linear response that was the same as would be expected from the linear sum of the response to the unimodal auditory and

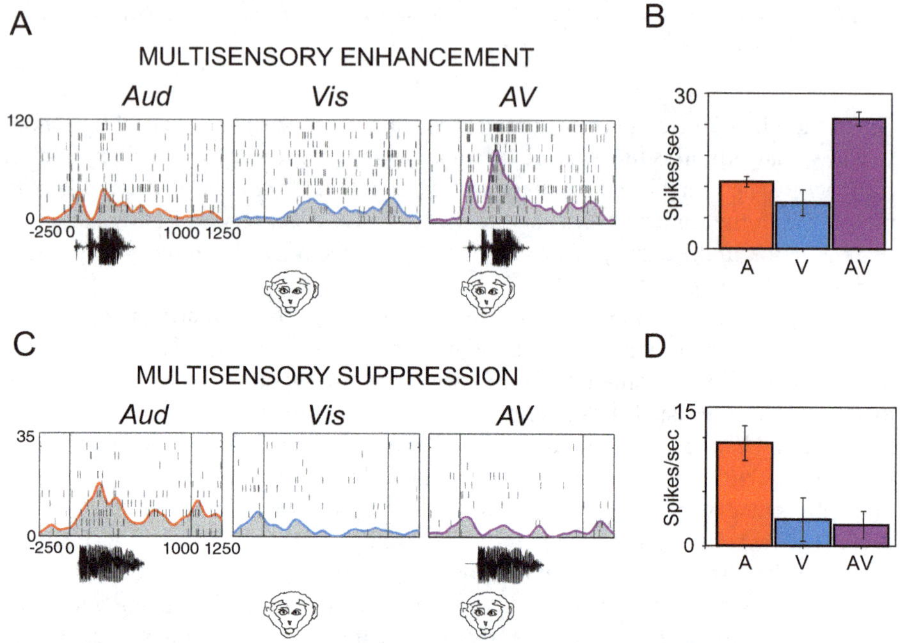

Fig. 7.4 Multisensory neuronal response enhancement and suppression in the VLPFC to dynamic face-vocalization stimuli presented individually (auditory [Aud; A]; visual [Vis; V]) and then combined (audiovisual [AV]). (**A, C**) Responses of two prefrontal cells to an auditory vocalization (*red*), a visual face (*blue*), and the face and vocalization played together (*purple*) are shown as raster/spike-density plots. (**B, D**) Bar graphs of the mean response to these stimuli. (**A**) Cell exhibited multisensory enhancement, in which the response to the AV stimulus was greater than the unimodal responses. (**C**) In contrast, the cell demonstrated multisensory suppression, in which the response to the AV stimulus was inhibited relative to the best unimodal response. Adapted from Romanski and Averbeck (2009), with permission

the unimodal visual response (i.e., an additive response). In contrast, other responses were nonlinear, in which case the response to the audiovisual stimulus was greater than the linear sum of the two unimodal responses. This is known as superadditivity and was seen in some VLPFC multisensory interactions (Fig. 7.4A, B).

The magnitude of these multisensory responses and whether they were enhanced or suppressed, relative to the unimodal responses, was entirely dependent on the particular pair of stimuli presented (Sugihara et al. 2006). One face-vocalization pair could evoke an enhanced interaction in one neuron and a different face-vocalization pair could evoke a suppressed interaction in the same cell. This suggests that any given VLPFC cell has the potential to be "multisensory responsive" given the presentation of the proper stimulus. For a given neuron, there was selectivity such that prefrontal neurons were selective and not all face-vocalization pairs evoked a multisensory response (Diehl and Romanski 2014). Demonstrating a multisensory response is conditional on the particular stimuli presented and does not appear to be a feature or quality of the individual neuron. Given the fact that the

VLPFC receives a dense projection from the multisensory region of the STS and also receives separate projections from the auditory association cortex, visual extrastriate regions and the somatosensory cortex, it may be possible to demonstrate multisensory responses across most of the VLPFC if a large enough stimulus repertoire is used.

7.3.3 Latency of Multisensory Responses in Ventrolateral Prefrontal Cortex Neurons

A number of studies have examined response times (i.e., latency) to unimodal and bimodal stimuli and typically show that faster responses occur for bimodal stimuli (Hershenson 1962; Posner et al. 1976). Typically, these reaction times are gathered via behavioral responses such as button presses or saccades to sensory targets. Whether such enhanced speed is seen in response to multisensory stimuli at the level of single-neuron latencies remained an open question. Neurophysiological studies have compared responses to unimodal and multisensory stimuli in several cortical and subcortical regions and have suggested a faster latency for the multisensory condition in some of these brain regions (Meredith and Stein 1986).

The VLPFC region sits at the very front of the brain, and latencies to sensory stimuli are typically longer in this cortical region than in sensory cortices or subcortical structures such as the superior colliculus. To examine the neuronal response latency to complex visual and auditory stimuli, neurons were tested for their responses to a static face taken from a vocalization movie and to the corresponding vocalization (Romanski and Hwang 2012). A static face that showed the prototypical facial expression corresponding to the vocalization was used so that the onset latency could be measured without any artifacts of motion-evoking responses or for the complication of a specific expression affecting the response latency. In most multisensory VLPFC neurons, the fastest response latency was elicited by the auditory vocalization while the longest latency or the slowest response occurred in response to the face stimuli. When presented simultaneously, the static face-vocalization combination elicited a response time that was between the fast vocalization and the slow face response (Fig. 7.5A; Romanski and Hwang 2012). When the vocalization movie is used instead of a static face, the latencies are similar. Despite the fact that the VLPFC is many synapses away from the sensory periphery, some of the unimodal response latencies were quite rapid with a range of latencies from 40 to 400 ms. (Romanski and Hwang 2012). The large range of response latencies suggests that auditory responses in different neurons might be explained by the selectivity of neurons to different components of the time-varying vocal stimulus or innervation from the early auditory cortex versus later association regions. Auditory afferents reach the VLPFC from as early as the lateral belt and from as late as the STS (Plakke and Romanski 2014). This could dictate the large variability in the response latencies of VLPFC cells and also the complexity of the features to which they respond.

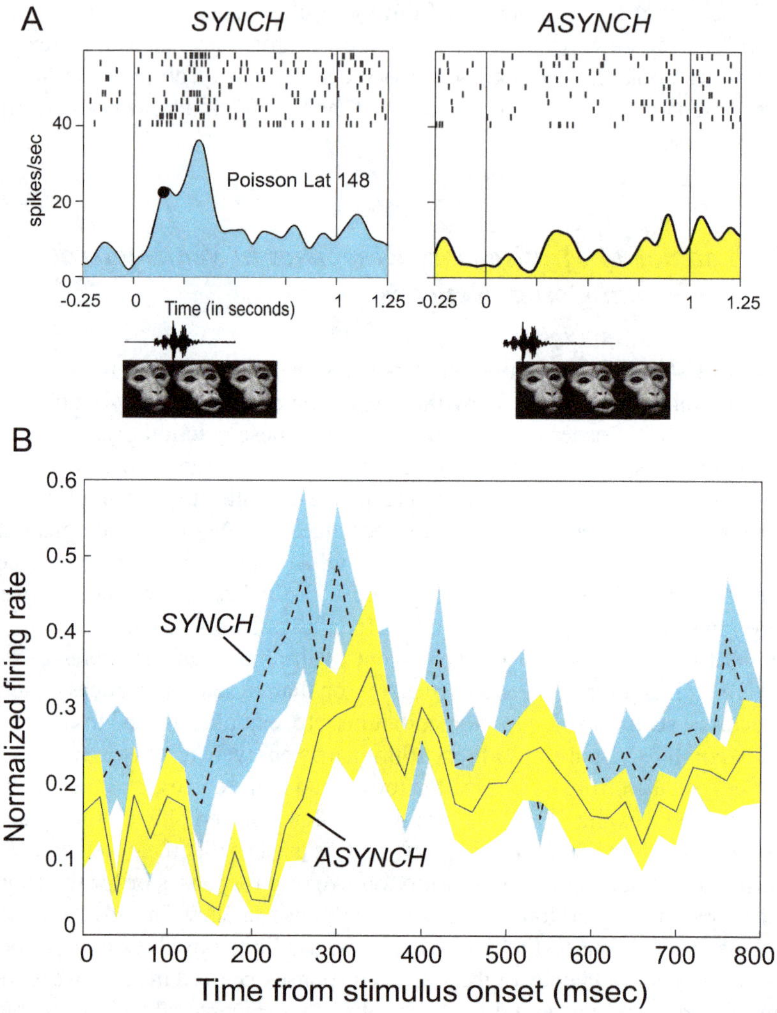

Fig. 7.5 A single neuronal example and the population response when a face-vocalization movie is played normally (synchronous [SYNCH]) and when the same movie with the auditory sound track played earlier than the face movie (asynchronous [ASYNCH]) compared with the original audiovisual movie stimulus. The latency was estimated as 148 ms in the SYNCH condition (*solid circle* on the SDF curve, Poisson latency [Lat] 148). (**A**) SYNCH movie stimulus elicited a robust response from a single neuron while the ASYNCH stimulus significantly suppressed the neuronal response. (**B**) Averaged population neuronal response to SYNCH (*blue*) and ASYNCH (*yellow*) audiovisual stimuli is shown as the normalized and averaged SDF ± SE from the responses of 18 multisensory neurons. The overall ASYNCH response was decreased in magnitude and had a longer latency relative to the SYNCH response. Adapted from Romanski and Hwang (2012), with permission

7.4 Factors That Affect Audiovisual Integration in Ventral Prefrontal Neurons

Prefrontal multisensory interactions are determined by the type, timing, and congruence of the stimuli that are paired. Although sensory stimuli converge in many regions throughout the brain, it is assumed that prefrontal neurons may have some specialization. The fact that the inferior frontal lobe is involved in communication in the human brain suggests that VLPFC neurons might be specialized for social communication stimuli in nonhuman primates. This is supported by the fact that "face cells" (O'Scalaidhe et al. 1997) and vocalization-responsive neurons are found in the VLPFC while responses to simple stimuli, such as pure tones, are not observed frequently in neurons of this region (Romanski et al. 2005; Romanski and Diehl 2011). In fact, direct testing with social and nonsocial stimuli showed that multisensory responses in VLPFC neurons are more frequently evoked by face-vocalization stimuli than by nonface-nonvocalization stimuli (Sugihara et al. 2006). This adds support to the notion that the VLPFC may be specialized for integrating face and vocalization information during communication rather than general auditory and visual stimuli and setting it apart from other brain regions that integrate sensory stimuli in a more general sense. Several studies have highlighted the importance of familiarity in stimulus preference for face-voice combinations in macaques (Adachi and Hampton 2011; Habbershon et al. 2013), indicating the importance of social interaction in the face-voice system. The importance of familiarity to multisensory interactions in the VLPFC has not yet been determined.

7.4.1 Processing of Congruent and Incongruent Stimuli in the Ventrolateral Prefrontal Cortex

An important aspect of social stimuli is the semantic relationship between the visual and auditory stimuli. During communication, congruent face and vocal stimuli, such as an angry expression and an angry voice, will reinforce the emotional tone of a message. Additionally, the congruent combination of social audiovisual stimuli can clarify and strengthen the information received (Laurienti et al. 2003; Molholm et al. 2004). In contrast, mismatching faces and vocal stimuli will provide confusing, conflicting information that can decrease or alter the information received. For example, an angry face paired with a happy voice is perceived as less angry (Watson et al. 2013). Another example is illustrated by the McGurk effect where an incongruent facial gesture and vocal stimulus may converge to elicit an entirely new percept (McGurk and MacDonald 1976; also see Lee, Maddox, and Bizley, Chap. 4, for a review). Single neurons recorded in primate VLPFC during the presentation of congruent and incongruent face-vocalization movies show significantly different responses, including significant differences in response dynamics. For example, a population of neurons in the VLPFC demonstrated neuronal suppression to an incongruent face-vocalization pair during the early part of the stimulus presentation period and enhancement occurred during the late part of the stimulus period (Diehl and Romanski 2014). Responses to incongruent pairs of stimuli occurred in both bimodal multisensory neurons and nonlinear multisensory-responsive neurons. The type of multisensory response did not predict a response to incongruent stimuli (Diehl and Romanski 2014). Nonetheless, incongruent responses were significantly different from responses to congruent stimuli in many VLPFC cells. This could portend a role for VLPFC in identity processing or comprehension as well as in the clarification of an audiovisual message.

7.4.2 Ventral Prefrontal Neurons Discriminate Asynchronous Audiovisual Stimuli

The temporal relationship of a paired auditory and visual stimulus is especially important for communication. During speech, the onset of the audible vocal stimulus typically lags behind the onset of the dynamic visual stimulus so that a speech sound is heard after the vocal apparatus begins to move. Changes to this expected temporal relationship are quite obvious and distracting, such as when watching a poorly synchronized movie. A functional magnetic resonance imaging (fMRI) study has shown that the human inferior frontal gyrus is one of several brain regions that changes its activity during asynchronous compared with synchronous audiovisual speech (Miller and D'Esposito 2005). Consequently, it may not be surprising that single neurons in nonhuman primate VLPFC, an area thought to be homologous with speech and language areas of the human inferior frontal gyrus, are sensitive to asynchronous communication stimuli (Romanski and Hwang 2012). When macaques were presented with normal audiovisual movies and the same movie where the auditory track had been offset in time relative to the video track, multisensory VLPFC neurons showed dramatic changes in latency and in response magnitude (Fig. 7.5B). These physiological studies show that the VLPFC, which is involved in the processing of communication information, is affected by temporal interactions of audiovisual stimuli.

These temporal interactions are also important in audio-vocal responses. When speaking, one unconsciously modulates one's vocal output while listening to the feedback of one's voice. This process is especially important in learning language. The VLPFC may play a role in this auditory-vocal feedback. Pitch-shift experiments in animals show that both auditory cortex and frontal lobe regions alter their activity during speech in response to altered audio feedback (Eliades and Wang 2012). It is essential during the learning of language that the timing of auditory, motor, and visual stimuli are represented with good fidelity. A delay along any pathway could have deleterious effects on the reception, comprehension, and production of speech. Recent studies have suggested that individuals with autism spectrum disorders may integrate audiovisual speech cues over larger temporal intervals and that this wider "window" can lead to deficits in perceiving audiovisual speech (Baum et al. 2015; see Baum Miller, and Wallace, Chap. 12, for a review). This wider integration window may be due to a transmission delay of stimuli. Delays in the transmission of information to PFC could lead to audiovisual or audio-motor timing aberrations that interfere with the early development of speech and language processing. Although it is difficult to study potential sources of communication deficits in animal models, recent studies in macaques have demonstrated some similarities in their vocal output system. For example, neurophysiological recordings have demonstrated modulatory changes in premotor and VLPFC neurons before self-initiated vocalizations (Hage and Nieder 2013). Furthermore, some VLPFC neurons respond to both hearing a vocalization and before a self-initiated vocalization (Hage and Nieder 2015). These "audio-vocal" cells may be crucial in modulating vocal output, a process that occurs during language learning in humans. Hence, this nonhuman primate system may help us understand the neural basis of communication disorders better.

7.5 Prefrontal Activity During Auditory and Visual Working Memory

As previously discussed in Sect. 7.1, multisensory cues are important for communication, a complex process that relies on working memory, and historically, the PFC has been implicated in both decision making and working memory. Despite the long history of neurophysiology studies linking activity with visual memory, decision making, and discrimination paradigms, lesion studies differ in their results where nonspatial working memory is concerned. Some studies showed visual working memory deficits after lesions of the lateral PFC (Passingham 1975; Mishkin and Manning 1978) while others do not. Recent investigations instead argue that the VLPFC assists in rule learning (Bussey et al. 2002; Rygula et al. 2010), stimulus selection (Rushworth et al. 1997; Passingham et al. 2000), and aspects of decision making (Baxter et al. 2009). Because visual afferents to the PFC originate in extrastriate visual areas, multisensory areas in the STS, the lateral parietal cortex, the perirhinal, and other regions, a diverse mix of visual functional attributes may be conferred to lateral prefrontal regions, providing the substrate for activity in complex cognitive abilities, such as decision making and discrimination. Clearly, more work is needed to delineate the role of the VLPFC in visual memory and decision processes.

Only a handful of studies have examined the role of the ventral frontal lobe during auditory working memory. Lesions of the frontal lobe suggested that it was essential in auditory memory. These lesion studies utilized large ablations of the lateral frontal lobe, including the VLPFC, and demonstrated impairments in some forms of auditory discrimination (Gross and Weiskrantz 1962; Iversen and Mishkin 1973). However, these studies included both dorsal and ventral regions and could not distinguish between behavioral effects such as perseveration or specific effects of auditory memory processing. In a similar manner, a recent neurophysiology study (Plakke et al. 2013a) involved recordings over a large frontal area including both the dorsal and ventral PFC and demonstrated neurons with task-related changes during auditory working memory and discrimination. Using a traditional match-to-sample paradigm, Plakke et al. (2013a) showed that the activity of lateral prefrontal neurons was modulated during key time periods, including when the animal encoded a sound, had to remember the sound, or discriminate and respond to a matching sound. Traditional memory markers such as delay activity and match suppression were also demonstrated in these prefrontal neurons during auditory working memory. Additional investigations of the lateral PFC showed that prefrontal neurons are active in auditory decision making and category operations where activity has been shown to correlate with behavioral choices and decisions in these paradigms (Russ et al. 2008; Cohen et al. 2009).

Although there have been studies examining working memory in the auditory or visual domain separately, there are very few studies that have attempted to examine cross-modal memory (Fuster et al. 2000). Thus, the same questions and experimental paradigms addressing unimodal stimulus processing can be applied to multisensory processes. For example, is the VLPFC involved when face and voice information must be remembered? This is an ability humans depend on during recognition and communication. Recent evidence from neurophysiological and inactivation studies indicates that the VLPFC not only processes communication stimuli but is essential for remembering them.

7.5.1 Encoding of Faces and Vocalizations During Audiovisual Working Memory

Hwang and Romanski (2015) examined the activity of VLPFC neurons in macaque monkeys during audiovisual working memory. Nonhuman primates were required to remember both the dynamic face and vocalization presented as a short movie clip and were required to detect when either the face or vocalization changed (Fig. 7.6).

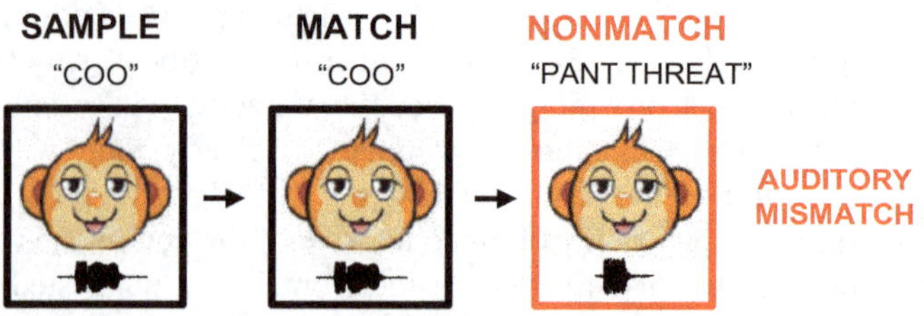

Fig. 7.6 (**A**) Nonmatch-to-sample task is illustrated where a vocalization movie (with an audio and video track) was presented as the sample stimulus. The subject was required to remember both the auditory and visual components (vocalization and accompanying facial gesture) and then to detect a change of either the face or vocalization component in subsequent stimulus presentations with a button press. Because the nonmatching stimulus can occur as either the second or third stimulus, it is unpredictable and subjects must rely on remembering the face and vocalization to detect any nonmatching stimuli. In the task example shown here, the vocalization is mismatched to this face so that the trial is an auditory nonmatch

Subjects performed well on both auditory and visual nonmatches, demonstrating that they were correctly remembering both the face and the vocalization component of the movie clip. VLPFC neurons increased their modulation during the sample phase, delay phase, and nonmatch phases of the task. These neurons were also active in other aspects of audiovisual working memory, including responding in a selective manner to particular face-vocalization combinations that were shown. Neurons also demonstrated unique context-dependent responses to nonmatching faces and vocalizations (Fig. 7.7A). Analyses of neuronal responses of VLPFC neurons revealed that even during a working memory paradigm, some neurons were modulated by the change of one modality, whereas others appeared to monitor the status of both the face and vocal stimulus in memory, an example of multisensory

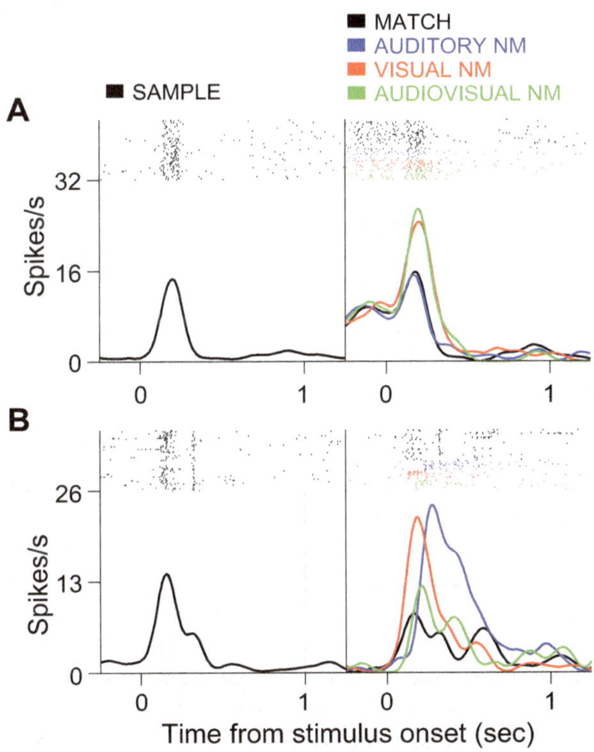

Fig. 7.7 Two example neurons during the audiovisual working memory task are shown. *Left*, response elicited during the sample presentation; *right*, different types of neural responses that occurred with the different types of nonmatch (NM) or match presentations. The example cells were both responsive during the nonmatch period. (**A**) Cell had enhanced firing rates when the visual stimulus was nonmatching (*red*) and when both the auditory and visual (audiovisual; *green*) stimuli were nonmatching. (**B**) Cell had enhanced firing to a change in either the auditory or visual component (*blue* or *red*, respectively), but only when one component changed at a time. This cell was considered a multisensory responsive neuron. Each panel depicts an SDF plot with rasters above the SDF for 10–12 trials

working memory. The activity of VLPFC neurons has proven to be quite complex and is modulated both by the specific face-vocalization stimulus and by their multisensory context as seen in Fig. 7.7B, where a cell exhibited an enhanced firing rate for an auditory or visual change but only when one modality changed at a time. These multisensory working-memory neurons were mostly located in and around the inferior prefrontal dimple, an area known for its strong auditory and multisensory responses (Sugihara et al. 2006; Diehl and Romanski 2014).

7.5.2 Neural Signatures of Memory During Audiovisual Working Memory

Further evidence that VLPFC neurons maintain audiovisual information online is evident when responses during other epochs of the audiovisual task are examined. When a stimulus is repeated as a match, it can lead to phenomena known as match enhancement or match suppression, where neurons show increased or decreased responses to "remembered" repeated stimuli. Hwang and Romanski (2015) examined responses of VLPFC neurons during the match repetition and found that the neurons demonstrated match suppression, showing that the neuron "remembers" the repeated stimulus with a decrease in response. These effects are typically described as evidence of working memory and have been found for both visual working memory (Miller et al. 1991, 1996) and auditory working memory (Plakke et al. 2013a) within the PFC. Hwang and Romanski (2015) also found evidence of match enhancement in the VLPFC. This is the first demonstration of these cognitive effects in the VLPFC during cross-modal working memory in nonhuman primates.

Neurons in the VLPFC show an increase in activity in the delay period of an audiovisual task in the absence of the stimulus when the stimulus needs to be remembered. These changes in firing rate are similar to previous reports of delay activity reported for visual working memory (Goldman-Rakic 1987; Asaad et al. 2000) and are viewed as neural markers of memory. In addition to general activity during the delay, some cells were responsive to a particular stimulus and maintained that selectivity throughout the delay period (Hwang and Romanski 2015), a finding similar to results described for faces in a visual memory paradigm (Wilson et al. 1993). This selectivity for particular faces or vocalizations across the delay is strong evidence supporting the idea that the VLPFC is not only involved in rule learning or attention but also contributes to working memory processing. Furthermore, in Hwang and Romanski (2015), when the decision to respond had already been made but the subject was waiting to respond, there was a decrease in neural activity. This supports the notion that delay activity is important for maintenance, and when maintenance is no longer necessary and a response is imminent, a corresponding decrease in delay activity occurs.

7.5.3 *Inactivation of the Ventrolateral Prefrontal Cortex Impairs Auditory and Audiovisual Working Memory*

The results from Hwang and Romanski (2015) clearly demonstrate that neurons in the VLPFC are actively involved in the encoding of face and vocal information and in maintaining this information in memory, i.e., audiovisual working memory. However, it does not tell us whether the VLPFC is essential in auditory and audiovisual working memory. Previous lesion studies have suggested that auditory working memory may depend on some lateral prefrontal areas, but which of these areas is necessary is not clear. In these studies, large lesions that included a number of prefrontal domains were used (Gross and Weiskrantz 1962; Iversen and Mishkin 1973). Furthermore, no studies have examined multisensory integration after prefrontal lesions. To investigate these issues, Plakke et al. (2015) reversibly inactivated the VLPFC with cooling while macaques performed an audiovisual working memory task. The animals performed several variations of the nonmatch-to-sample working-memory paradigm previously used in Hwang and Romanski (2015). This included an auditory-only memory task, a visual-only memory task, and an audiovisual memory version. While animal subjects performed each one of these three tasks, the VLPFC was temporarily inactivated by cooling. When the VLPFC was cooled during an audiovisual working memory task in which subjects were required to remember *both the face and vocalization*, animals were significantly impaired when detecting changes on both auditory and visual trials (Fig. 7.8). These results suggest a definitive role for the VLPFC during audiovisual working memory, when it is necessary for information from both modalities to be held in memory as it is during communication.

It is certainly possible that remembering two complex items, a face and its associated vocalization, is difficult and that this increased memory load may be dependent on a functioning VLPFC. This possibility was tested by having subjects perform the same task but only requiring them remember either the auditory or the visual stimulus in separate tests. In testing the role of the VLPFC in auditory-only working memory, a face-vocalization movie was presented, but subjects only had to remember the vocalization component. The results showed that when the VLPFC was cooled, performance was significantly impaired, with subjects unable to correctly remember and detect the vocalization, which suggests the VLPFC is essential for auditory working memory (Plakke et al. 2015). This finding is supported by neuroimaging studies where activation has been found within the inferior frontal gyrus during auditory voice recognition (Rama et al. 2004; Rama and Courtney 2005), verbal working memory (Schumacher et al. 1996; Crottaz-Herbette et al. 2004), and phonological maintenance (Strand et al. 2008).

In a third series of experiments, the effect of transient inactivation of the VLPFC on visual-only working memory was studied. As before, a face-vocalization movie was shown as the sample, but in this version of the task, only the facial stimulus was required to be remembered. Interestingly, when subjects were asked to remember

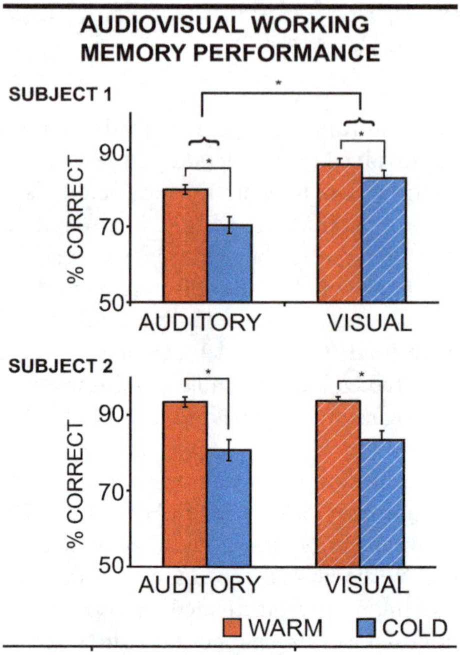

Fig. 7.8 Inactivation of the VLPFC during working memory in two subjects. VLPFC inactivation with cortical cooling significantly impaired overall performance accuracy on the audiovisual nonmatch-to-sample task. *Solid bars*, accuracy on auditory nonmatch trials, where a different vocalization was substituted; *striped bars*, accuracy on visual nonmatch trials, where a different face was used in the stimulus video. Both subjects showed a significant decrease in response accuracy during cooling (COLD; *blue*) compared with the control period (WARM; *red*). Additionally, subject 1 was significantly worse on auditory nonmatch trials compared with visual nonmatch trials. Error bars indicate SE. $*P < 0.05$

only the face, performance was *not* impaired when the VLPFC was cooled. This lack of an impairment during visual memory for the face alone indicates that inactivation of the VLPFC does not affect all cognitive processes or cause a general impairment that could explain the deficit during the auditory and audiovisual memory tasks. It was hypothesized that the auditory working memory specifically requires the VLPFC and that when this process is made even more difficult by requiring subjects to also remember a face stimulus during the audiovisual memory task, the larger memory load leads to a decrease in performance accuracy in both auditory nonmatch and visual nonmatch trials (Plakke et al. 2015). Similar load effects in human subjects have occurred during auditory and visual detection (Yu et al. 2014). Neuroimaging studies have also found activation of the ventral prefrontal region or inferior frontal gyrus when subjects are remembering face and voice information (Rama and Courtney 2005). These data suggest that the VLPFC is necessary for audiovisual working memory and supports the previous neurophysiological findings. Taken together, the recent neurophysiological and inactivation studies suggest that the VLPFC is a multisensory region with complex neural responses that play an important role in audiovisual processing. Single neurons encode specific face-vocalization combinations, respond to particular combinations of those faces and vocalizations, demonstrate match suppression/enhancement, and are selectively active during the delay period. Collectively, these characteristics are all essential functions in support of audiovisual working memory.

7.6 Summary

Scientific discoveries that led to a clearer understanding of the anatomical and physiological organization of the primate auditory cortex were a necessary prerequisite in localizing auditory functional zones in connected prefrontal cortical regions. Studies that have examined these auditory-recipient prefrontal zones demonstrated that many auditory-responsive neurons were, in fact, multisensory. Many VLPFC cells respond to face and vocalization stimuli and exhibit multisensory enhancement or suppression when face-vocalization stimuli are combined. The demonstration of audiovisual responsive regions in the ventral PFC provides the neural architecture to explain neuroimaging and behavioral findings, demonstrating the integration of speech sounds and gestures during communication. The demonstration of a nonhuman primate region that is involved in integration and remembering communication-relevant face and vocal stimuli suggests that this region may have some basic functional homologies to the human frontal lobe language areas.

Acknowledgments The authors acknowledge the hard work and assistance of Theodore Lincoln and Skye Kevelson in writing, editing, and figure creation. We also thank Adrian K. C. Lee, Mark Wallace, Allison Coffin, and Arthur N. Popper for all their helpful comments and suggestions in editing this chapter.

Compliance with Ethics Requirements Lizabeth M. Romanski declares that she has no conflict of interest.

Bethany Plakke declares that she has no conflict of interest.

References

Adachi, I., & Hampton, R. (2011). Rhesus monkeys see who they hear: Spontaneous cross-modal memory for familiar conspecifics. *PLoS One, 6*(8), e23345.

Asaad, W. F., Rainer, G., & Miller, E. K. (2000). Task-specific neural activity in the primate prefrontal cortex. *Journal of Neurophysiology, 84*, 451–459.

Averbeck, B. B., & Romanski, L. M. (2006). Probabilistic encoding of vocalizations in macaque ventral lateral prefrontal cortex. *The Journal of Neuroscience, 26*, 11023–11033.

Barbas, H. (1992). Architecture and cortical connections of the prefrontal cortex in the rhesus monkey. *Advances in Neurology, 57*, 91–115.

Barraclough, N. E., Xiao, D., Baker, C. I., Oram, M. W., & Perrett, D. I. (2005). Integration of visual and auditory information by superior temporal sulcus neurons responsive to the sight of actions. *Journal of Cognitive Neuroscience, 17*(3), 377–391.

Baum, S. H., Stevenson, R. A., & Wallace, M. T. (2015). Behavioral, perceptual, and neural alterations in sensory and multisensory function in autism spectrum disorder. *Progress in Neurobiology, 134*, 140–160.

Baxter, M. G., Gaffan, D., Kyriazis, D. A., & Mitchell, A. S. (2009). Ventrolateral prefrontal cortex is required for performance of a strategy implementation task but not reinforcer devaluation effects in rhesus monkeys. *European Journal of Neuroscience, 29*(10), 2049–2059.

Bodner, M., Kroger, J., & Fuster, J. M. (1996). Auditory memory cells in dorsolateral prefrontal cortex. *Neuroreport, 7*, 1905–1908.

Brody, C. D., Hernandez, A., Zainos, A., & Romo, R. (2003). Timing and neural encoding of somatosensory parametric working memory in macaque prefrontal cortex. *Cerebral Cortex, 13*(11), 1196–1207.

Bussey, T. J., Wise, S. P., & Murray, E. A. (2002). Interaction of ventral and orbital prefrontal cortex with inferotemporal cortex in conditional visuomotor learning. *Behavioral Neuroscience, 116*(4), 703–715.

Cohen, Y. E., Russ, B. E., Davis, S. J., Baker, A. E., Ackelson, A. L., & Nitecki, R. (2009). A functional role for the ventrolateral prefrontal cortex in non-spatial auditory cognition. *Proceedings of the National Academy of Sciences of the United States of America, 106*(47), 20045–20050.

Crottaz-Herbette, S., Anagnoson, R. T., & Menon, V. (2004). Modality effects in verbal working memory: Differential prefrontal and parietal responses to auditory and visual stimuli. *NeuroImage, 21*(1), 340–351.

Diehl, M. M., & Romanski, L. M. (2014). Responses of prefrontal multisensory neurons to mismatching faces and vocalizations. *The Journal of Neuroscience, 34*(34), 11233–11243.

Eliades, S. J., & Wang, X. (2012). Neural correlates of the Lombard effect in primate auditory cortex. *The Journal of Neuroscience, 32*(31), 10737–10748.

Fuster, J. M., Bodner, M., & Kroger, J. K. (2000). Cross-modal and cross-temporal association in neurons of frontal cortex. *Nature, 405*(6784), 347–351.

Ghazanfar, A. A., Maier, J. X., Hoffman, K. L., & Logothetis, N. K. (2005). Multisensory integration of dynamic faces and voices in rhesus monkey auditory cortex. *The Journal of Neuroscience, 25*(20), 5004–5012.

Ghazanfar, A. A., Morrill, R. J., & Kayser, C. (2013). Monkeys are perceptually tuned to facial expressions that exhibit a theta-like speech rhythm. *Proceedings of the National Academy of Sciences of the United States of America, 110*(5), 1959–1963.

Gifford, G. W., III, Maclean, K. A., Hauser, M. D., & Cohen, Y. E. (2005). The neurophysiology of functionally meaningful categories: Macaque ventrolateral prefrontal cortex plays a critical role in spontaneous categorization of species-specific vocalizations. *Journal of Cognitive Neuroscience, 17*, 1471–1482.

Goldman-Rakic, P. S. (1987). Circuitry of primate prefrontal cortex and regulation of behavior by representational memory. In F. Plum (Ed.), *Handbook of physiology, Vol. V: Higher functions of the brain: The nervous system, Sect. 1* (pp. 373–418). Bethesda: American Physiological Society.

Gross, C. G., & Weiskrantz, L. (1962). Evidence for dissociation of impairment on auditory discrimination and delayed response following lateral frontal lesions in monkeys. *Experimental Neurology, 5*, 453–476.

Habbershon, H. M., Ahmed, S. Z., & Cohen, Y. E. (2013). Rhesus macaques recognize unique multimodal face-voice relations of familiar individuals and not of unfamiliar ones. *Brain, Behavior and Evolution, 81*(4), 219–225.

Hackett, T. A., Stepniewska, I., & Kaas, J. H. (1998). Subdivisions of auditory cortex and ipsilateral cortical connections of the parabelt auditory cortex in macaque monkeys. *Journal of Comparative Neurology, 394*, 475–495.

Hackett, T. A., Stepniewska, I., & Kaas, J. H. (1999). Prefrontal connections of the parabelt auditory cortex in macaque monkeys. *Brain Research, 817*, 45–58.

Hackett, T. A., de la Mothe, L. A., Camalier, C. R., Falchier, A., Lakatos, P., Kajikawa, Y., & Schroeder, C. E. (2014). Feedforward and feedback projections of caudal belt and parabelt areas of auditory cortex: Refining the hierarchical model. *Frontiers in Neuroscience, 8*, 72.

Hage, S. R., & Nieder, A. (2013). Single neurons in monkey prefrontal cortex encode volitional initiation of vocalizations. *Nature Communications, 4*, 2409.

Hage, S. R., & Nieder, A. (2015). Audio-vocal interaction in single neurons of the monkey ventrolateral prefrontal cortex. *The Journal of Neuroscience, 35*, 7030–7040.

Hershenson, M. (1962). Reaction time as a measure of intersensory facilitation. *Journal of Experimental Psychology, 63*, 289–293.

Hwang, J., & Romanski, L. M. (2015). Prefrontal neuronal responses during audiovisual nmemonic processing. *The Journal of Neuroscience, 35*, 960–971.

Iversen, S. D., & Mishkin, M. (1973). Comparison of superior temporal and inferior prefrontal lesions on auditory and non-auditory tasks in rhesus monkeys. *Brain Research, 55*(2), 355–367.

Jones, E. G., Dell'Anna, M. E., Molinari, M., Rausell, E., & Hashikawa, T. (1995). Subdivisions of macaque monkey auditory cortex revealed by calcium-binding protein immunoreactivity. *Journal of Comparative Neurology, 362*, 153–170.

Jurgens, U. (2009). The neural control of vocalization in mammals: A review. *Journal of Voice, 23*(1), 1–10.

Kosaki, H., Hashikawa, T., He, J., & Jones, E. G. (1997). Tonotopic organization of auditory cortical fields delineated by parvalbumin immunoreactivity in macaque monkeys. *Journal of Comparative Neurology, 386*(2), 304–316.

Laurienti, P. J., Wallace, M. T., Maldjian, J. A., Susi, C. M., Stein, B. E., & Burdette, J. H. (2003). Cross-modal sensory processing in the anterior cingulate and medial prefrontal cortices. *Human Brain Mapping, 19*(4), 213–223.

McGurk, H., & MacDonald, J. (1976). Hearing lips and seeing voices. *Nature, 264*, 746–748.

Medalla, M., & Barbas, H. (2014). Specialized prefrontal "auditory fields": Organization of primate prefrontal-temporal pathways. *Frontiers in Neuroscience, 8*, 77.

Meredith, M. A., & Stein, B. E. (1986). Visual, auditory, and somatosensory convergence on cells in superior colliculus results in multisensory integration. *Journal of Neurophysiology, 56*(3), 640–662.

Miller, L. M., & D'Esposito, M. (2005). Perceptual fusion and stimulus coincidence in the crossmodal integration of speech. *The Journal of Neuroscience, 25*, 5884–5893.

Miller, E. K., Li, L., & Desimone, R. (1991). A neural mechanism for working and recognition memory in inferior temporal cortex. *Science, 254*(5036), 1377–1379.

Miller, E. K., Erickson, C. A., & Desimone, R. (1996). Neural mechanisms of visual working memory in prefrontal cortex of the macaque. *The Journal of Neuroscience, 16*, 5154–5167.

Mishkin, M., & Manning, F. J. (1978). Non-spatial memory after selective prefrontal lesions in monkeys. *Brain Research, 143*, 313–323.

Molholm, S., Ritter, W., Javitt, D. C., & Foxe, J. J. (2004). Multisensory visual-auditory object recognition in humans: A high-density electrical mapping study. *Cerebral Cortex, 14*(4), 452–465.

O'Scalaidhe, S. P., Wilson, F. A., & Goldman-Rakic, P. S. (1997). Areal segregation of face-processing neurons in prefrontal cortex. *Science, 278*, 1135–1138.

O'Scalaidhe, S. P. O., Wilson, F. A. W., & Goldman-Rakic, P. G. R. (1999). Face-selective neurons during passive viewing and working memory performance of rhesus monkeys: Evidence for intrinsic specialization of neuronal coding. *Cerebral Cortex, 9*, 459–475.

Passingham, R. (1975). Delayed matching after selective prefrontal lesions in monkeys (*Macaca mulatta*). *Brain Research, 92*, 89–102.

Passingham, R. E., Toni, I., & Rushworth, M. F. (2000). Specialisation within the prefrontal cortex: The ventral prefrontal cortex and associative learning. *Experimental Brain Research, 133*(1), 103–113.

Petrides, M., & Pandya, D. N. (2002). Comparative cytoarchitectonic analysis of the human and the macaque ventrolateral prefrontal cortex and corticocortical connection patterns in the monkey. *European Journal of Neuroscience, 16*(2), 291–310.

Plakke, B., & Romanski, L. M. (2014). Auditory connections and functions of prefrontal cortex. *Frontiers in Neuroscience, 8*, 199.
Plakke, B., Diltz, M. D., & Romanski, L. M. (2013a). Coding of vocalizations by single neurons in ventrolateral prefrontal cortex. *Hearing Research, 305*, 135–143.
Plakke, B., Ng, C. W., & Poremba, A. (2013b). Neural correlates of auditory recognition memory in primate lateral prefrontal cortex. *Neuroscience, 244*, 62–76.
Plakke, B., Hwang, J., & Romanski, L. M. (2015). Inactivation of primate prefrontal cortex impairs auditory and audiovisual working memory. *The Journal of Neuroscience, 35*, 9666–9675.
Posner, M. I., Nissen, M. J., & Klein, R. M. (1976). Visual dominance: An information-processing account of its origins and significance. *Psychological Review, 83*(2), 157–171.
Price, C. J. (2012). A review and synthesis of the first 20 years of PET and fMRI studies of heard speech, spoken language and reading. *NeuroImage, 62*(2), 816–847.
Rama, P., & Courtney, S. M. (2005). Functional topography of working memory for face or voice identity. *NeuroImage, 24*(1), 224–234.
Rama, P., Poremba, A., Sala, J. B., Yee, L., Malloy, M., Mishkin, M., & Courtney, S. M. (2004). Dissociable functional cortical topographies for working memory maintenance of voice identity and location. *Cerebral Cortex, 14*, 768–780.
Rauschecker, J. P., Tian, B., & Hauser, M. (1995). Processing of complex sounds in the macaque nonprimary auditory cortex. *Science, 268*(5207), 111–114.
Romanski, L. M. (2012). Integration of faces and vocalizations in ventral prefrontal cortex: Implications for the evolution of audiovisual speech. *Proceedings of the National Academy of Sciences of the United States of America, 109*(Suppl 1), 10717–10724.
Romanski, L. M., & Averbeck, B. B. (2009). The primate cortical auditory system and neural representation of conspecific vocalizations. *Annual Review of Neuroscience, 32*, 315–346.
Romanski, L. M., & Diehl, M. M. (2011). Neurons responsive to face-view in the primate ventrolateral prefrontal cortex. *Neuroscience, 189*, 223–235.
Romanski, L. M., & Goldman-Rakic, P. S. (2002). An auditory domain in primate prefrontal cortex. *Nature Neuroscience, 5*, 15–16.
Romanski, L. M., & Hwang, J. (2012). Timing of audiovisual inputs to the prefrontal cortex and multisensory integration. *Neuroscience, 214*, 36–48.
Romanski, L. M., Bates, J. F., & Goldman-Rakic, P. S. (1999a). Auditory belt and parabelt projections to the prefrontal cortex in the rhesus monkey. *Journal of Comparative Neurology, 403*, 141–157.
Romanski, L. M., Tian, B., Fritz, J., Mishkin, M., Goldman-Rakic, P. S., & Rauschecker, J. P. (1999b). Dual streams of auditory afferents target multiple domains in the primate prefrontal cortex. *Nature Neuroscience, 2*(12), 1131–1136.
Romanski, L. M., Averbeck, B. B., & Diltz, M. (2005). Neural representation of vocalizations in the primate ventrolateral prefrontal cortex. *Journal of Neurophysiology, 93*(2), 734–747.
Rushworth, M. F., Nixon, P. D., Eacott, M. J., & Passingham, R. E. (1997). Ventral prefrontal cortex is not essential for working memory. *The Journal of Neuroscience, 17*(12), 4829–4838.
Russ, B. E., Orr, L. E., & Cohen, Y. E. (2008). Prefrontal neurons predict choices during an auditory same-different task. *Current Biology, 18*(19), 1483–1488.
Rygula, R., Walker, S. C., Clarke, H. F., Robbins, T. W., & Roberts, A. C. (2010). Differential contributions of the primate ventrolateral prefrontal and orbitofrontal cortex to serial reversal learning. *The Journal of Neuroscience, 30*(43), 14552–14559.
Schumacher, E. H., Lauber, E., Awh, E., Jonides, J., Smith, E. E., & Koeppe, R. A. (1996). PET evidence for an amodal verbal working memory system. *NeuroImage, 3*(2), 79–88.
Strand, F., Forssberg, H., Klingberg, T., & Norrelgen, F. (2008). Phonological working memory with auditory presentation of pseudo-words—An event related fMRI study. *Brain Research, 1212*, 48–54.
Sugihara, T., Diltz, M. D., Averbeck, B. B., & Romanski, L. M. (2006). Integration of auditory and visual communication information in the primate ventrolateral prefrontal cortex. *The Journal of Neuroscience, 26*, 11138–11147.

Tanila, H., Carlson, S., Linnankoski, I., Lindroos, F., & Kahila, H. (1992). Functional properties of dorsolateral prefrontal cortical neurons in awake monkey. *Behavioral Brain Research, 47*, 169–180.

Tanila, H., Carlson, S., Linnankoski, I., & Kahila, H. (1993). Regional distribution of functions in dorsolateral prefrontal cortex of the monkey. *Behavioral Brain Research, 53*, 63–71.

Vaadia, E., Benson, D. A., Hienz, R. D., & Goldstein, M. H., Jr. (1986). Unit study of monkey frontal cortex: Active localization of auditory and of visual stimuli. *Journal of Neurophysiology, 56*, 934–952.

Watson, R., Latinus, M., Noguchi, T., Garrod, O., Crabbe, F., & Belin, P. (2013). Dissociating task difficulty from incongruence in face-voice emotion integration. *Frontiers in Human Neuroscience, 7*, 744.

West, R. A., & Larson, C. R. (1995). Neurons of the anterior mesial cortex related to faciovocal activity in the awake monkey. *Journal of Neurophysiology, 74*(5), 1856–1869.

Wilson, F. A., O'Scalaidhe, S. P., & Goldman-Rakic, P. S. (1993). Dissociation of object and spatial processing domains in primate prefrontal cortex. *Science, 260*, 1955–1958.

Yu, J. C., Chang, T. Y., & Yang, C. T. (2014). Individual differences in working memory capacity and workload capacity. *Frontiers in Psychology, 5*, 1465.

Chapter 8
Using Multisensory Integration to Understand the Human Auditory Cortex

Michael S. Beauchamp

Abstract Accurate and meaningful parcellation of the human cortex is an essential endeavor to facilitate the collective understanding of brain functions across sensory and cognitive domains. Unlike in the visual cortex, the details of anatomical and functional mapping associated with the earliest stages of auditory processing in the cortex are still a topic of active debate. Interestingly, aspects of multisensory processing may provide a unique window to meaningfully subdivide the auditory sensory areas by exploring different functional properties other than the traditional tonotopic approach. In this chapter, a tour of the auditory cortical areas is first provided, starting from its core area, Heschl's gyrus, then moving onto surrounding areas. Evidence from different sources, including postmortem studies of the human auditory cortex, resting-state functional connectivity derived from the Human Connectome Project, and electrocorticographic studies, is presented to better understand how different subdivisions of the human auditory cortex and its surrounding areas are involved in auditory and multisensory processing. The chapter concludes with the remaining challenges to account for individual variability in functional anatomy, particularly pertaining to multisensory processing.

Keywords Auditory cortex · Cross-modal · Electrocorticography · Functional anatomy · Functional connectivity · Heschl's gyrus · Human Connectome Project · Sensory integration · Superior temporal sulcus · Temporal cortex

8.1 Introduction

The human cerebrum is divided into sensory cortices specialized for the processing of a specific sensory modality, with the visual cortex located in the occipital lobe and the auditory cortex centered on Heschl's gyrus on the plane of the superior

M. S. Beauchamp (✉)
Department of Neurosurgery and Core for Advanced MRI, Baylor College of Medicine, Houston, TX, USA
e-mail: michael.beauchamp@bcm.edu

A. K. C. Lee et al. (eds.), *Multisensory Processes*, Springer Handbook of Auditory Research 68, https://doi.org/10.1007/978-3-030-10461-0_8

temporal gyrus. The visual cortex and auditory cortex may be further subdivided into multiple cortical areas, each specialized for performing a specific computation on the incoming sensory data. The best-known example is Felleman and Van Essen's (1991) subdivision of the macaque visual cortex into 32 areas in a 10-level hierarchy. More than 25 years later, neuroscientists are still struggling to develop a similarly detailed description of the auditory cortex. Even at the earliest stages of cortical auditory processing, the number of areas, their anatomical layout, and their nomenclature are topics of active research and debate. One reason for this slow progress is the difficulty in finding functional properties that allow the auditory cortex to be subdivided. In this chapter, the possibility is explored that consideration multisensory integration processes, here the integration of both auditory and nonauditory stimuli, may lead to a better understanding of the human auditory cortex. The organization of the human auditory cortex is presented first (Sect. 8.2) and is framed around the general division into core, belt, and parabelt regions. Next, several subdivisions of the human parabelt cortex are examined from different perspectives (Sects. 8.3 and 8.4). Finally, the chapter concludes by using the auditory cortex as an example of the challenges that face functional brain mapping from the perspective of incorporating individual variability into the process of drawing meaningful functional distinctions between brain regions (Sect. 8.5).

8.2 Organization of the Human Auditory Cortex

The auditory cortex is located in the temporal lobe of the human neocortex (Fig. 8.1). Moving from dorsal to ventral, the temporal cortex encompasses the superior temporal gyrus (STG), the superior temporal sulcus (STS), the middle temporal gyrus (MTG), and the inferior temporal sulcus (ITS). Heschl's gyrus (HG) is a short gyrus engraved on the superior surface of the STG that begins on the lateral convexity and runs lateral-anterior to medial-posterior before terminating in the insula.

HG is the location of so-called "core" auditory areas. The use of the term core instead of "primary" denotes the idea that there are multiple areas that coexist at the

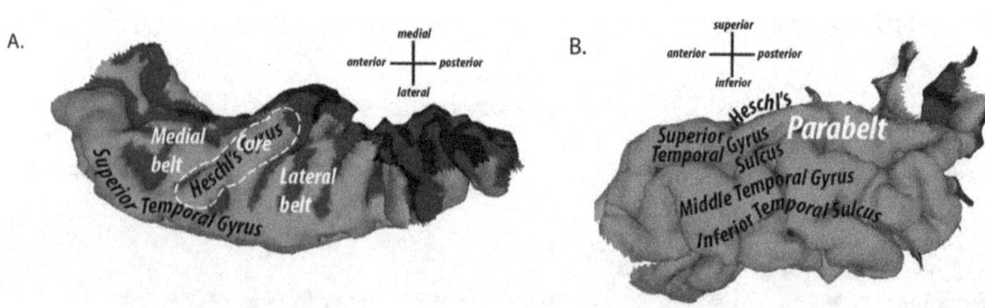

Fig. 8.1 Human auditory cortex. (**A**) Virtual dissection of the human temporal lobe, viewed from above. *Black labels*, anatomically defined structures; *white labels*, functionally defined regions. *Lighter grays*, gyri; *darker grays*, sulci. (**B**) Lateral view of virtual dissection of the pial surface of the temporal lobe

first stage of cortical processing. Each core area contains a tonotopic representation or map. In these maps, there is a gradual change in the preferred auditory frequency of the neurons across the area, with neighboring neurons having similar frequency tuning, and the entire range of perceptible frequencies is represented. At the boundaries between areas, the gradient of change in preferred frequency reverses so that adjacent areas have mirror-symmetrical tonotopic representations. Although the organization of these tonotopic maps in the core auditory cortex has been well established in animal models (Kass et al. 1999), the organization in humans has remained elusive. Data from ultrahigh-field 7-tesla functional magnetic resonance imaging (fMRI) of the blood oxygen level-dependent (BOLD) signal in human subjects led to a proposal by Moerel et al. (2014) that low frequencies are represented along the posterior edge of HG, whereas high frequencies are represented along the anterior edge, and that the core auditory cortex consists of three complete tonotopic maps. These three maps are named based on the conventions applied from studies in nonhuman primate models and consist of area A1 at the medial-posterior edge of HG, the rostrotemporal area at the lateral-anterior edge of HG, and the rostral area in the center of HG. Although this classification scheme is still a subject of debate, it provides a view into the current state of understanding of the functional architecture of the human core auditory cortex.

The cortex anterior and posterior to HG on the superior surface of the STG is the location of "belt" areas of the auditory cortex, so-called because of their anatomical location encircling HG. The areas anterior and medial to HG are referred to as the medial belt areas. Posterior and lateral to HG is a triangular patch of cortex termed the temporal plane (*planum temporale*), the location of the lateral belt areas. Data from tonotopic mapping at 7 tesla was used by Moerel et al. (2014) to divide the belt areas into six subdivisions, with anatomical labels derived from the nomenclature developed in physiological studies of nonhuman primate models. Moving from lateral to medial across the cortex, these areas are rostromedial, mediomedial, and caudomedial within the medial belt and anterolateral, mediolateral, and caudolateral within the lateral belt.

Responses to auditory stimuli extend laterally and posteriorly from the lateral belt areas onto the lateral surface of the STG and into the STS. Collectively, this region is termed the auditory cortex "parabelt." Although the auditory parabelt is larger than the core and belt areas, it fails to show a robust tonotopic organization, making functional parcellation based on frequency tuning impossible. However, as detailed in Sect. 8.3, substantial effort has been made to better delimit the functional organization of the parabelt areas.

8.3 Subdivisions of the Human Parabelt Cortex

8.3.1 Postmortem and In Vivo Magnetic Resonance Imaging

Classic and more contemporary cytoarchitectonic studies derived from human postmortem tissue may shed some light on the functional organization of the parabelt regions. Although Brodmann in his atlas (1909) classified the entire posterior two-thirds of the STG and STS, extending posteriorly all the way to the inferior parietal lobe, as a single cytoarchitectonic region (i.e., area 22), more recent studies suggest additional subdivisions (Fig. 8.2A; Morosan et al. 2005). Using multimodal architectonic mapping contrasting neuronal cell types, neuronal packing density, columnar organization, and neurotransmitter receptor distributions, the Jülich group identified the existence of a distinct area, labeled Te3, on the lateral bulge of the STG that does not extend onto the dorsal or ventral banks of the STG (Morosan et al. 2005).

A second valuable source of evidence about the functional organization of the human parabelt cortex is the multimodal MRI dataset derived from 210 subjects as part of the Human Connectome Project (HCP; Van Essen et al. 2013). Using this dataset, Glasser et al. (2016) subdivided the cerebral cortex into 180 areas in each hemisphere, including 10 distinct areas located in parabelt cortex (Fig. 8.2B).

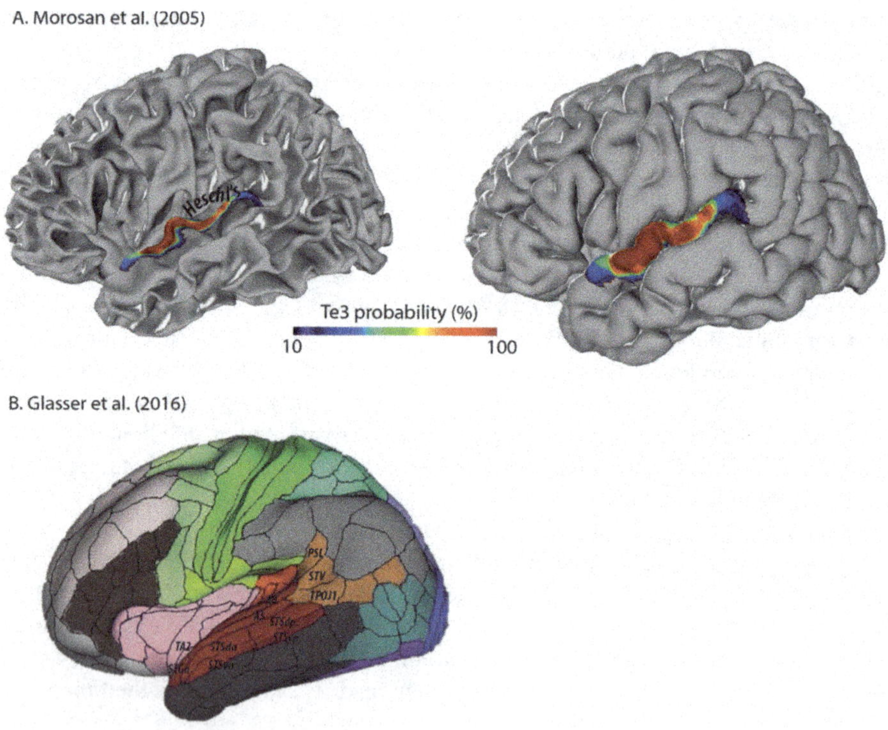

Fig. 8.2 Anatomical-functional subdivisions in lateral temporal cortex. (**A**) Morosan et al. (2005) described a cytoarchitectonic region termed Te3. Ten postmortem human brains were examined. The color scale shows the probability of the brain area containing Te3, visualized in a white matter cortical surface (*left*) and a pial cortical surface (*right*). *Dashed white line*, Heschl's gyrus; *red line*, cutting plane defined by Heschl's gyrus. (**B**) Inflated cortical surface model showing the Human Connectome Project (HCP) 1.0 brain parcellation. *Labeled areas*, parabelt cortex. *A4*, auditory 4 complex; *A5*, auditory 5 complex; *PSL*, perisylvian language area; *STG/S*, superior temporal gyrus/sulcus; *STV*, superior temporal visual area; *TPOJ1*, temporo-parieto-occipital junction 1; *TA2*, area TA2; *d*, dorsal; *v*, ventral; *a*, anterior; *p*, posterior

To consider this parcellation of the parabelt cortex in more detail, it is first necessary to briefly review its composition, which consists of four different measurements of brain structure and function. First, functional connectivity or resting-state fMRI measures of BOLD signal fluctuations can be obtained as the subject lays in the scanner without a task. Areas that show synchronous fluctuations in the BOLD signal are presumed to be functionally connected. If adjacent regions of the cortex show very different patterns of connectivity, this is taken as evidence for an areal boundary. Second, task-based fMRI measures the activity in brain areas in response to different stimuli and task conditions. Of particular relevance for parabelt delimitation in the HCP data is the language story condition, in which subjects listened to a story, and the language math condition, in which subjects listened to a math problem. Subtracting the math from the story condition task (story vs. math contrast) reveals areas specific for higher order language processes. Third, MRI pulse sequences and analysis techniques have been developed to measure correlates of the cytoarchitecture, notably myelin content (Glasser and Van Essen 2011; Glasser et al. 2014). Using this method, gradients in myelin content can also be used to distinguish between areas. Fourth, structural MRI can be used to determine cortical thickness and cortical folding patterns. The assumption is that changes in cortical thickness, or differences relative to cortical landmarks such as the fundus of a sulcus, represent areal boundaries.

Use of these four dimensions resulted in the division of the parabelt into ten distinct divisions. The two most anterior parabelt areas are small areas lateral to the medial belt areas and are labeled areas TA2 and STGa. The next four areas tile most of the STS. STSda and STSdp tile the upper bank of the STS in the anterior-to-posterior direction, whereas STSva and STSvp tile the lower bank of the STS in the same direction. Interestingly, the functional connectivity gradient is strongest along the fundus of the STS, providing strong evidence for an important functional distinction between the upper and lower bank areas. Medial to the STS, two areas tile the crown of the STG, area A4 more medially and area A5 more laterally. Finally, the most posterior of the parabelt regions are the superior temporal visual area (STV), made up of the most posterior section of STG before it angles up into the parietal lobe, and the temporo-parieto-occipital junction 1 area, which is the most posterior section of STS (Fig. 8.2B).

8.3.2 Electrocortigraphic Evidence That Multisensory Integration in the Auditory Cortex Provides Valuable Functional Information

The aforementioned atlases (see Sect. 8.3.1) derived from postmortem histology and group MRI data clearly suggest the existence of functionally specialized areas within the auditory cortex that are located anterior to posterior along the STG and STS. However, these group atlases are of limited value in understanding the organization of a particular individual brain. For instance, the 180 areas in the HCP atlas were defined by assuming that areas existed with sharp boundaries between them. A natural question is whether the parabelt cortex is best described as a series of areas with sharply defined boundaries or as a broad region of the cortex with gradual transitions between zones with different functional properties.

The ability of a technique such as BOLD fMRI to answer this question is limited because it does not directly measure neural activity. Any property of the cortical map, whether it is a sharp boundary or a gradual transition, could be ascribed to the properties of the cerebral vasculature in the region rather than the functional properties of the underlying neurons. For instance, if the upper and lower banks of the STS are drained by different venous beds, this could create a sharp boundary in BOLD fMRI between the two regions that does not necessarily reflect a functional distinction.

Another method to examine functional specialization in the parabelt cortex is intracranial encephalography (iEEG), also known as electrocorticography (ECoG), a technique to record activity directly from the cortex of awake human subjects (generally patients with conditions such as intractable epilepsy). Relative to BOLD fMRI, iEEG has the advantage that it directly measures neural activity without imposing a blurry hemodynamic filter.

A study used iEEG to probe the functional organization of auditory cortical regions by examining responses to audiovisual speech presented within varying levels of auditory noise (Ozker et al. 2017). The key observation motivating this study is that humans can use the visual mouth movements observed in the face of the talker to better understand the talker's voice and that these improvements grow larger as the auditory speech signal becomes noisier. The hypothesis behind the study was that parabelt areas involved in speech processing would be differentiated into those that process exclusively auditory information and those that integrate visual and auditory speech information.

Posterior portions of the STS/STG are multisensory in that they respond not only to auditory but also to visual and somatosensory stimuli in both humans (Beauchamp et al. 2004a, b) and nonhuman primates (Bruce et al. 1981). Therefore, for the purposes of this study, STG was divided into a posterior section and an anterior section (Fig. 8.3).

The responses to clear versus noisy audiovisual speech were strikingly different between the anterior and posterior STS/STG. Whereas in the anterior half, noisy speech greatly decreased the amplitude of the response when compared with clear speech, in the posterior half there was no decrease in the amplitude of the response. This effect was highly consistent; all anterior electrodes showed larger responses for the stimuli consisting of clear audiovisual speech, whereas all posterior STG electrodes showed similar responses for stimuli consisting of clear or noisy audiovisual speech.

Because iEEG directly measures neural activity from a small region of the cortex, activity in each electrode can be confidently assigned with anatomical precision. To examine the effect of anatomical location on the response to clear and noisy audiovisual speech with more detail than a simple division of the STS/STG into

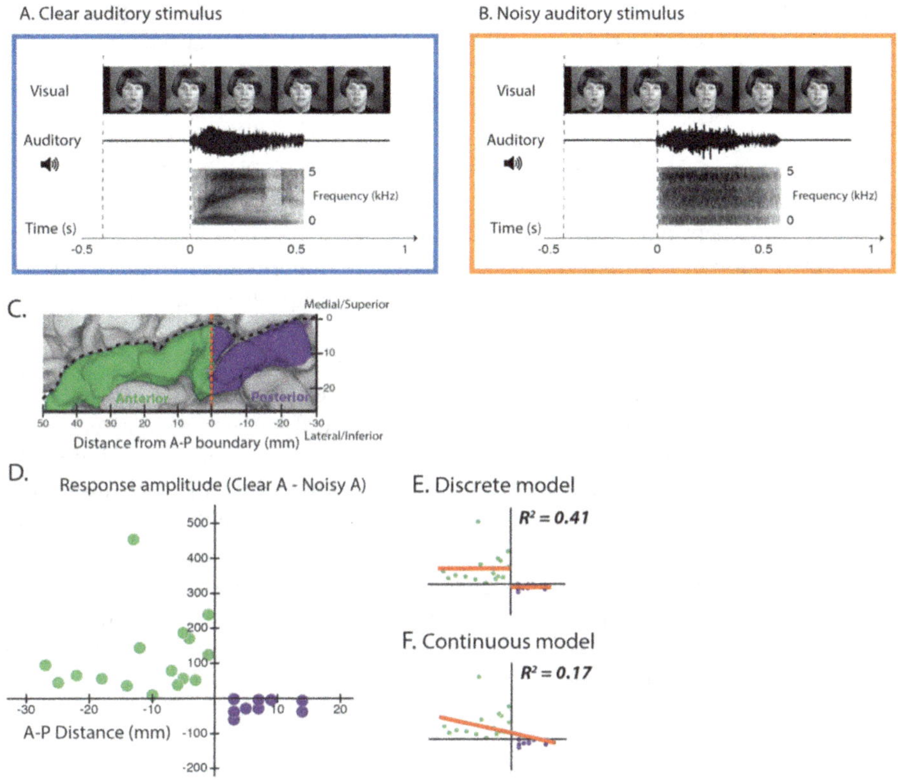

Fig. 8.3 Using electrocorticography to parcellate the parabelt cortex. (**A**) Example stimulus consisting of clear auditory speech (Clear A) and a movie of a talking face. Spectrogram shows clear formant bands in auditory speech. (**B**) Example stimulus consisting of noisy auditory speech (Noisy A) and a movie of a talking face. Spectrogram shows a lack of formant bands. (**C**) Lateral view of a cortical surface model of the temporal lobe showing anterior STG (*green*) and posterior STG (*purple*). Heschl's gyrus (not visible on the superior face of temporal lobe) extends from anterior-lateral to posterior-medial. The posterior most point of Heschl's gyrus is used to define an origin (*red dashed line*). All points anterior to this origin are classified as anterior and given a positive value corresponding to their distance from the origin (values on x-axis). The distance from the origin in the inferior-to-superior direction is shown on the y-axis.*Black dashed line*, distance from medial/superior border of STG. (**D**) Response to Clear A versus Noisy A speech for each individual electrode. *Green circles*, each anterior electrode; *purple circles*, each posterior electrode. The response amplitude is the mean percent change in high-gamma power (70–110 Hz) in the 0- to 500-ms time window relative to the prestimulus baseline (−500 to −100 ms). (**E**) Discrete model: constant values were fit separately to the anterior and posterior electrode data in **B** ($y = a$ for all electrodes with $x > 0$ and $y = b$ for all electrodes with $x < 0$) and the correlation with the data was calculated. (**F**) Continuous model: a linear model with two parameters was fit to both anterior and posterior electrodes ($y = mx + b$, where m is the slope and b is the constant term). Adapted from Ozker et al. (2017)

anterior and posterior segments, the location of each electrode was plotted in a functional reference frame defined by the responses to auditory speech. In creating this reference frame, the location of each electrode was plotted against its preference for clear compared with noisy audiovisual speech. Consistent with the first analysis, anterior electrodes showed greater responses for clear speech, whereas posterior electrodes showed similar or smaller responses for clear compared with noisy speech. Most importantly, when examined on an electrode-by-electrode basis, a sharp transition was found between anterior and posterior electrodes. This observation received quantitative support from a comparison of two Bayesian models, one of which posited a discrete transition (Fig. 8.3E) and one of which posited a gradual transition (Fig. 8.3F). The discrete model was more than 100 times more likely to explain the observed data.

Hence, using iEEG, an electrophysiological method with excellent spatial resolution, it was demonstrated that in the parabelt cortex of individual subjects there is a sharp functional boundary between the anterior and posterior STG, paralleling the findings from group maps created using postmortem anatomy or multimodal MRI. Critically, this functional difference was only evident with the use of multisensory stimuli (i.e., audiovisual speech) because both the anterior and posterior STG respond to unisensory auditory speech (either clear or noisy). It is postulated that the application of such multisensory approaches may allow for the differentiation of additional functionally distinct regions in the parabelt cortex and to other regions beyond the auditory cortex.

8.4 Posterior Boundary of the Parabelt Cortex

Anatomically, the posterior STS/STG is situated between the visual cortex and the auditory cortex, a finding consistent with the general organizational observation that multisensory zones exist at the borders between unisensory cortices (Wallace et al. 2004). If lateral temporal (STG/STS) regions that respond to auditory stimulation are considered as part of the parabelt cortex, the question arises: What is the posterior boundary of the parabelt cortex or where does the auditory cortex end and the visual cortex begin?

There is substantial evidence from fMRI that the inferotemporal sulcus (ITS) is a reasonable boundary for the transition from the visual cortex to the multisensory cortex. Two visual areas are situated along the ITS, area MT, which is typically located on the posterior bank of the ascending limb of the ITS, and area MST, which is typically located on the anterior bank of the ITS. These two areas, although both highly responsive to visual motion, have markedly different response properties. In macaque monkeys, single neurons in area MST are multisensory, responding to both visual and vestibular stimuli, potentially reflecting the role of this area in spatial navigation and postural control (Takahashi et al. 2007). In contrast, similar recordings from area MT in macaques reveal this area to be almost exclusively responsive to visual motion stimuli. A second difference between these areas is that whereas area MT appears to only respond to visual stimuli in the contralateral visual field, area MST responds to both contralateral and ipsilateral visual stimuli (Huk et al. 2002). When using fMRI to measure activation patterns to simple vibrotactile stimuli, area MT was found to respond only to visual stimuli, whereas area MST was found to respond to both visual and somatosensory stimuli (Fig. 8.4; Beauchamp et al. 2007). Therefore, the fundus of the inferotemporal sulcus (which typically corresponds to the border between areas MT and MST) appears to represent the functional boundary between the multisensory cortex (area MST) and visual cortex (area MT; Jiang et al. 2015).

One potential objection to this schema is the claim in the literature that MT in humans is, in fact, multisensory. A number of prior studies have claimed that MT responds to tactile motion (as well as to visual motion), such as an (unseen) brush stroking the arm (Hagen et al. 2002; van Kemenade et al. 2014). These results have been interpreted to mean that in humans, area MT is multisensory and, more generally, serves as a cross-modal motion-processing module (Pascual-Leone and Hamilton 2001; Ricciardi et al. 2007). However, a recent attempt to replicate these results (Hagen et al. 2002) found that in any individual subject, there was no overlap between visual and tactile motion activations in and around area MT (Fig. 8.4A).

So how can these seemingly contradictory results be reconciled? First, some of these studies did not actually locate area MT in individual subjects, instead relying on stereotactic coordinate values (Matteau et al. 2010). This is problematic because atlas values are blind to anatomical or functional landmarks, and it is known that the location of area MT in any individual can vary by a centimeter or more. Thus, multisensory activity on one side of the ITS can easily be confused with unisensory visual activity on the other bank of the ITS. Other studies rely on group-average activation maps to compare the location of tactile and visual motion activations (Ricciardi et al. 2007; Summers et al. 2009). The problem with this approach is illustrated in Fig. 8.4B, in which the absence of overlap between tactile and visual motion in any individual subject can result in an overlapping group activation map. Once again, this misleading activation pattern in the group map can be attributed to variability in the spatial location of activity across individual subjects. Averaging across individuals to create a group map acts as a blurring filter, taking two distinct tactile and visual motion peaks and merging them together. A simple illustration of this effect is shown in Fig. 8.4C. Although in any individual automobile, the front seats and back seats are in completely separate locations along the anterior-to-posterior axis of the auto, a map of the average seat locations across vehicles shows substantial overlap. However, the inference that the front and back seats are in the same location is obviously false. By extension, the use of group activation maps and meta-analysis techniques such as activation-likelihood estimation (Eickhoff et al. 2012) that creates what is, in effect, group maps by combining data from multiple studies must be used with extreme caution when making inferences about the anatomical organization multisensory responses.

A final set of observations relevant to the multisensory character of area MT is the fact that work in the blind and those with some degree of sight restoration have suggested that this area can support the processing of auditory motion in the absence

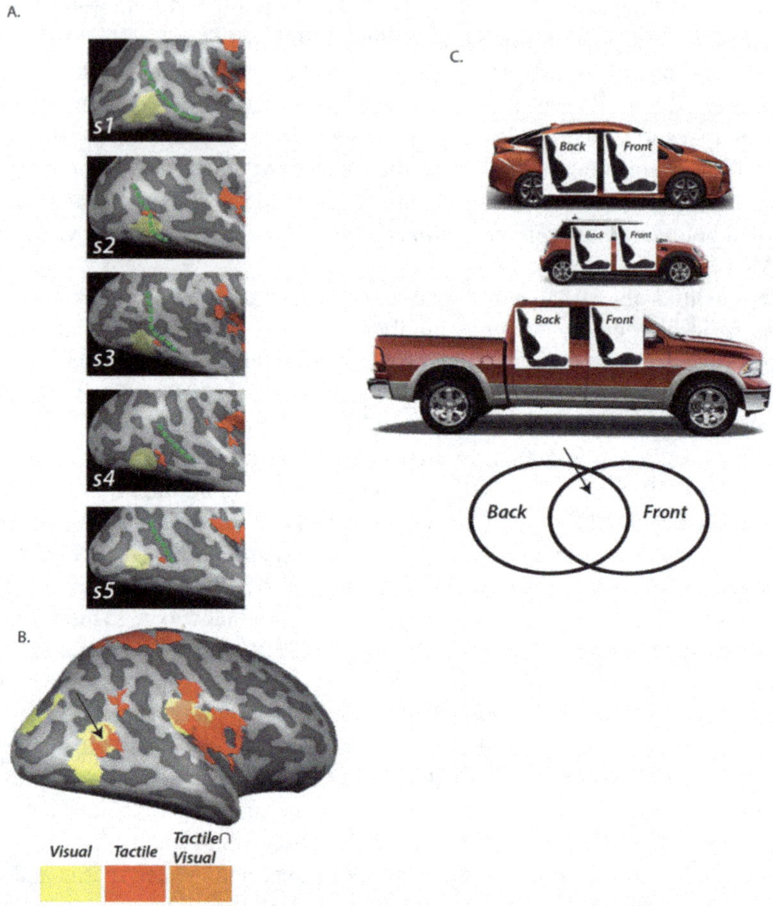

Fig. 8.4 Posterior boundary of parabelt cortex. (**A**) Functional magnetic resonance imaging (fMRI) activation maps from 5 subjects (s1 to s5) show the lateral views of inflated cortical surface models of the gray-white matter boundary for the right hemisphere. *Yellow*, areas with significantly greater blood oxygen level-dependent (BOLD) signal ($t > 2$ uncorrected [t statistic]) for a visual stimulus of moving dots compared with the fixation baseline in the vicinity of human area MT; *red*, areas with significantly greater BOLD signal ($t > 2$ uncorrected) for auditory sounds compared with the fixation baseline; *orange*, areas with significant activity for both visual and tactile stimuli. Adapted from Jiang et al. (2015). (**B**) Group map constructed from the individual subjects shown in **A**. Note that the group map shows overlapping tactile and visual activity in the vicinity of area MT (*black arrow*) even though this overlap is not present in any individual subject. Adapted from Jiang et al. (2015). (**C**) Example illustrating how average position maps can lead to incorrect inferences. In different vehicles, front and back seats are always in different spatial locations. A group map showing the average location of front and backs seats shows overlap between their positions (*black arrow*) even though this overlap is not present in any individual vehicle

of normal visual input (Saenz et al. 2008; Jiang et al. 2016). Although these examples are evidence for cross-modal plasticity, they highlight that area MT has some degree of multisensory or "supramodal" character that may contribute to the confusion as to whether it is truly a visual or a multisensory area.

Although this example of MT and MST illustrates some of the difficulties in drawing distinctions between multisensory and unisensory brain regions, the same challenges and issues are likely to apply to a host of brain regions that have been characterized as multisensory using methods such as fMRI and in which the spatial resolution is sufficiently coarse to result in the blurring of true functional distinctions. Thus, similar arguments can be applied to parietal cortical regions that lie along the intraparietal sulcus (IPS), and that is interposed between more posterior occipital (visual) cortical regions and more ventral temporal (auditory) cortical regions. Although invasive physiological studies in animal models have established the multisensory character of a number of the divisions of the IPS and the important role these areas play in saccadic eye and reach movements and spatial attention (Snyder et al. 1997; Grefkes and Fink 2005), a number of human imaging studies focused on the IPS are subject to the same concerns as articulated in Fig. 8.4C in regard to spatial blurring and the potential overestimation of true regions of multisensory convergence and integration.

8.5 Difficulties with the Use of Task-Based Functional Magnetic Resonance Imaging to Demarcate Area Boundaries

In addition to the concerns about spatial blurring in cross-subject analyses, there are other are fundamental difficulties with using task-based fMRI on its own to define area boundaries. Defining a given cortical area using task-based fMRI requires the use of statistical criteria, with different criteria giving different results (Beauchamp 2005). Even if the statistical criterion is fixed across subjects, a ubiquitous observation is that there is remarkable interindividual variability in fMRI activation maps. For instance, Fig. 8.5 shows the activation patterns in six healthy subjects during the presentation of unisensory auditory and visual stimuli, with multisensory activations generated using a simple conjunction analysis criterion ($t > 2$ for auditory stimulation and $t > 2$ for visual stimulation [t statistic]). Not surprisingly, visual activations are concentrated in the occipital lobe, whereas auditory activations are concentrated in the superior temporal gyrus. Regions responsive to both visual and auditory stimuli are found in the posterior superior temporal sulcus and gyrus (pSTS)/STG, with most activations located between the boundaries of HG and the ITS. However, in general, the activation in these areas is patchy, making it difficult to delineate sharp boundaries between unisensory and multisensory cortices.

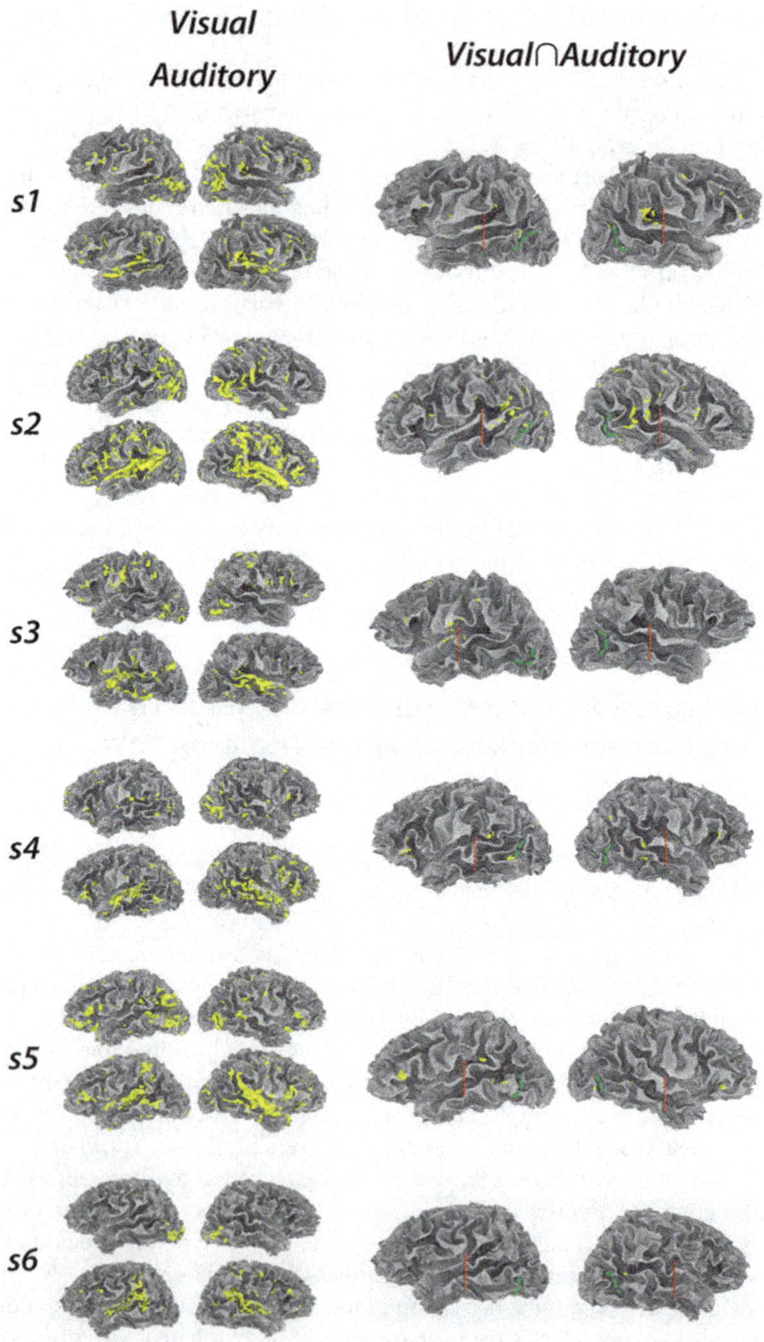

Fig. 8.5 Auditory and visual single subject activation maps. fMRI activation maps from 6 subjects (1 subject per row, s1 to s6) show the lateral views of surface models of the gray-white matter

A related problem is that the definition of a “significant” response in fMRI is strongly dependent on the amount of data collected. Gonzalez-Castillo et al. (2012) measured brain activity with fMRI while subjects viewed a simple visual stimulus consisting of a flashing checkerboard. As opposed to traditional designs for such studies, which would entail 4–6 “runs” or blocks of data collection, the authors carried out 100 runs per subject. Their remarkable observation was that the number of active voxels continued to increase as more data were collected, with no plateau. Using all 100 runs, about 96% of the entire brain was active in response to the simple visual stimulus. Although this is an impractical amount of data to collect under most circumstances, increases in the signal-to-noise ratio with improved scanner hardware, pulse sequences, and cardiac and respiratory noise removal means that similar results could soon be achieved with much less data. Similar results arise from the use of multivoxel pattern analysis methods that attempt to effectively “decode” the presence (or identity) of a given stimulus within a given brain region (cluster of voxels). Increasingly, such studies are illustrating the capacity of unisensory cortices to decode stimuli presented in other modalities; for instance, the visual cortex can decode the identity of auditory stimuli above chance. As a natural extension of these univariate and multivariate analyses, with sufficient data, it is very clear that the entire brain can ultimately be labeled “multisensory.” Indeed, an influential review (Ghazanfar and Schroeder 2006) had the provocative title “Is Neocortex Essentially Multisensory” and highlighted the growing number of studies that were illustrating some degree of multisensory influence even in low-level (i.e., primary) sensory cortices. Although this work has changed the traditional views of the sensory cortex, it also serves to unintentionally blur the true functional distinctions between primary sensory cortices (which can be influenced by other senses but which are still largely unisensory) and multisensory brain regions (those with active convergence and integration of information from multiple sensory modalities). Indeed, as highlighted in earlier chapters (see Willet, Groh, and Maddox, Chap. 5; King, Hammond-Kenny, and Nodal, Chap. 6), the evidence for multisensory convergence (as well as some forms of integration) now extends to a number of subcortical loci beyond the classic multisensory structure, the superior colliculus, in which many of the initial formative observations about the behavior of multisensory neurons were first made (Stein and Meredith 1993).

Fig. 8.5 (continued) boundary for the left (*left columns*) and right (*right columns*) hemispheres for three conditions. In the visual condition (*top rows in 2 left columns*), *yellow* shows areas with significantly greater BOLD signal ($t > 2$ uncorrected) for a visual stimulus of moving dots compared with the fixation baseline. In the auditory condition (*bottom rows in 2 left columns*), *yellow* shows areas with significantly greater BOLD signal ($t > 2$ uncorrected) for auditory sounds compared with the fixation baseline. In the conjunction condition (*2 right columns*), *yellow* shows areas with significant activity for both visual and auditory stimuli. *Red dashed line*, Heschl’s gyrus landmark; *green dashed line*, inferior temporal sulcus boundary

8.6 Summary and Future Directions

Although a great deal of focus has been placed on understanding the structural and functional organization of the auditory cortex, this work has (not surprisingly) had a strong emphasis on the processing of auditory signals. This chapter posits that a greater understanding of the functional role played by the auditory cortex can also be gained through bringing a multisensory lens to studies of these brain regions. This approach becomes increasingly important as one moves outward from the predominantly auditory core regions into the increasingly complex and multisensory belt and parabelt regions, where influences from other sensory modalities become increasingly prevalent. One of the best illustrations of this comes in the context of naturalistic speech, which is generally encountered in an audiovisual manner in which the spoken signal is accompanied by correlated visual signals largely associated with mouth movements (see Grant and Bernstein, Chap. 3). Indeed, using such naturalistic speech in both quiet and noisy settings reveals a functional distinction in parabelt regions not evident in auditory signals alone.

Given the spatial limitations of fMRI (even at high field) and the enormous amount of temporal information available in other electrophysiological approaches that can be applied to human subjects (i.e., EEG, iEEG, and magnetoencephalography [MEG]), future work that employs a conjunctive set of approaches toward questions of the auditory and multisensory cortical processes are likely to reveal additional insights into the functional organization of this complex set of cortical regions.

Compliance with Ethics Requirement Michael S. Beauchamp declares that he has no conflict of interest.

References

Beauchamp, M. S. (2005). Statistical criteria in FMRI studies of multisensory integration. *Neuroinformatics, 3*(2), 93–113.

Beauchamp, M. S., Lee, K. E., Argall, B. D., & Martin, A. (2004a). Integration of auditory and visual information about objects in superior temporal sulcus. *Neuron, 41*(5), 809–823.

Beauchamp, M. S., Argall, B. D., Bodurka, J., Duyn, J. H., & Martin, A. (2004b). Unraveling multisensory integration: Patchy organization within human STS multisensory cortex. *Nature Neuroscience, 7*(11), 1190–1192.

Beauchamp, M. S., Yasar, N. E., Kishan, N., & Ro, T. (2007). Human MST but not MT responds to tactile stimulation. *The Journal of Neuroscience, 27*(31), 8261–8267.

Brodmann, K. (1994). Vergleichende Lokalisationslehre der Grosshirnrinde [Brodmann's Localization in the Cerebral Cortex] (L. J. Garey, Trans.) Leipzig: Barth. London: Smith Gordon (Original work published in 1909).

Bruce, C., Desimone, R., & Gross, C. G. (1981). Visual properties of neurons in a polysensory area in superior temporal sulcus of the macaque. *Journal of Neurophysiology, 46*(2), 369–384.

Eickhoff, S. B., Bzdok, D., Laird, A. R., Kurth, F., & Fox, P. T. (2012). Activation likelihood estimation meta-analysis revisited. *NeuroImage, 59*(3), 2349–2361.

Felleman, D. J., & Van Essen, D. C. (1991). Distributed hierarchical processing in the primate cerebral cortex. *Cerebral Cortex, 1*(1), 1–47.

Ghazanfar, A. A., & Schroeder, C. E. (2006). Is neocortex essentially multisensory? *Trends in Cognitive Sciences, 10*(6), 278–285.

Glasser, M. F., & Van Essen, D. C. (2011). Mapping human cortical areas in vivo based on myelin content as revealed by T1- and T2-weighted MRI. *The Journal of Neuroscience, 31*(32), 11597–11616.

Glasser, M. F., Goyal, M. S., Preuss, T. M., Raichle, M. E., & Van Essen, D. C. (2014). Trends and properties of human cerebral cortex: Correlations with cortical myelin content. *NeuroImage, 93*, 165–175.

Glasser, M. F., Coalson, T. S., Robinson, E. C., Hacker, C. D., Harwell, J., Yacoub, E., Ugurbil, K., Andersson, J., Beckmann, C. F., Jenkinson, M., & Smith, S. M. (2016). A multi-modal parcellation of human cerebral cortex. *Nature, 536*(7), 171–178.

Gonzalez-Castillo, J., Saad, Z. S., Handwerker, D. A., Inati, S. J., Brenowitz, N., & Bandettini, P. A. (2012). Whole-brain, time-locked activation with simple tasks revealed using massive averaging and model-free analysis. *Proceedings of the National Academy of Sciences of the United States of America, 109*(14), 5487–5492.

Grefkes, C., & Fink, G. R. (2005). The functional organization of the intraparietal sulcus in humans and monkeys. *Journal of Anatomy, 207*(1), 3–17.

Hagen, M. C., Franzen, O., McGlone, F., Essick, G., Dancer, C., & Pardo, J. V. (2002). Tactile motion activates the human middle temporal/V5 (MT/V5) complex. *European Journal of Neuroscience, 16*(5), 957–964.

Huk, A. C., Dougherty, R. F., & Heeger, D. J. (2002). Retinotopy and functional subdivision of human areas MT and MST. *The Journal of Neuroscience, 22*(16), 7195–7205.

Jiang, F., Beauchamp, M. S., & Fine, I. (2015). Re-examining overlap between tactile and visual motion responses within hMT+ and STS. *NeuroImage, 119*, 187–196.

Jiang, F., Stecker, G. C., Boynton, G. M., & Fine, I. (2016). Early blindness results in developmental plasticity for auditory motion processing within auditory and occipital cortex. *Frontiers in Human Neuroscience, 10*, 324.

Kaas, J. H., Hackett, T. A., & Tramo, M. J. (1999). Auditory processing in primate cerebral cortex. *Current Opinion in Neurobiology, 9*(2), 164–170.

Matteau, I., Kupers, R., Ricciardi, E., Pietrini, P., & Ptito, M. (2010). Beyond visual, aural and haptic movement perception: hMT+ is activated by electrotactile motion stimulation of the tongue in sighted and in congenitally blind individuals. *Brain Research Bulletin, 82*(5–6), 264–270.

Moerel, M., De Martino, F., & Formisano, E. (2014). An anatomical and functional topography of human auditory cortical areas. *Frontiers in Neuroscience, 8*, 225.

Morosan, P., Schleicher, A., Amunts, K., & Zilles, K. (2005). Multimodal architectonic mapping of human superior temporal gyrus. *Anatomy and Embryology, 210*(5–6), 401–406.

Ozker, M., Schepers, I. M., Magnotti, J. F., Yoshor, D., & Beauchamp, M. S. (2017). A double dissociation between anterior and posterior superior temporal gyrus for processing audiovisual speech demonstrated by electrocorticography. *Journal of Cognitive Neuroscience, 29*(6), 1044–1060.

Pascual-Leone, A., & Hamilton, R. (2001). The metamodal organization of the brain. *Progress in Brain Research, 134*, 427–445.

Ricciardi, E., Vanello, N., Sani, L., Gentili, C., Scilingo, E. P., Landini, L., Guazzelli, M., Bicchi, A., Haxby, J. V., & Pietrini, P. (2007). The effect of visual experience on the development of functional architecture in hMT+. *Cerebral Cortex, 17*(12), 2933–2939.

Saenz, M., Lewis, L. B., Huth, A. G., Fine, I., & Koch, C. (2008). Visual motion area MT+/V5 responds to auditory motion in human sight-recovery subjects. *The Journal of Neuroscience, 28*(20), 5141–5148.

Snyder, L. H., Batista, A. P., & Andersen, R. A. (1997). Coding of intention in the posterior parietal cortex. *Nature, 386*(6621), 167–170.

Stein, B. E., & Meredith, M. A. (1993). *The merging of the senses*. Cambridge, MA: The MIT Press.

Summers, I. R., Francis, S. T., Bowtell, R. W., McGlone, F. P., & Clemence, M. (2009). A functional-magnetic-resonance-imaging investigation of cortical activation from moving vibrotactile stimuli on the fingertip. *The Journal of the Acoustical Society of America, 125*(2), 1033–1039.

Takahashi, K., Gu, Y., May, P. J., Newlands, S. D., Deangelis, G. C., & Angelaki, D. E. (2007). Multimodal coding of three-dimensional rotation and translation in area MSTd: Comparison of visual and vestibular selectivity. *The Journal of Neuroscience, 27*(36), 9742–9756.

Van Essen, D. C., Smith, S. M., Barch, D. M., Behrens, T. E. J., Yacoub, E., Ugurbil, K., & Wu-Minn HCP Consortium. (2013). The WU-Minn Human Connectome Project: An overview. *NeuroImage, 80*, 62–79.

van Kemenade, B. M., Seymour, K., Wacker, E., Spitzer, B., Blankenburg, F., & Sterzer, P. (2014). Tactile and visual motion direction processing in hMT+/V5. *NeuroImage, 84*, 420–427.

Wallace, M. T., Ramachandran, R., & Stein, B. E. (2004). A revised view of sensory cortical parcellation. *Proceedings of the National Academy of Sciences of the United States of America, 101*(7), 2167–2172.

Chapter 9
Combining Voice and Face Content in the Primate Temporal Lobe

Catherine Perrodin and Christopher I. Petkov

Abstract The interactions of many social animals critically depend on identifying other individuals to approach or avoid. Recognizing specific individuals requires extracting and integrating cross-sensory indexical cues from richly informative communication signals such as voice and face content. Knowledge on how the brain processes faces and voices as unisensory or multisensory signals has grown; neurobiological insights are now available not only from human neuroimaging data but also from comparative neuroimaging studies in nonhuman animals, which together identify the correspondences that can be made between brain processes in humans and other species. These advances have also had the added benefit of establishing animal models in which neuronal processes and pathways are interrogated at finer neurobiological scales than possible in humans. This chapter overviews the latest insights on neuronal representations of voice and face content, including information on sensory convergence sites and pathways that combine multisensory signals in the primate temporal lobe. The information synthesized here leads to a conceptual model whereby the sensory integration of voice and face content depends on temporal lobe convergence sites, which are a midway processing stage and a conduit between audiovisual sensory-processing streams and the frontal cortex.

Keywords Auditory · Communication · Comparative · Face areas · Functional magnetic resonance imaging · Neurons · Neurophysiology · Oscillations · Spikes · Superior-temporal sulcus · Visual · Voice areas

---

C. Perrodin
Institute of Behavioural Neuroscience, University College London, London, UK
e-mail: c.perrodin@ucl.ac.uk

C. I. Petkov (✉)
Institute of Neuroscience, Newcastle University Medical School, Newcastle upon Tyne, UK
e-mail: chris.petkov@ncl.ac.uk

A. K. C. Lee et al. (eds.), *Multisensory Processes*, Springer Handbook of Auditory Research 68, https://doi.org/10.1007/978-3-030-10461-0_9

9.1 Neurobiological Processing of Voice and Face Content in Communication Signals

Far from being redundant, information from different sensory inputs is complementary and expedites behavioral recognition of an object or entity. Yet how the brain achieves sensory integration remains a challenging question to answer, in part because it has become evident that multisensory neural interactions are distributed, taking place between multiple sites throughout the brain. Although, for brevity, a handful of multisensory association areas are often emphasized in summaries and reviews (Schroeder and Foxe 2002; Stein and Stanford 2008), it is well accepted that multisensory influences abound from early cortical and subcortical sensory processing stages and beyond (Damasio 1989; Ghazanfar and Schroeder 2006).

Thus the task for neuroscientists has been steadily shifting away from a focus on particular sensory convergence sites toward an emphasis on identifying the neural multisensory influences and transformations that occur between sites along particular processing pathways (Yau et al. 2015; Bizley et al. 2016). Moreover, comparing the forms of multisensory convergence seen at different brain sites can pinpoint common principles of multisensory integration and identify how specializations in neural multisensory interactions may occur, such as duplication with differentiation. In this chapter, evidence on neuronal representations and multisensory interactions along pathways involved in processing voice and face content are considered. Finally, this chapter concludes with the identification of epistemic gaps that inspired readers might be encouraged to empirically shrink by helping to further advance neurobiological knowledge.

Initial insights into how the brain processes identity-related information were obtained in the visual modality. Neurons that respond stronger to faces than to other nonface objects were first identified in the monkey inferior temporal (IT) cortex (Bruce et al. 1981; Perrett et al. 1982). Subsequently, human neuroimaging studies identified face-category-preferring regions in the fusiform gyrus, occipital cortex, and adjacent visual areas (Sergent et al. 1992; Kanwisher et al. 1997). Shortly thereafter, functionally homologous face-sensitive regions in the monkey inferior bank and fundus of the superior-temporal sulcus (STS) were identified (Logothetis et al. 1999; Tsao et al. 2006). We next consider complementary information from the auditory modality that has recently become available (see Perrodin et al. 2015b for a review). Together, these developments have opened pathways for understanding how multisensory (voice and face) content is combined in the brain (see Fig. 9.1; also see Plakke and Romanski, Chap. 7).

9.1.1 Voice-Sensitive Brain Regions in Humans, Monkeys, and Other Mammals

Human neuroimaging studies aiming to shed light on the processing of auditory communication signals tend to focus on the neurobiology of speech and language, which is a fundamental aspect of human communication (Hickok and Poeppel 2007;

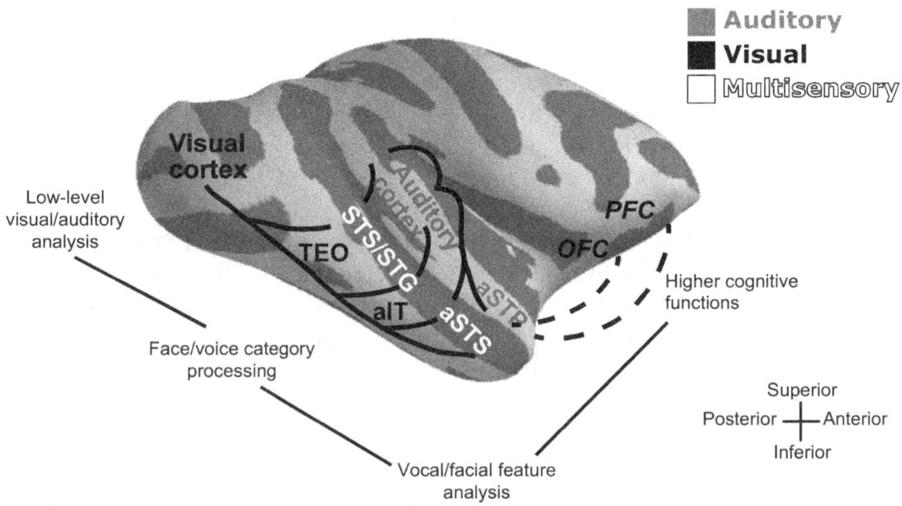

Fig. 9.1 Model of ascending auditory and visual cortical streams of auditory voice- and visual face-processing pathways in the primate brain rendered on a rhesus macaque brain. It features early sensory cortices, processing stages extracting face content in visual areas of the inferior temporal lobe (TEO and aIT), and auditory regions of the anterior superior-temporal plane/gyrus (aSTP/STG) extracting voice-related content. Multisensory interactions are possible between voice- and face-processing regions including by way of association areas along the superior-temporal sulcus (STS) and frontal cortex (prefrontal cortex [PFC]; orbitofrontal cortex [OFC]). The cortical regions are interconnected via bidirectional pathways of interregional projections, including feedforward and feedback projections to the auditory- and visual-processing streams (dashed and solid lines, respectively). Reproduced from Perrodin et al. (2015b)

Binder et al. 2009). In parallel, work in carnivore, rodent, and primate models aims to unravel the neurobiological substrates for referential social communication (i.e., "what" was vocalized), a likely evolutionary precursor on which human vocal communication evolved (Ghazanfar and Takahashi 2014; Seyfarth and Cheney 2014).

More recently, investigators focusing on understanding identity-related content ("who" vocalized) have identified voice-sensitive regions in the human brain using functional magnetic resonance imaging (fMRI). The approach of comparing how the brain responds to voice versus nonvoice content in communication signals is analogous to neurobiological studies in the visual domain comparing responses to face versus nonface objects (Belin et al. 2004). These and other studies have identified the presence of several voice-sensitive clusters in the human temporal lobe, including in the superior-temporal gyrus/sulcus (Belin et al. 2000; von Kriegstein et al. 2003).

However, it is known that human voice regions also strongly respond to speech (Fecteau et al. 2004) and that both speech and voice content can be decoded from largely overlapping areas in the superior portions of the human temporal lobe (Formisano et al. 2008). These observations left open the possibility that human voice and speech processes are so functionally intertwined that human brain specializations for voice processing may have occurred alongside those for speech, raising the question whether "voice regions" would be evident in nonhuman animals.

This question of whether nonhuman animals have voice-sensitive regions as humans do was answered in the affirmative initially in rhesus macaques (*Macaca mulatta*), an Old World monkey species, with evidence in other primate species and mammals following shortly thereafter. The macaque monkey fMRI study identified temporal lobe voice-sensitive regions that are more strongly activated by voice than by nonvoice sounds (Petkov et al. 2008). Moreover, of the several fMRI-identified voice-sensitive clusters in the monkey superior temporal lobe, the most anterior one was found to be particularly sensitive to who vocalized rather than what was vocalized, forging a more direct link to human fMRI studies on voice identity-sensitive processes in the anterior temporal lobe (Belin and Zatorre 2003; McLaren et al. 2009). More recently, Andics et al. (2014) imaged domesticated dogs with fMRI to reveal voice-preferring regions in the temporal lobe of these carnivores, broadening the evolutionary picture. Relatedly, a fMRI study in marmosets (*Callithrix jacchus*, a New World monkey species) identified temporal lobe regions that respond more strongly to conspecific vocalizations than to other categories of sounds (Sadagopan et al. 2015), which in the future could be interrogated for voice-content sensitivity.

In laboratory animals that are established neurobiological models, the fMRI identified voice-sensitive clusters that can be targeted for neurophysiological study at a fundamental scale of neural processing, such as at the level of single neurons. Moreover, the identification of voice-sensitive regions in nonhuman animals helps to forge links to analogous processes in the visual system.

9.1.2 Voice-Sensitive Neurons in the Ventral Auditory-Processing Stream

The anterior voice-sensitive fMRI cluster in rhesus macaques is located in hierarchically higher neuroanatomically delineated cortical regions (Galaburda and Pandya 1983; Saleem and Logothetis 2007). These areas reside in the anterior portion of the superior-temporal plane (aSTP; the dorsal and anterior surface of the temporal lobe; see Fig. 9.1). The anterior temporal lobe voice-sensitive cluster falls somewhere between the fourth or fifth stage of processing in the auditory cortical hierarchy, rostral to the tonotopically organized core (first), "belt" (second), and parabelt (third) areas (Rauschecker 1998; Kaas and Hackett 2000). The anatomical localization of the voice area in the aSTP places it at an intermediate level in the ventral auditory "object"-processing stream (Rauschecker and Tian 2000; Romanski 2012). Other downstream cortical regions interconnected with the aSTP include the superior-temporal gyrus, sulcus, and frontal cortex (see Fig. 9.1; Perrodin et al. 2011).

Although the fMRI results on voice identity-sensitive processing and the corresponding anatomical findings identify the anterior aSTP region as a higher order cortex; the auditory feature selectivity displayed by neurons in this region was unknown before electrophysiological recordings from the fMRI-identified clusters in a neurobiological model, in this case macaques. In the initial neuronal recording studies from the anterior voice-sensitive cluster, neuronal-spiking responses were modulated by differences in the vocal features of the auditory stimuli, such as call type, caller identity, and caller species (Perrodin et al. 2014). In particular, the results revealed a distinct subpopulation of voice-sensitive neurons, which accounted for much of the observed sensitivity to caller identity features (who vocalized).

Thus, the neuronal recordings showed that the responses of neurons in voice-sensitive clusters can simultaneously be sensitive to the category of voices over non-voice stimuli and to auditory features of individual stimuli within the voice category. This dual sensitivity in the recordings is paralleled at a very different spatiotemporal scale by the human and monkey fMRI results, which in turn show that voice-sensitive clusters are sensitive to both the categorical distinction between voice versus nonvoice stimuli *and* the specific conspecific voices within the category of voices (Belin and Zatorre 2003; Petkov et al. 2008).

The neuronal sensitivity to auditory vocal features in the voice area was compared with that in a different part of the anterior temporal lobe, the anterior upper bank of the superior temporal sulcus (aSTS; see Fig. 9.1), which has long been considered to be multisensory because it belongs to a higher order association cortex (Stein and Stanford 2008). More posterior regions of the STS were known to contain both auditory and visually responsive clusters of neurons (Beauchamp et al. 2004; Dahl et al. 2009; also see Beauchamp, Chap. 8). The neuronal recordings in the aSTS confirmed that a substantial proportion of neurons in this area are driven by sounds, but the results also showed that neurons in this area are not very sensitive to auditory vocal features, unlike the auditory-feature sensitive neurons in the aSTP voice-sensitive cluster (Perrodin et al. 2014).

In comparison to these observations from neural recordings in the aSTP and aSTS, neurons in the ventrolateral prefrontal cortex, which are hierarchically further along the ventral processing stream, show a sensitivity to the complex acoustical features of vocalizations, such as call type (Gifford et al. 2005) and caller identity (Plakke et al. 2013). Why certain areas in the processing pathways to the frontal cortex show less auditory feature specificity than others is a topic that is visited later in this chapter, after considering more of the available information.

Converging evidence from the visual and auditory modalities in humans and monkeys points to anterior subregions of the temporal lobe being involved in the processing of identity-related features. In the visual domain, face regions in the anterior inferior-temporal lobe (aIT; see Fig. 9.1) are particularly sensitive to identity-related content in humans (Kriegeskorte et al. 2007; Tsao and Livingstone 2008) and monkeys (Freiwald and Tsao 2010; Morin et al. 2014). Likewise in the auditory modality, more anterior temporal lobe areas are sensitive to identity-related content in communication sounds in both humans (e.g., Belin and Zatorre 2003; von Kriegstein et al. 2003) and monkeys (Petkov et al. 2008). A number of theoretical models also highlight the anterior temporal lobe as a region containing sites sensitive to voice or face identity-related content (Bruce and Young 1986; Campanella and Belin 2007). However, because anterior temporal lobe sites are nodes in a broader network processing voice and face content (Fecteau et al. 2005; Tsao et al. 2008), other more posterior voice- or face-sensitive sites in the temporal lobe are undoubtedly involved in ways that need to be better understood. For instance, more posterior superior temporal lobe regions can also be sensitive to identity-related information regardless of the sensory modality (Chan et al. 2011; Watson et al. 2014).

In summary, the results from neuronal recordings in the voice-sensitive aSTP specify the auditory-response characteristics of neurons in this region of the ventral processing stream and distinguish these characteristics in relation to those from neurons in the adjacent association cortex of the aSTS. However, despite the surface resemblance, sensory processing of voices and faces in the auditory and visual modalities, respectively, does not seem to be identical. We consider that in Sect. 9.1.3 where we ask, Do voice cells exist and, if so, are they direct analogs of visual face-sensitive neurons?

9.1.3 Do Voice-Sensitive Regions Contain "Voice Cells," and, If So, How Do Their Responses Compare with "Face Cells" in the Visual System?

An initial question while interrogating neuronal responses in the voice-sensitive cortex, given the evidence for face-sensitive cells in the visual system, is do" voice cells" exist? Arguably, at the cellular level, the auditory system tends to show relatively less tangible organizational properties than those seen in the visual and somatosensory systems for a host of fundamental sensory-processing features (Griffiths et al. 2004; King and Nelken 2009). The better established view of auditory cortical processing is that many auditory functions are supported by neuronal processes that are distributed across populations of auditory neurons and do not require topographical maps or individual cells with high feature selectivity (Bizley et al. 2009; Mizrahi et al. 2014).

Thus another open question was whether fMRI-identified voice clusters contain highly voice content-sensitive neurons, or voice cells. Voice cells could be defined as neurons that exhibit twofold greater responses to voice versus nonvoice stimuli, in direct analogy to how face cells have been defined: This was the approach of Perrodin et al. (2011) in searching for voice cells in the anterior voice identity-sensitive fMRI cluster in macaque monkeys. They first used an auditory voice localizer borrowed from the earlier monkey fMRI study (Petkov et al. 2008), which allowed them to identify neurons within the fMRI voice clusters that were preferentially sensitive to the voice category of stimuli. The voice localizer stimulus set included a collection of macaque voices from many individuals (many voices) and two comparison categories containing either animal vocalizations and voices from other species or a set of natural sounds. All stimuli were subsampled from a larger stimulus set so that the selected sounds from each category were matched in multiple low-level acoustical features. Using these sounds as stimuli, the researchers observed a modest proportion (~25% of the sample) of neurons within the aSTP that could be defined as voice cells.

Yet, comparisons of the proportions and response characteristics of the voice cells in relation to what is known about face cells suggest that voice cells may not be direct analogs to face cells. For instance, visual studies of face clusters identified a high density of face cells in monkey face-sensitive fMRI regions (Tsao et al. 2006; Aparicio et al. 2016). This very high clustering (>90%) of face cells in these fMRI face clusters is in stark contrast to the much more modest (~25%) clustering of voice cells (Perrodin et al. 2011). Furthermore, the voice-sensitive cells in the anterior temporal lobe fMRI cluster are remarkably stimulus selective, responding to only a small proportion or just a few of the voices within the category of stimuli (Perrodin et al. 2011). This high neuronal selectivity of voice cells seems to diverge from the functional encoding properties that have been reported for face cells, whereby face cells typically respond more broadly to the majority of faces in the stimulus set (Hasselmo et al. 1989; Tsao et al. 2006).

The high stimulus selectivity of voice cells is on par with the level of selectivity measured in neurons responding to conspecific vocalizations in the ventrolateral prefrontal cortex (Gifford et al. 2005; Romanski et al. 2005); both temporal and frontal sites show higher stimulus selectivity than that reported for neurons in and around the primary auditory cortex (Tian et al. 2001; Recanzone 2008), in an auditory region in the insula (Remedios et al. 2009), or as parts of the superior temporal gyrus (Russ et al. 2008). Thus, in relation to reports on face cell selectivity, so far the available evidence raises the intriguing possibility that voice cells are not direct auditory analogs of face cells, which may reflect specialization under different evolutionary pressures in the auditory versus visual domains (Miller and Cohen 2010; Perrodin et al. 2011).

9.2 How Multisensory Is the Anterior Voice-Sensitive Temporal Cortex?

We and the other authors involved in the initial monkey neuroimaging and electrophysiological studies on voice regions and voice cells were rather bold in our initial claims identifying the anterior temporal fMRI-identified cluster as a voice-sensitive area. In the initial monkey neuronal recording study (Perrodin et al. 2011), multisensory stimulation conditions were not used to rule out or rule in that the region is multisensory rather than auditory. Moreover, human fMRI studies had already shown evidence for both functional crosstalk and direct structural connections between voice- and face-sensitive regions (von Kriegstein et al. 2005; Blank et al. 2011), which suggests that the neuroanatomical pathways for the exchange of visual face and auditory voice information are in place. Other potential sources of visual input into the auditory STP include corticocortical projections from the visual areas

(Bizley et al. 2007; Blank et al. 2011) as well as feedback projections from higher association areas such as the inferior frontal cortex (Romanski et al. 1999a, b), and the upper bank of the STS (Pandya et al. 1969; Cappe and Barone 2005). Multisensory projections with subcortical origins could also directly or indirectly influence cross-modal interactions, such as those from the suprageniculate nucleus of the thalamus or the superior colliculus.

A number of electrophysiological studies have directly evaluated the multisensory influences of face input on voice processing in nonhuman primates at a number of cortical sites, including posterior auditory regions closer to the primary and adjacent auditory cortices (Ghazanfar et al. 2005; Kayser et al. 2008) as well as higher order association regions such as the STS (Chandrasekaran and Ghazanfar 2009; Dahl et al. 2009) or the ventrolateral prefrontal cortex (Sugihara et al. 2006; Romanski 2007; also see Plakke and Romanski, Chap. 7). However, whether the aSTP could be classified as an auditory or association/multisensory cortex had been ambiguous based on its neuroanatomy (Galaburda and Pandya 1983; Kaas and Hackett 1998), begging the question whether multisensory interactions in the voice-sensitive aSTP are comparable to those in the early auditory cortex or, alternatively, whether they are more like those seen in multisensory association areas.

9.2.1 How Multisensory Are Neurons in the Anterior Voice Identity-Sensitive Functional Magnetic Resonance Imaging Cluster?

To directly study whether and how auditory responses to voices are affected by simultaneously presented visual facial information, neuronal recordings were performed from the aSTP voice cluster during auditory, visual, or audiovisual presentation of dynamic face and voice stimuli. As might be expected of a predominantly auditory area, neurons in the voice-sensitive cortex primarily respond to auditory stimuli, whereas silent visual stimuli are mostly ineffective in eliciting neuronal firing (see Fig. 9.2B; Perrodin et al. 2014). Yet, comparing the amplitudes of spiking

Fig. 9.2 (continued) incongruent audiovisual (AVi) pairs. *Dark horizontal line*, duration of the auditory stimulus; *light shaded box*, 400-ms response window in which the response amplitude was computed. *Bar plots*, response amplitudes (means ± SE) in the 400-ms response window. Symbols refer to significantly nonlinear audiovisual interactions, defined by comparing the audiovisual response with all possible summations of auditory and visual responses: AVc vs. (A + Vv) and AVi vs. (A + Vi), *z*-test, $*P < 0.05$; n.s., not significant. (**E**) Summary of the congruency specificity of visually modulated units in the anterior voice-sensitive cortex (aSTP; *left*) and the anterior superior-temporal sulcus (*right*). Bar plots indicate the percentage of auditory responsive units that showed significant nonadditive audiovisual interactions in response to the congruent pair only (*black*), the incongruent pair only (*light gray*), or both the congruent and the incongruent stimuli (*dark gray*). $**P < 0.01$, χ^2-test comparing the number of visually modulated units for each of the three categories to a uniform distribution. (**A**) and (**C–F**) reproduced from Perrodin et al. (2014), with permission from the Society for Neuroscience; (**B**) reproduced from Perrodin et al. (2015b)

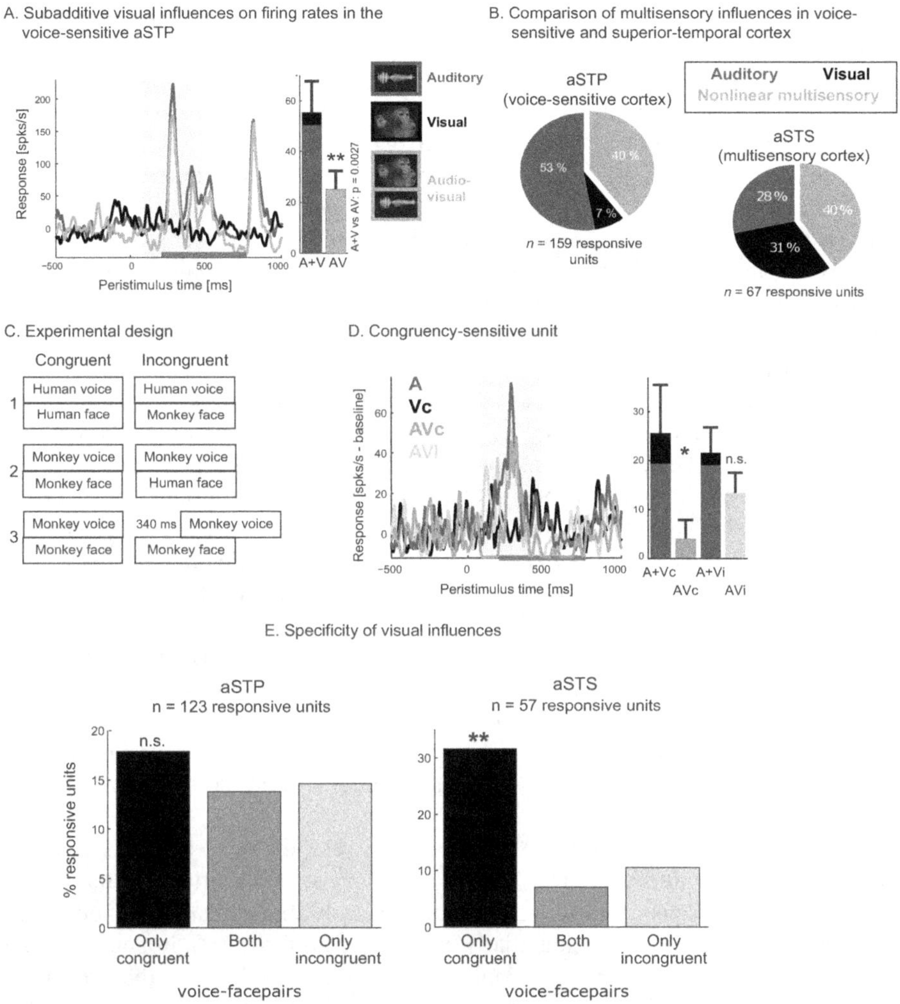

Fig. 9.2 Neuronal multisensory influences and effect of voice-face congruency in voice-sensitive and superior-temporal cortex. (**A**) Example spiking response of a unit in the anterior voice-sensitive fMRI cluster on the superior-temporal plane showing nonlinear (subadditive) visual modulation of auditory activity. Firing rates in response to combined audiovisual stimulation (AV; voice and face) are significantly lower than the sum of the responses to the unimodal stimuli: AV vs. (A + V), *z*-test, $^{**}P < 0.01$. *Dark horizontal line*, duration of the auditory stimulus; *light shaded box*, 400-ms peak-centered response window. *Bar plots*, response amplitudes (means ± SE) in the 400-ms response window. (**B**) Neuronal multisensory influences are prominent in the voice-sensitive cortex (aSTP) but are qualitatively different from those in the anterior superior-temporal sulcus (aSTS). For example, aSTS neurons more often display bimodal responses (Perrodin et al. 2014). (**B**) Reproduced from Perrodin et al. (2015b). (**C**) Illustration of the stimulus set containing three congruency violations in primate voice/face pairs. (**D**) Example response of a unit with congruency-specific visual influences: a congruent, but not an incongruent, visual stimulus significantly modulated the auditory response. Plot shows spiking activity in response to the auditory stimulus alone (A), the congruent visual stimulus alone (Vc), and the congruent audiovisual (AVc) and

responses to unimodal (auditory alone) versus bimodal (audiovisual) stimulation conditions revealed clear nonlinear influences (subadditive or superadditive) on the responses of auditory neurons (Fig. 9.2A, B). This provided evidence for robust visual modulation of auditory neuronal responses at the anterior voice-sensitive fMRI cluster in the aSTP (Perrodin et al. 2014). From here, a comparison can be made between these multisensory influences in the anterior voice area and those seen in earlier auditory areas. Interestingly, similar proportions and types of multisensory influences have been reported for neurons in the posterior core/belt auditory areas (Ghazanfar et al. 2005; Kayser et al. 2008), suggesting qualitatively comparable multisensory interactions throughout the auditory cortex, from earlier auditory cortical processing stages to the anterior voice cluster.

Beyond the proportions of modulated neurons, and the types of multisensory interactions, cross-modal influences are also known to differ in their specificity to behaviorally relevant multisensory combinations used for stimulation (Werner and Noppeney 2010). The neuronal sensitivity of visual influences to speaker congruency was investigated using a set of congruent and incongruent audiovisual stimulus conditions, in which a voice was paired with a mismatched face (Fig. 9.2C). The neuronal responses to these conditions showed that multisensory influences on aSTP units were relatively insensitive to speaker congruency and were not strongly affected by the mismatching audiovisual stimulus conditions, such as when a monkey voice was paired with a human face (see Fig. 9.2E; Perrodin et al. 2014). The relative lack of specificity of visual influences in the aSTP is consistent with the notion that the anterior voice cluster shows more general cross-modal influences belonging to a relatively early stage of audiovisual processing, which includes the primary auditory cortex and surrounding auditory areas (Schroeder et al. 2003; Ghazanfar and Schroeder 2006).

The impressions given by these observations is that there are clear visual influences on many auditory neurons in the anterior voice-sensitive cluster in the aSTP. These multisensory influences are qualitatively more like those reported in early auditory cortical fields, potentially differing from those seen in neurons from the multisensory association cortex in the aSTS (see Sect. 9.2.3). Thereby, the anterior voice identity-sensitive cluster in monkeys is primarily sensitive to auditory features, with the multisensory influences seen in this region being of a more general modulatory form.

9.2.2 Natural Asynchronies in Audiovisual Communication Signals and Their Impact on Neuronal Excitability

As we have considered, neurons in the anterior voice-sensitive cluster in the aSTP, although predominantly auditory, do show certain types of cross-modal influences from faces on their auditory spiking responses to voices. During audiovisual communication, a caller is perceived to produce a vocalization while the facial expression changes. However, although these multisensory signals are often perceived to emanate simultaneously, in natural communication signals, there is a considerable level of temporal asynchrony between the onset of informative content in one modality relative to the other. For instance, visual orofacial movements can precede the vocalization by tens to hundreds of milliseconds (see Fig. 9.3A–C; Ghazanfar et al. 2005; Chandrasekaran et al. 2009). Although a coherent multisensory percept can be maintained across a wide range of spatial and temporal discrepancies (McGrath and Summerfield 1985; Slutsky and Recanzone 2001), these subtle-to-moderate temporal misalignments have the potential to drastically impact on neuronal excitability and population dynamics. Yet because neurons in the voice-sensitive cortex lack the temporal response fidelity of neurons in and around primary auditory cortical or subcortical regions (Creutzfeldt et al. 1980; Bendor and Wang 2007), it was uncertain whether such stimulus asynchronies affect audiovisual influences on neurons in the aSTP voice-sensitive cortex.

Relevant studies have shown that the temporal dynamics of sensory streams, such as those typical for processing human speech or other natural stimuli, can shape and synchronize cortical oscillations through entrainment (Ghitza 2011; Giraud and Poeppel 2012). More generally, neuronal oscillations are thought to reflect the state-dependent excitability of local networks to subsequent incoming sensory inputs (Schroeder et al. 2008; Thut et al. 2012). These oscillatory responses are considered to reflect neuronal population mechanisms for routing information to downstream stages and prioritizing the processing at sensory nodes in the brain network (Bastos et al. 2015). It is also known that cortical oscillations are influenced by rhythmic multisensory input (Thorne and Debener 2014; van Atteveldt et al. 2014).

The impact of cross-modal stimulus asynchronies on neuronal responses and cortical oscillations in the voice-sensitive cortex of the primate aSTP was assessed using a set of dynamic face and voice combinations spanning a broad range of naturally occurring audiovisual asynchronies (Fig. 9.3A–C). The results of this study revealed that the prevalence of two key forms of audiovisual interactions in neuronal spiking responses (multisensory enhancement or suppression) varied according to the degree of asynchrony between the onsets of informative communication content in either sensory input stream (see Fig. 9.3D; Perrodin et al. 2015a). Time-frequency analyses of the local-field potential signal in the aSTP showed that this cross-modal asynchrony selectively affects low-frequency neuronal oscillations (Perrodin et al. 2015a). By aligning and transiently synchronizing the phase of ongoing low-frequency cortical oscillations, the visual input cyclically influences the excitability of auditory neuronal responses in the aSTP. Thus, whether the majority of neurons show enhancement or suppression in their multisensory responses depends to a large extent on the visual-to-auditory stimulus onset delay present in natural communication signals. These effects on neuronal excitability span several hundreds of milliseconds or the full range of asynchronies observed in audiovisual communication signals (Chandrasekaran et al. 2009; Perrodin et al. 2015a).

The functional role of cortical oscillations and how they modulate sensory perception is the topic of ongoing research. In comparison, comparable cross-modal phase resetting in local-field potentials is also seen in early auditory and visual cortical areas (Lakatos et al. 2007; Mercier et al. 2013). In the primary auditory cortex,

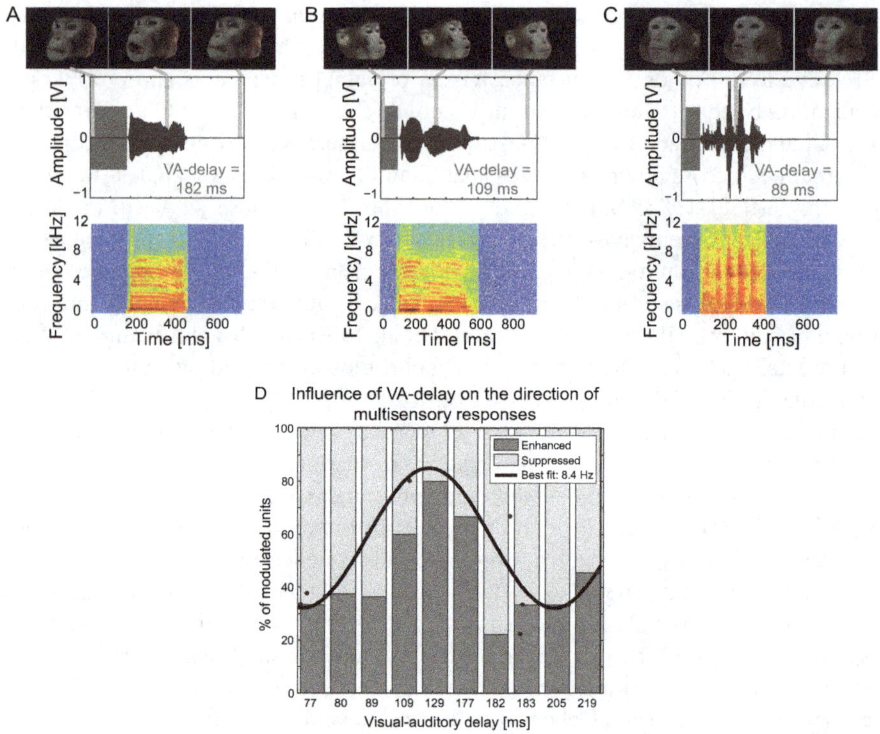

Fig. 9.3 Audiovisual primate vocalizations, visual-auditory onset delays, and the direction (sign) of multisensory interactions. (**A**–**C**) Examples of audiovisual rhesus macaque "coo" (**A** and **B**) and "grunt" (**C**) vocalizations used for stimulation and their respective natural visual-to-auditory onset asynchronies/delays (time interval between the onset of mouth movement and the onset of the vocalization; *shaded areas*). *Top row:* video starts at the onset of mouth movement, with the first frame showing a neutral facial expression, followed by mouth movements associated with the vocalization. *Vertical lines*, temporal position of the representative video frames. *Center row:* amplitude waveforms; *bottom row:* spectrograms of the corresponding auditory vocalization. (**D**) Proportions of enhanced and suppressed multisensory units by stimulus, arranged as a function of increasing visual-to-auditory onset delays (VA-delay; $n = 81$ units). The bars are spaced at equidistant intervals for display purposes. *Solid circles*, proportion of enhanced units for each VA-delay value while respecting the real relative positions of the VA-delay values; *black line*, sinusoid with the best-fitting frequency (8.4 Hz; adjusted $R^2 = 0.58$). Reproduced from Perrodin et al. (2015a)

another study shifting somatosensory nerve stimulation combined with pure-tone stimuli found a comparable alternating pattern of multisensory enhancement and suppression of multiunit activity for different relative stimulus onset asynchronies (Lakatos et al. 2007). Other studies on visual influences in the primary auditory cortex have reported comparable neural response dependencies on temporal stimulus alignment (Ghazanfar et al. 2005; Bizley et al. 2007). Thus, in combination with the naturally occuring timing differences in multisensory streams, cross-modal resetting of ongoing oscillations allows the leading visual input to shape or "window" subsequent auditory responses.

One hypothesis proposes that these temporal relationships in natural communicative situations segment sensory input into an appropriate temporal granularity (Giraud and Poeppel 2012; Gross et al. 2013). For instance, neurons in the monkey STS show specific patterns of slow oscillatory activity and spike timing that reflect visual category-specific information in faces versus other objects (Turesson et al. 2012). Anterior voice area neurons seem to be comparably involved in oscillatory responses whereby the neuronal spiking responses display different types of multisensory interactions (enhancement vs. suppression) depending on the phase alignment of low-frequency oscillatory responses. Taken together, these findings suggest an interplay between neuronal firing and the surrounding oscillatory context that needs to be better explored in terms of the causal interactions underlying auditory and audiovisual transformations of neural responses between brain areas. The potential behavioral relevance of these oscillatory phenomena for stimulus identification and detection will also need to be described. These issues are currently being investigated in humans (Henry and Obleser 2012; Keil et al. 2014) for a host of perceptual processes (Strauss et al. 2015; Ten Oever and Sack 2015; also see Keil and Senkowski, Chap. 10). These efforts in humans could, in turn, benefit from insights obtained at the neuronal level in animal models to better understand the perceptual correlates (Chen et al. 2016), especially from subjects participating in active tasks (Fetsch et al. 2012; Osmanski and Wang 2015).

9.2.3 How Do Visual Interactions at Voice Clusters Compare with Those in Multisensory Areas of the Temporal Lobe?

Extracellular recordings of neuronal activity in the anterior upper bank of the STS in response to the same voice and face stimulus set described in Sect. 9.2.1 revealed a comparable proportion of nonlinear audiovisual interactions as in aSTP neurons (Fig. 9.2B). However, in agreement with previous electrophysiological studies (Benevento et al. 1977; Dahl et al. 2009), evidence for a greater level of cross-modal convergence was prevalent in the STS, where neurons showed a balance of both auditory and visual responses alongside modulatory multisensory influences (Fig. 9.2B). These observations are consistent with studies highlighting the STS as a cortical association area and a prominent target for both auditory and visual afferents in the temporal lobe (Seltzer and Pandya 1994; Beauchamp et al. 2004).

The presentation of incongruent audiovisual stimuli revealed that, in contrast to the generic visual influences in voice-sensitive neurons, those modulating the auditory responses of STS neurons occurred with greater specificity; multisensory interactions were sensitive to the congruency of the presented voice-face pairing, and nonlinear multisensory responses (both super- and subadditive) occurred more frequently in response to matching compared with mismatching audiovisual stimuli (e.g., were more likely to be disrupted by incongruent stimulation; see Fig. 9.2D, E). Dahl et al. (2010) similarly reported congruency-sensitive auditory influences on visual responses in the monkey lower-bank STS. These observations are consistent with the evidence for integrative multisensory processes in the human and monkey STSs (Beauchamp et al. 2004; Dahl et al. 2009), potentially at the cost of decreased specificity for representing unisensory features (see Sect. 9.1.2; Werner and Noppeney 2010; Perrodin et al. 2014). More generally, this increased audiovisual feature specificity in STS neurons, a classically defined multisensory region, is in agreement with current models of audiovisual processing and the important role of the STS in multisensory integration (Beauchamp et al. 2004; Stein and Stanford 2008).

Thereby, neurons in the anterior voice-sensitive cluster in the aSTP show a double dissociation in functional properties with respect to neurons in the aSTS; aSTP neurons, maybe because they are primarily engaged in sensory analysis in the unisensory (auditory) modality, show little specificity in their cross-sensory influences. In contrast, neurons in the STS show more specific multisensory influences but display a lack of fidelity in their unisensory representations (see Sect. 9.1.2). Together, these observations suggest that a high level of specificity is not retained in both the unisensory and the multisensory domains. As such, the results are consistent with the notion of reversed gradients of functional specificity in the unisensory versus multisensory pathways, whereby unisensory stimulus-response fidelity decreases along the sensory-processing hierarchy as multisensory feature sensitivity and specificity increase. These comparisons of results across different brain areas suggest an intermediate functional role of voice-sensitive neurons in the auditory and audiovisual processing hierarchies relative to early auditory fields and the multisensory STS, which is of relevance for building better neurobiologically informed models of multisensory integration (for reviews, see, e.g., Ghazanfar and Schroeder 2006; Stein and Stanford 2008).

9.3 Multisensory Pathways to the Primate Prefrontal Cortex

Sections 9.2.1 to 9.2.3 reviewed some of the evidence for visual influences on the neuronal processing of voices at voice-sensitive and association regions in the temporal lobe. However, these findings do not address whether and how voice regions in the primate temporal lobe are interconnected. Previous neuroanatomical studies have identified pathways for visual and auditory input to the frontal lobe, including dense projections from the second stage of auditory cortical processing, the auditory belt (Romanski et al. 1999b; Plakke and Romanski 2014; also see Plakke and Romanski, Chap. 7). Projections to the frontal cortex from the association cortex in the anterior superior-temporal gyrus are considerable (Petrides and Pandya 1988; Seltzer and Pandya 1989). Yet, the strength and functional impact of the connections between the aSTP and frontal cortex were unclear.

Insights into the effective connectivity between some of these regions were recently provided using combined microstimulation and fMRI in monkeys. This approach allows charting the directional connectivity of a specific pathway, in this

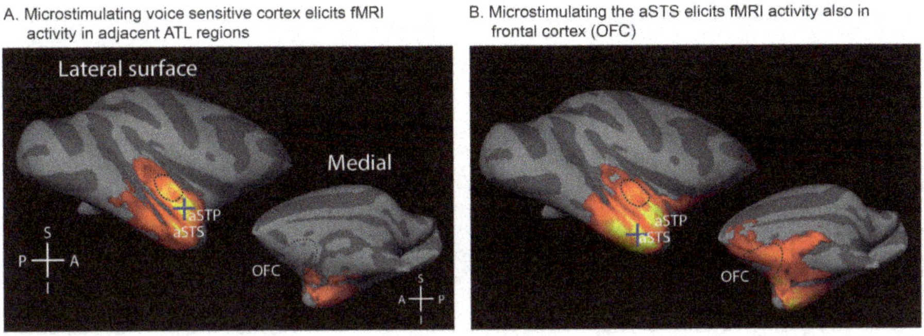

Fig. 9.4 Effective functional connectivity between voice-sensitive and frontal cortices. (**A**) Study of effective functional connectivity using combined microstimulation and functional magnetic resonance imaging (fMRI) shows that stimulating the voice-sensitive cortex (*dark cross*) on the aSTP tends to elicit fMRI activity in neighboring regions of the anterior temporal lobe (Petkov et al. 2015). (**B**) In contrast, stimulating the (aSTS also elicits fMRI activity in the frontal cortex, in particular the OFC. *A*, anterior, *P*, posterior, *S*, superior, *I*, inferior. Reproduced from Perrodin et al. (2015b)

case between the temporal and prefrontal brain regions. Namely, electrically stimulating a specific cortical brain region and using fMRI to assess which regions are activated in response can reveal synaptic targets of the stimulated site, a presumption supported by the fact that target regions activated by stimulation are often consistent with those identified using neuronal anterograde tractography (e.g., Matsui et al. 2011; Petkov et al. 2015). Surprisingly, microstimulating voice identity-sensitive cortex did not strongly activate the prefrontal cortex, unlike stimulation of downstream multisensory areas in the STS and upstream auditory cortical areas in the lateral belt (Petkov et al. 2015). The voice-sensitive cortex in the primate aSTP seemed to interact primarily with a local multisensory network in the temporal lobe, including the upper bank of the aSTS and regions around the temporal pole (Fig. 9.4A). By contrast, stimulating the aSTS resulted in significantly stronger frontal fMRI activation, particularly in orbital frontal cortex (Fig. 9.4B). These observations complement those on interregional connectivity (Frey et al. 2004; Plakke and Romanski 2014) and information on neuronal properties in some of these regions (Kikuchi et al. 2010; Perrodin et al. 2014), which altogether suggest that multisensory voice/face processes are initially integrated in a network of anterior temporal lobe regions, only parts of which have direct access to the frontal cortex.

9.4 Voice- and Face-Processing Pathways: Comparative Perspective

Much of this review has thus far focused on studies in human and nonhuman primates. However, it is important to at least briefly consider the benefits of pursuing a broader evolutionary perspective for advancing the understanding of the

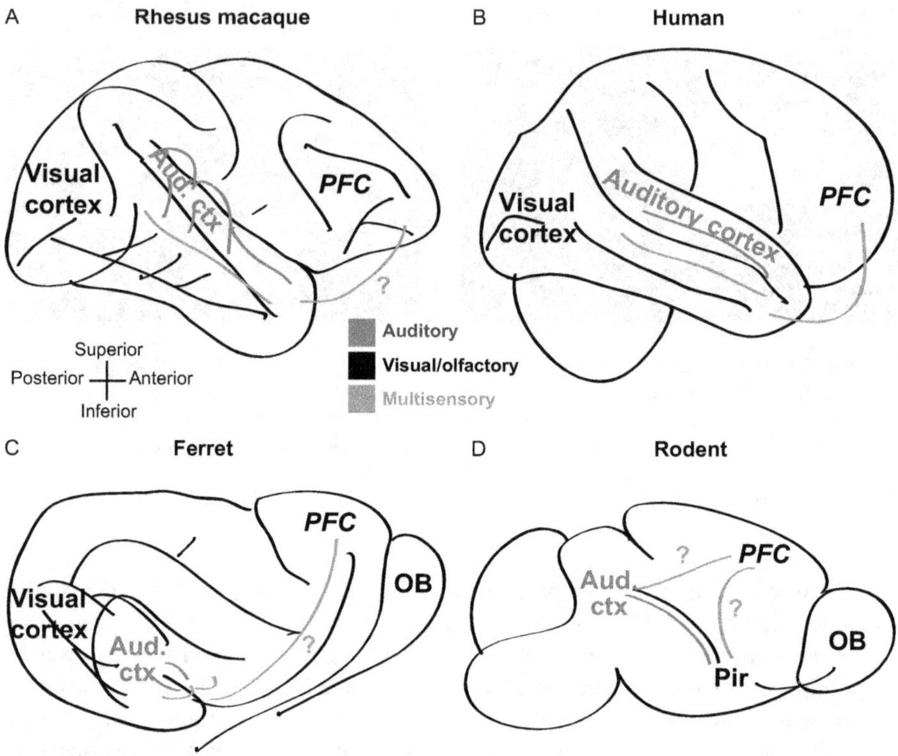

Fig. 9.5 A comparative view of ascending auditory and visual or olfactory cortical streams of the sensory processing pathways in the right hemisphere of mammalian brains supporting social communication across several species. (**A**) rhesus macaque monkey; (**B**) human; (**C**) ferret; (**D**) rodent. Unisensory neuronal representations of communication signals (visual: face or facial expressions in primates; auditory: voices or vocalizations; olfactory: odor or pheromonal cues) become progressively more selective in relation to primary sensory cortices. *Medium gray lines*, auditory projections; *dark gray lines*, visual and/or olfactory. *OB*, olfactory bulb; *Pir*, piriform (olfactory) cortex. Bidirectional anatomical and functional crosstalk occurs at multiple levels throughout the sensory streams, for example, visual projections into the auditory cortices (Aud. Ctx; Bizley et al. 2007) or auditory projections from the primary auditory cortex into the olfactory cortex (Budinger et al. 2006). There are also feedforward and feedback projections between cortical areas, including multisensory influences from high-order association areas in the frontal lobes (PFC) onto sensory processing streams (Hackett et al. 1998; Romanski et al. 1999a). Directions for future study include better understanding the nature and dynamics of the bidirectional functional links between higher level unisensory and frontal cortices and how these mediate the transformation/abstractions of multisensory neuronal representations

neurobiology of communication signal processing and integration (Fig. 9.5). Some animal models will allow teasing apart circuit mechanisms and processes that remain difficult or not possible to achieve in primate models.

Although a number of nonprimate species rely less on voices and faces for social interactions than other forms of communication, relevant ethologically suitable paradigms can be found. The ferret (*Mustela putorius*) is an animal in which both the auditory and multisensory cortical representations and pathways are actively being studied. For instance, studies in ferrets are relied on to reveal the neuronal coding principles supporting the representation of multiple simultaneous auditory features, such as pitch and the timbre of resonant sources (Bizley et al. 2009; Walker et al. 2011). These auditory features, although more generally found in many natural sounds, nevertheless are related to the processing of voice content, given that prominent indexical cues for identifying an individual by voice are provided by formants, with the vocal folds as the source and vocal tract as the filter (Fitch 2000; Smith and Patterson 2005). Multisensory interactions between auditory and visual stimuli have also been well studied, both anatomically and functionally, in ferrets (Bizley et al. 2007).

Many rodents, including mice (*Mus musculus*), rats (*Rattus norvegicus*), and gerbils (*Meriones unguiculatus*), strongly rely on olfactory/pheromonal and auditory information for social interactions, and these animals can readily identify each other by odor (Brennan 2004). Information about odor identity is represented in the olfactory piriform cortex (Kadohisa and Wilson 2006; Gire et al. 2013) and can synergistically interact with vocalization sounds to influence maternal behavior in mice (Okabe et al. 2013). There appear to be multisensory interactions between the rodent olfactory- and auditory-processing systems associated with improved maternal behavior (Budinger et al. 2006; Cohen et al. 2011). A broader comparative approach will clarify evolutionary relationships and better define the function of behaviorally relevant uni- and multisensory pathways.

9.5 Summary, Conclusions, and Look Ahead

This chapter has reviewed the current state of neuroscientific knowledge on the neural representation of voice and face content in communication signals, focusing in particular on some of the processing sites in the anterior temporal lobe in primates. Guided by neuroimaging results in humans and rhesus macaques and the resulting functional correspondences across the species, invasive electrophysiological recordings in the nonhuman primates revealed evidence for voice cells and characterized their basic functional properties, including how these relate to information on face cell characteristics in the visual system. Neuronal processing in the aSTP voice cluster was found to be sensitive to voice identity and very acoustically stimulus selective in relation to upstream auditory areas. Additionally, a double dissociation in the auditory feature sensitivity versus the specificity of multisensory interactions was identified between, on the one hand, neurons in the anterior voice-sensitive cluster on the supratemporal plane and, on the other, adjacent regions in the temporal association cortex. Insights into the directed functional connectivity have also been obtained, providing information on interregional connectivity to complement that on neuronal response characteristics. Together, these initial forays into the neurobiological substrates of voice processing in the temporal lobe raise a number of new questions: What are the perceptual and behavioral correlates of the observed neuronal and oscillatory responses and multisensory interactions? What are the transformations and causal interactions that occur between brain regions involved in voice and face processing as well as multisensory integration for identifying individuals and other entities? Pursuing answers to these questions will be essential for solidifying the next set of advances in understanding how the brain integrates sensory information to guide behavior.

Acknowledgments We thank Christoph Kayser and Nikos Logothetis for useful discussion, support, and encouragement. This work was supported by the Max Planck Society, Wellcome Trust Investigator Award WT092606AIA to Christopher I. Petkov, Sir Henry Wellcome Fellowship 110238/Z/15/Z to Catherine Perrodin, the European Research Council to Christopher I. Petkov (MECHIDENT), and Biotechnology and Biological Sciences Research Council BB/J009849/1 to Christopher I. Petkov jointly with Q. Vuong.

Compliance with Ethics Requirements Catherine Perrodin declares that she has no conflict of interest.

Christopher I. Petkov declares that he has no conflict of interest.

References

Andics, A., Gácsi, M., Faragó, T., Kis, A., & Miklósi, Á. (2014). Voice-sensitive regions in the dog and human brain are revealed by comparative fMRI. *Current Biology, 24*(5), 574–578.

Aparicio, P. L., Issa, E. B., & DiCarlo, J. J. (2016). Neurophysiological organization of the middle face patch in macaque inferior temporal cortex. *The Journal of Neuroscience, 36*(50), 12729–12745.

Bastos, A. M., Vezoli, J., Bosman, C. A., Schoffelen, J.-M., Oostenveld, R., Dowdall, J. R., De Weerd, P., Kennedy, H., & Fries, P. (2015). Visual areas exert feedforward and feedback influences through distinct frequency channels. *Neuron, 85*(2), 390–401.

Beauchamp, M. S., Argall, B. D., Bodurka, J., Duyn, J. H., & Martin, A. (2004). Unraveling multisensory integration: Patchy organization within human STS multisensory cortex. *Nature Neuroscience, 7*(11), 1190–1192.

Belin, P., & Zatorre, R. J. (2003). Adaptation to speaker's voice in right anterior temporal lobe. *Neuroreport, 14*(16), 2105–2109.

Belin, P., Zatorre, R. J., Lafaille, P., Ahad, P., & Pike, B. (2000). Voice-selective areas in human auditory cortex. *Nature, 403*(6767), 309–312.

Belin, P., Fecteau, S., & Bedard, C. (2004). Thinking the voice: Neural correlates of voice perception. *Trends in Cognitive Sciences, 8*(3), 129–135.

Bendor, D., & Wang, X. (2007). Differential neural coding of acoustic flutter within primate auditory cortex. *Nature Neuroscience, 10*(6), 763–771.

Benevento, L. A., Fallon, J., Davis, B. J., & Rezak, M. (1977). Auditory-visual interaction in single cells in the cortex of the superior temporal sulcus and the orbital frontal cortex of the macaque monkey. *Experimental Neurology, 57*(3), 849–872.

Binder, J. R., Desai, R. H., Graves, W. W., & Conant, L. L. (2009). Where is the semantic system? A critical review and meta-analysis of 120 functional neuroimaging studies. *Cerebral Cortex, 19*(12), 2767–2796.

Bizley, J. K., Nodal, F. R., Bajo, V. M., Nelken, I., & King, A. J. (2007). Physiological and anatomical evidence for multisensory interactions in auditory cortex. *Cerebral Cortex, 17*(9), 2172–2189.

Bizley, J. K., Walker, K. M. M., Silverman, B. W., King, A. J., & Schnupp, J. W. H. (2009). Interdependent encoding of pitch, timbre, and spatial location in auditory cortex. *The Journal of Neuroscience, 29*(7), 2064–2075.

Bizley, J. K., Jones, G. P., & Town, S. M. (2016). Where are multisensory signals combined for perceptual decision-making? *Current Opinion in Neurobiology, 40*, 31–37.

Blank, H., Anwander, A., & von Kriegstein, K. (2011). Direct structural connections between voice-and face-recognition areas. *The Journal of Neuroscience, 31*(36), 12906–12915.

Brennan, P. A. (2004). The nose knows who's who: Chemosensory individuality and mate recognition in mice. *Hormones and Behavior, 46*(3), 231–240.

Bruce, V., & Young, A. (1986). Understanding face recognition. *British Journal of Psychology, 77*, 305–327.

Bruce, C., Desimone, R., & Gross, C. G. (1981). Visual properties of neurons in a polysensory area in superior temporal sulcus of the macaque. *Journal of Neurophysiology, 46*(2), 369–384.

Budinger, E., Heil, P., Hess, A., & Scheich, H. (2006). Multisensory processing via early cortical stages: Connections of the primary auditory cortical field with other sensory systems. *Neuroscience, 143*(4), 1065–1083.

Campanella, S., & Belin, P. (2007). Integrating face and voice in person perception. *Trends in Cognitive Sciences, 11*(12), 535–543.

Cappe, C., & Barone, P. (2005). Heteromodal connections supporting multisensory integration at low levels of cortical processing in the monkey. *European Journal of Neuroscience, 22*(11), 2886–2902.

Chan, A. M., Baker, J. M., Eskandar, E., Schomer, D., Ulbert, I., Marinkovic, K., Cash, S. S., & Halgren, E. (2011). First-pass selectivity for semantic categories in human anteroventral temporal lobe. *The Journal of Neuroscience, 31*(49), 18119–18129.

Chandrasekaran, C., & Ghazanfar, A. A. (2009). Different neural frequency bands integrate faces and voices differently in the superior temporal sulcus. *Journal of Neurophysiology, 101*(2), 773–788.

Chandrasekaran, C., Trubanova, A., Stillittano, S., Caplier, A., & Ghazanfar, A. A. (2009). The natural statistics of audiovisual speech. *PLoS Computational Biology, 5*(7), e1000436.

Chen, A., Gu, Y., Liu, S., DeAngelis, G. C., & Angelaki, D. E. (2016). Evidence for a causal contribution of macaque vestibular, but not intraparietal, cortex to heading perception. *The Journal of Neuroscience, 36*(13), 3789–3798.

Cohen, L., Rothschild, G., & Mizrahi, A. (2011). Multisensory integration of natural odors and sounds in the auditory cortex. *Neuron, 72*(2), 357–369.

Creutzfeldt, O., Hellweg, F. C., & Schreiner, C. (1980). Thalamocortical transformation of responses to complex auditory stimuli. *Experimental Brain Research, 39*(1), 87–104.

Dahl, C. D., Logothetis, N. K., & Kayser, C. (2009). Spatial organization of multisensory responses in temporal association cortex. *The Journal of Neuroscience, 29*(38), 11924–11932.

Dahl, C. D., Logothetis, N. K., & Kayser, C. (2010). Modulation of visual responses in the superior temporal sulcus by audio-visual congruency. *Frontiers in Integrative Neuroscience, 4*, 10.

Damasio, A. R. (1989). The brain binds entities and events by multiregional activation from convergence zones. *Neural Computation, 1*(1), 123–132.

Fecteau, S., Armony, J. L., Joanette, Y., & Belin, P. (2004). Is voice processing species-specific in human auditory cortex? An fMRI study. *NeuroImage, 23*(3), 840–848.

Fecteau, S., Armony, J. L., Joanette, Y., & Belin, P. (2005). Sensitivity to voice in human prefrontal cortex. *Journal of Neurophysiology, 94*(3), 2251–2254.

Fetsch, C. R., Pouget, A., DeAngelis, G. C., & Angelaki, D. E. (2012). Neural correlates of reliability-based cue weighting during multisensory integration. *Nature Neuroscience, 15*(1), 146–154.

Fitch, W. T. (2000). The evolution of speech: A comparative review. *Trends in Cognitive Sciences, 4*(7), 258–267.

Formisano, E., De Martino, F., Bonte, M., & Goebel, R. (2008). "Who" is saying "what"? Brain-based decoding of human voice and speech. *Science, 322*(5903), 970–973.

Freiwald, W. A., & Tsao, D. Y. (2010). Functional compartmentalization and viewpoint generalization within the macaque face-processing system. *Science, 330*(6005), 845–851.

Frey, S., Kostopoulos, P., & Petrides, M. (2004). Orbitofrontal contribution to auditory encoding. *NeuroImage, 22*(3), 1384–1389.

Galaburda, A. M., & Pandya, D. N. (1983). The intrinsic architectonic and connectional organization of the superior temporal region of the rhesus monkey. *Journal of Comparative Neurology, 221*(2), 169–184.

Ghazanfar, A. A., & Schroeder, C. E. (2006). Is neocortex essentially multisensory? *Trends in Cognitive Sciences, 10*(6), 278–285.

Ghazanfar, A. A., & Takahashi, D. Y. (2014). The evolution of speech: Vision, rhythm, cooperation. *Trends in Cognitive Sciences, 18*(10), 543–553.

Ghazanfar, A. A., Maier, J. X., Hoffman, K. L., & Logothetis, N. K. (2005). Multisensory integration of dynamic faces and voices in rhesus monkey auditory cortex. *The Journal of Neuroscience, 25*(20), 5004–5012.

Ghitza, O. (2011). Linking speech perception and neurophysiology: Speech decoding guided by cascaded oscillators locked to the input rhythm. *Frontiers in Psychology, 2*, 130.

Gifford, G. W., 3rd, MacLean, K. A., Hauser, M. D., & Cohen, Y. E. (2005). The neurophysiology of functionally meaningful categories: Macaque ventrolateral prefrontal cortex plays a critical role in spontaneous categorization of species-specific vocalizations. *Journal of Cognitive Neuroscience, 17*(9), 1471–1482.

Giraud, A. L., & Poeppel, D. (2012). Cortical oscillations and speech processing: Emerging computational principles and operations. *Nature Neuroscience, 15*(4), 511–517.

Gire, D. H., Whitesell, J. D., Doucette, W., & Restrepo, D. (2013). Information for decision-making and stimulus identification is multiplexed in sensory cortex. *Nature Neuroscience, 16*(8), 991–993.

Griffiths, T. D., Warren, J. D., Scott, S. K., Nelken, I., & King, A. J. (2004). Cortical processing of complex sound: A way forward? *Trends in Neurosciences, 27*(4), 181–185.

Gross, J., Hoogenboom, N., Thut, G., Schyns, P., Panzeri, S., Belin, P., & Garrod, S. (2013). Speech rhythms and multiplexed oscillatory sensory coding in the human brain. *PLoS Biology, 11*(12), e1001752.

Hackett, T. A., Stepniewska, I., & Kaas, J. H. (1998). Subdivisions of auditory cortex and ipsilateral cortical connections of the parabelt auditory cortex in macaque monkeys. *Journal of Comparative Neurology, 394*(4), 475–495.

Hasselmo, M. E., Rolls, E. T., & Baylis, G. C. (1989). The role of expression and identity in the face-selective responses of neurons in the temporal visual cortex of the monkey. *Behavioural Brain Research, 32*(3), 203–218.

Henry, M. J., & Obleser, J. (2012). Frequency modulation entrains slow neural oscillations and optimizes human listening behavior. *Proceedings of the National Academy of Sciences of the United States of America, 109*(49), 20095–20100.

Hickok, G., & Poeppel, D. (2007). The cortical organization of speech processing. *Nature Reviews Neuroscience, 8*(5), 393–402.

Kaas, J. H., & Hackett, T. A. (1998). Subdivisions of auditory cortex and levels of processing in primates. *Audiology and Neuro-Otology, 3*(2-3), 73–85.

Kaas, J. H., & Hackett, T. A. (2000). Subdivisions of auditory cortex and processing streams in primates. *Proceedings of the National Academy of Sciences of the United States of America, 97*(22), 11793–11799.

Kadohisa, M., & Wilson, D. A. (2006). Separate encoding of identity and similarity of complex familiar odors in piriform cortex. *Proceedings of the National Academy of Sciences of the United States of America, 103*(41), 15206–15211.

Kanwisher, N., McDermott, J., & Chun, M. M. (1997). The fusiform face area: A module in human extrastriate cortex specialized for face perception. *The Journal of Neuroscience, 17*(11), 4302–4311.

Kayser, C., Petkov, C. I., & Logothetis, N. K. (2008). Visual modulation of neurons in auditory cortex. *Cerebral Cortex, 18*(7), 1560–1574.

Keil, J., Muller, N., Hartmann, T., & Weisz, N. (2014). Prestimulus beta power and phase synchrony influence the sound-induced flash illusion. *Cerebral Cortex, 24*(5), 1278–1288.

Kikuchi, Y., Horwitz, B., & Mishkin, M. (2010). Hierarchical auditory processing directed rostrally along the monkey's supratemporal plane. *The Journal of Neuroscience, 30*(39), 13021–13030.

King, A. J., & Nelken, I. (2009). Unraveling the principles of auditory cortical processing: Can we learn from the visual system? *Nature Neuroscience, 12*(6), 698–701.

Kriegeskorte, N., Formisano, E., Sorger, B., & Goebel, R. (2007). Individual faces elicit distinct response patterns in human anterior temporal cortex. *Proceedings of the National Academy of Sciences of the United States of America, 104*(51), 20600–20605.

Lakatos, P., Chen, C. M., O'Connell, M. N., Mills, A., & Schroeder, C. E. (2007). Neuronal oscillations and multisensory interaction in primary auditory cortex. *Neuron, 53*(2), 279–292.

Logothetis, N. K., Guggenberger, H., Peled, S., & Pauls, J. (1999). Functional imaging of the monkey brain. *Nature Neuroscience, 2*(6), 555–562.

Matsui, T., Tamura, K., Koyano, K. W., Takeuchi, D., Adachi, Y., Osada, T., & Miyashita, Y. (2011). Direct comparison of spontaneous functional connectivity and effective connectivity measured by intracortical microstimulation: An fMRI study in macaque monkeys. *Cerebral Cortex, 21*(10), 2348–2356.

McGrath, M., & Summerfield, Q. (1985). Intermodal timing relations and audio-visual speech recognition by normal-hearing adults. *The Journal of the Acoustical Society of America, 77*(2), 678–685.

McLaren, D. G., Kosmatka, K. J., Oakes, T. R., Kroenke, C. D., Kohama, S. G., Matochik, J. A., Ingram, D. K., & Johnson, S. C. (2009). A population-average MRI-based atlas collection of the rhesus macaque. *NeuroImage, 45*(1), 52–59.

Mercier, M. R., Foxe, J. J., Fiebelkorn, I. C., Butler, J. S., Schwartz, T. H., & Molholm, S. (2013). Auditory-driven phase reset in visual cortex: Human electrocorticography reveals mechanisms of early multisensory integration. *NeuroImage, 79*, 19–29.

Miller, C. T., & Cohen, Y. E. (2010). Vocalizations as auditory objects: Behavior and neurophysiology. In M. L. Platt & A. A. Ghazanfar (Eds.), *Primate neuroethology* (pp. 237–255). New York: Oxford University Press.

Mizrahi, A., Shalev, A., & Nelken, I. (2014). Single neuron and population coding of natural sounds in auditory cortex. *Current Opinion in Neurobiology, 24*, 103–110.

Morin, E. L., Hadj-Bouziane, F., Stokes, M., Ungerleider, L. G., & Bell, A. H. (2014). Hierarchical encoding of social cues in primate inferior temporal cortex. *Cerebral Cortex, 25*(9), 3036–3045.

Okabe, S., Nagasawa, M., Kihara, T., Kato, M., Harada, T., Koshida, N., Mogi, K., & Kikusui, T. (2013). Pup odor and ultrasonic vocalizations synergistically stimulate maternal attention in mice. *Behavioural Neuroscience, 127*(3), 432–438.

Osmanski, M. S., & Wang, X. (2015). Behavioral dependence of auditory cortical responses. *Brain Topography, 28*(3), 365–378.

Pandya, D. N., Hallett, M., & Kmukherjee, S. K. (1969). Intra- and interhemispheric connections of the neocortical auditory system in the rhesus monkey. *Brain Research, 14*(1), 49–65.

Perrett, D. I., Rolls, E. T., & Caan, W. (1982). Visual neurones responsive to faces in the monkey temporal cortex. *Experimental Brain Research, 47*(3), 329–342.

Perrodin, C., Kayser, C., Logothetis, N. K., & Petkov, C. I. (2011). Voice cells in the primate temporal lobe. *Current Biology, 21*(16), 1408–1415.

Perrodin, C., Kayser, C., Logothetis, N. K., & Petkov, C. I. (2014). Auditory and visual modulation of temporal lobe neurons in voice-sensitive and association cortices. *The Journal of Neuroscience, 34*(7), 2524–2537.

Perrodin, C., Kayser, C., Logothetis, N. K., & Petkov, C. I. (2015a). Natural asynchronies in audiovisual communication signals regulate neuronal multisensory interactions in voice-sensitive cortex. *Proceedings of the National Academy of Sciences of the United States of America, 112*(1), 273–278.

Perrodin, C., Kayser, C., Abel, T. J., Logothetis, N. K., & Petkov, C. I. (2015b). Who is that? Brain networks and mechanisms for identifying individuals. *Trends in Cognitive Sciences, 19*(12), 783–796.

Petkov, C. I., Kayser, C., Steudel, T., Whittingstall, K., Augath, M., & Logothetis, N. K. (2008). A voice region in the monkey brain. *Nature Neuroscience, 11*(3), 367–374.

Petkov, C. I., Kikuchi, Y., Milne, A. E., Mishkin, M., Rauschecker, J. P., & Logothetis, N. K. (2015). Different forms of effective connectivity in primate frontotemporal pathways. *Nature Communications, 6*. https://doi.org/10.1038/ncomms7000.

Petrides, M., & Pandya, D. N. (1988). Association fiber pathways to the frontal cortex from the superior temporal region in the rhesus monkey. *Journal of Comparative Neurology, 273*(1), 52–66.

Plakke, B., & Romanski, L. M. (2014). Auditory connections and functions of prefrontal cortex. *Frontiers in Neuroscience, 8*, 199.

Plakke, B., Diltz, M. D., & Romanski, L. M. (2013). Coding of vocalizations by single neurons in ventrolateral prefrontal cortex. *Hearing Research, 305*, 135–143.

Rauschecker, J. P. (1998). Parallel processing in the auditory cortex of primates. *Audiology and Neurotology, 3*(2-3), 86–103.

Rauschecker, J. P., & Tian, B. (2000). Mechanisms and streams for processing of "what" and "where" in auditory cortex. *Proceedings of the National Academy of Sciences of the United States of America, 97*, 11800–11806.

Recanzone, G. H. (2008). Representation of con-specific vocalizations in the core and belt areas of the auditory cortex in the alert macaque monkey. *The Journal of Neuroscience, 28*(49), 13184–13193.

Remedios, R., Logothetis, N. K., & Kayser, C. (2009). An auditory region in the primate insular cortex responding preferentially to vocal communication sounds. *The Journal of Neuroscience, 29*(4), 1034–1045.

Romanski, L. M. (2007). Representation and integration of auditory and visual stimuli in the primate ventral lateral prefrontal cortex. *Cerebral Cortex, 17*(Suppl. 1), i61–i69.

Romanski, L. M. (2012). Integration of faces and vocalizations in ventral prefrontal cortex: Implications for the evolution of audiovisual speech. *Proceedings of the National Academy of Sciences of the United States of America, 109*(Suppl. 1), 10717–10724.

Romanski, L. M., Bates, J. F., & Goldman-Rakic, P. S. (1999a). Auditory belt and parabelt projections to the prefrontal cortex in the rhesus monkey. *Journal of Comparative Neurology, 403*(2), 141–157.

Romanski, L. M., Tian, B., Fritz, J., Mishkin, M., Goldman-Rakic, P. S., & Rauschecker, J. P. (1999b). Dual streams of auditory afferents target multiple domains in the primate prefrontal cortex. *Nature Neuroscience, 2*(12), 1131–1136.

Romanski, L. M., Averbeck, B. B., & Diltz, M. (2005). Neural representation of vocalizations in the primate ventrolateral prefrontal cortex. *Journal of Neurophysiology, 93*(2), 734–747.

Russ, B. E., Ackelson, A. L., Baker, A. E., & Cohen, Y. E. (2008). Coding of auditory-stimulus identity in the auditory non-spatial processing stream. *Journal of Neurophysiology, 99*(1), 87–95.

Sadagopan, S., Temiz-Karayol, N. Z., & Voss, H. U. (2015). High-field functional magnetic resonance imaging of vocalization processing in marmosets. *Scientific Reports, 5*, 10950.

Saleem, K. S., & Logothetis, N. K. (2007). *A combined MRI and histology: Atlas of the rhesus monkey brain in stereotaxic coordinates*. London: Academic.

Schroeder, C. E., & Foxe, J. J. (2002). The timing and laminar profile of converging inputs to multisensory areas of the macaque neocortex. *Cognitive Brain Research, 14*(1), 187–198.

Schroeder, C. E., Smiley, J., Fu, K. G., McGinnis, T., O'Connell, M. N., & Hackett, T. A. (2003). Anatomical mechanisms and functional implications of multisensory convergence in early cortical processing. *International Journal of Psychophysiology, 50*(1-2), 5–17.

Schroeder, C. E., Lakatos, P., Kajikawa, Y., Partan, S., & Puce, A. (2008). Neuronal oscillations and visual amplification of speech. *Trends in Cognitive Sciences, 12*(3), 106–113.

Seltzer, B., & Pandya, D. N. (1989). Frontal lobe connections of the superior temporal sulcus in the rhesus monkey. *Journal of Comparative Neurology, 281*(1), 97–113.

Seltzer, B., & Pandya, D. N. (1994). Parietal, temporal, and occipital projections to cortex of the superior temporal sulcus in the rhesus monkey: A retrograde tracer study. *Journal of Comparative Neurology, 343*(3), 445–463.

Sergent, J., Ohta, S., & MacDonald, B. (1992). Functional neuroanatomy of face and object processing. A positron emission tomography study. *Brain, 115*(1), 15–36.

Seyfarth, R. M., & Cheney, D. L. (2014). The evolution of language from social cognition. *Current Opinion in Neurobiology, 28*, 5–9.

Slutsky, D. A., & Recanzone, G. H. (2001). Temporal and spatial dependency of the ventriloquism effect. *Neuroreport, 12*(1), 7–10.

Smith, D. R., & Patterson, R. D. (2005). The interaction of glottal-pulse rate and vocal-tract length in judgements of speaker size, sex, and age. *The Journal of the Acoustical Society of America, 118*(5), 3177–3186.

Stein, B. E., & Stanford, T. R. (2008). Multisensory integration: Current issues from the perspective of the single neuron. *Nature Reviews Neuroscience, 9*(4), 255–266.

Strauss, A., Henry, M. J., Scharinger, M., & Obleser, J. (2015). Alpha phase determines successful lexical decision in noise. *The Journal of Neuroscience, 35*(7), 3256–3262.

Sugihara, T., Diltz, M. D., Averbeck, B. B., & Romanski, L. M. (2006). Integration of auditory and visual communication information in the primate ventrolateral prefrontal cortex. *The Journal of Neuroscience, 26*(43), 11138–11147.

Ten Oever, S., & Sack, A. T. (2015). Oscillatory phase shapes syllable perception. *Proceedings of the National Academy of Sciences of the United States of America, 112*(52), 15833–15837.

Thorne, J. D., & Debener, S. (2014). Look now and hear what's coming: On the functional role of cross-modal phase reset. *Hearing Research, 307*, 144–152.

Thut, G., Miniussi, C., & Gross, J. (2012). The functional importance of rhythmic activity in the brain. *Current Biology, 22*(16), R658–R663.

Tian, B., Reser, D., Durham, A., Kustov, A., & Rauschecker, J. P. (2001). Functional specialization in rhesus monkey auditory cortex. *Science, 292*(5515), 290–293.

Tsao, D. Y., & Livingstone, M. S. (2008). Mechanisms of face perception. *Annual Review of Neuroscience, 31*, 411–437.

Tsao, D. Y., Freiwald, W. A., Tootell, R. B., & Livingstone, M. S. (2006). A cortical region consisting entirely of face-selective cells. *Science, 311*(5761), 670–674.

Tsao, D. Y., Schweers, N., Moeller, S., & Freiwald, W. A. (2008). Patches of face-selective cortex in the macaque frontal lobe. *Nature Neuroscience, 11*(8), 877–879.

Turesson, H. K., Logothetis, N. K., & Hoffman, K. L. (2012). Category-selective phase coding in the superior temporal sulcus. *Proceedings of the National Academy of Sciences of the United States of America, 109*(47), 19438–19443.

van Atteveldt, N., Murray, M. M., Thut, G., & Schroeder, C. E. (2014). Multisensory integration: Flexible use of general operations. *Neuron, 81*(6), 1240–1253.

von Kriegstein, K., Eger, E., Kleinschmidt, A., & Giraud, A. L. (2003). Modulation of neural responses to speech by directing attention to voices or verbal content. *Cognitive Brain Research, 17*(1), 48–55.

von Kriegstein, K., Kleinschmidt, A., Sterzer, P., & Giraud, A. L. (2005). Interaction of face and voice areas during speaker recognition. *Journal of Cognitive Neuroscience, 17*(3), 367–376.

Walker, K. M., Bizley, J. K., King, A. J., & Schnupp, J. W. (2011). Multiplexed and robust representations of sound features in auditory cortex. *The Journal of Neuroscience, 31*(41), 14565–14576.

Watson, R., Latinus, M., Charest, I., Crabbe, F., & Belin, P. (2014). People-selectivity, audiovisual integration and heteromodality in the superior temporal sulcus. *Cortex, 50*, 125–136.

Werner, S., & Noppeney, U. (2010). Distinct functional contributions of primary sensory and association areas to audiovisual integration in object categorization. *The Journal of Neuroscience, 30*(7), 2662–2675.

Yau, J. M., Deangelis, G. C., & Angelaki, D. E. (2015). Dissecting neural circuits for multisensory integration and crossmodal processing. *Philosophical Transactions of the Royal Society B: Biological Sciences, 370*(1677), 20140203.

Chapter 10
Neural Network Dynamics and Audiovisual Integration

Julian Keil and Daniel Senkowski

Abstract Why does seeing a speaker's lip movements improve understanding speech in noisy environments? Why does simultaneous ringing and vibrating quicken answering a phone? These are questions of interest for researchers in the field of multisensory information processing. Electrophysiological approaches are suited to map the neural network dynamics underlying multisensory perception. Combining findings from behavioral, functional neuroimaging, and electrophysiological studies allows a comprehensive understanding of how information is integrated across the different senses. This chapter first provides an introduction on the relationships between neural network dynamics, as reflected in neural oscillations, and unisensory perception. Then, the relevance of neural network dynamics for multisensory perception is described, with a special focus on the auditory system. Moreover, the chapter provides an overview on how visual and auditory information can mutually influence each other and highlights the crucial role of ongoing neural network dynamics for upcoming multisensory perception. Finally, general principles of audiovisual integration are established, and open questions and future direction in the field of multisensory perception are discussed.

Keywords Alpha band · Amplitude · Beta band · Connectivity · Electroencephalography · Electrophysiology · Gamma band · Illusion · Magnetoencephalography · Multisensory integration · Oscillation · Phase · Speech

---

J. Keil (✉)
Biological Psychology, Christian-Albrechts-University Kiel, Kiel, Germany
e-mail: keil@psychologie.uni-kiel.de

D. Senkowski
Department of Psychiatry and Psychotherapy, Charité—Universitätsmedizin Berlin, Berlin, Germany
e-mail: daniel.senkowski@charite.de

A. K. C. Lee et al. (eds.), *Multisensory Processes*, Springer Handbook of Auditory Research 68, https://doi.org/10.1007/978-3-030-10461-0_10

10.1 Introduction

Why does seeing a speaker's lip movements improve understanding speech in noisy environments? Why does simultaneous ringing and vibrating quicken answering a phone? These are questions of interest for researchers in the field of multisensory information processing (Sumby and Polack 1954; Pomper et al. 2014). The study of multisensory integration at the behavioral level can provide valuable information about the conditions under which information from different senses interact. Moreover, functional neuroimaging approaches are well suited to study which cortical regions are involved in the perception and processing of multisensory information. Electrophysiological approaches, in particular, are suited to map the neural network dynamics underlying multisensory perception. The combined knowledge from behavioral, functional neuroimaging, and electrophysiological studies allows a comprehensive understanding of how information is integrated across the different senses.

This chapter first provides an introduction on the relationships between neural network dynamics, as reflected in neural oscillations, and perception (Sect. 10.1.1). Then, the relevance of neural network dynamics for multisensory perception is described, with a special focus on the auditory system (Sects. 10.2 and 10.3). Subsequently, the chapter describes how visual and auditory information can mutually influence each other (Sects. 10.4 and 10.5). This chapter also highlights the crucial role of ongoing neural network dynamics for upcoming perception (Sect. 10.6). Finally, general principles of audiovisual integration based on presented findings are established (Sect. 10.7), and open questions and future direction in the field of multisensory perception are discussed (Sect. 10.7.4).

10.1.1 Oscillatory Neural Activity Relates to Cognition and Perception

"Clocks tick, bridges vibrate, and neural networks oscillate" (Buzsáki and Draguhn 2004). Oscillatory neural activity recorded through electroencephalography (EEG) or magnetoencephalography (MEG) can be understood as the synchronous waxing and waning of summed postsynaptic activity of large neural populations in circumscribed brain regions (Lopes da Silva 1991; Wang 2010). The resulting waveform can be dissected into different frequency components with distinct amplitudes (also called power) and phases (Herrmann et al. 1999; Mitra and Pesaran 1999). In the frequency components, two types of oscillatory responses, which reflect different aspects of neural synchronization, can be distinguished: evoked and induced oscillations (Tallon-Baudry and Bertrand 1999). The former are closely related to the onset of an external event and are strictly phase and time locked to the stimulus onset. The phase locking of oscillatory responses can be quantified by intertrial coherence (ITC; Cheron et al. 2007) and the summation over trials of identical phase can result in event-related potentials (ERPs; Luck 2014). Induced oscillations can be elicited by stimulation but are also present independent of external stimulation. Induced oscillations do not have to be strictly phase and time locked to the onset of stimuli (Tallon-Baudry and Bertrand 1999). Evoked and induced oscillations can be modulated by cognitive processes. Moreover, functional connectivity, the interaction between oscillatory activities in different cortical regions, can be reflected in phase coherence. Neural oscillations of two brain regions are considered to be phase coherent when there is a constant relationship between the phases of the two signals over time (Fries 2005, 2015). Information processing as well as transfer and storage in the cortex has been hypothesized to rely on flexible cell assemblies, which are defined as transiently synchronized neural networks (Engel et al. 2001; Buzsáki and Draguhn 2004). The transient synchronization of cell assemblies by oscillatory activity depends on the coupling strength between neural populations as well as on the frequency distribution; as long as the frequencies of coupled cell assemblies are similar, the synchronization within the neural network can be sustained with weak synaptic links (Wang 2010). In general, the analysis of oscillatory cortical activity can provide valuable information on the temporal structure of local processes and network interactions underlying perception and cognition.

Neural networks in mammals exhibit oscillatory activity ranging between approximately 0.05 Hz and 350 Hz (Penttonen and Buzsáki 2003). In humans, oscillatory activity patterns were among the first signals recorded using EEG (Berger 1929; Bremer 1958). Within one neural network, neighboring frequency bands can compete with each other and can be associated with different cognitive and perceptual processes (Engel et al. 2001; Buzsáki and Draguhn 2004). Typically, multiple rhythms coexist at the same time, which result in complex waveforms consisting of high- and low-frequency oscillations (Steriade 2001). One way to organize the multiple rhythms is to divide the frequency spectrum into neighboring frequency bands (Buzsáki and Draguhn 2004). Oscillatory slow-wave activity (below 1 Hz) plays a prominent role in sleep and memory (Penttonen and Buzsáki 2003; Diekelmann and Born 2010), but also reflects changes in cortical excitability related to task performance (Birbaumer et al. 1990; Rockstroh et al. 1992). Above these slow-wave oscillations, Walter (1936) described the delta band, which comprises oscillatory activity below 4 Hz. In the frequency range of 4–7 Hz, Walter et al. (1966) identified the theta band. Both delta band and theta band activity have been related to memory processing (Klimesch 1999; Sauseng et al. 2005). More recently, theta band activity has been linked to cognitive control mechanisms such as attention and predictions (Cavanagh and Frank 2014). In his seminal article from the late 1920s, Berger described that the EEG is dominated by ongoing 8- to 12-Hz oscillations, which were later termed alpha band activity (Berger 1929). Of particular note was Berger's observation that alpha band activity changed with the participant's behavior: alpha band power increased when participants closed their eyes and decreased when they opened the eyes (Berger 1929). Ray and Cole (1985) proposed that oscillatory activity in different frequency bands reflects different cognitive processes. In two experiments, the authors established that alpha band activity relates to attentional processes and is increased if attention is not required. Additionally, ongoing alpha band oscillations influence subsequent perception (Lange et al. 2014). Recently, the alpha band has been ascribed an important role in attention and the routing of information (Jensen and Mazaheri 2010; Klimesch 2012). Above the alpha band, Berger (1929) identified the beta band (13–30 Hz), but its functional significance has only been studied many years later (Pfurtscheller 1992; Engel and Fries 2010). Recent studies have provided evidence that, besides motor functions, beta band activity relates to cognitive and emotional processing and that it might reflect cortical feedback processing (Keil et al. 2016; Michalareas et al. 2016). Cortical activity in frequencies above the beta band (i.e., >30 Hz) has been coined gamma band activity (Adrian 1942; Bressler 1990). It has been proposed that oscillatory activity in the gamma band may form a mechanism for feature representation of a given stimulus (Lopes da Silva 1991). Findings from the auditory and visual domains support this notion. For instance, using intracranial recordings from the calcarine region of the visual cortex in epileptic patients, Chatrian et al. (1960) described a rhythmic response to visual stimulation at a frequency of around 50 Hz. Moreover, in response to auditory stimuli, Pantev et al. (1991) described a transient oscillatory response at around 40 Hz.

Thus, oscillatory activity in different frequency bands relates to different perceptual and cognitive processes and reflects the functional states of neural networks (Lopes da Silva 1991). However, multiple neighboring frequency bands can be involved in a single process and multiple processes can relate to a single frequency. Moreover, the boundaries between different frequency bands can vary by task and recording technique (Buzsáki and Draguhn 2004). In summary, there is robust evidence that oscillatory activity in different frequency bands relates to various perceptual and cognitive functions (Table 10.1).

10.2 Role of Oscillatory Processes for Multisensory Integration and Perception

Multisensory perception requires processing in primary sensory areas as well as the formation of multimodal coherent percepts in distributed neural networks. In an early EEG study on oscillatory activity and multisensory processing, Sakowitz et al. (2001) found increased gamma band power in response to audiovisual stimuli

Table 10.1 Overview of the classical frequency bands found in human electrophysiological data and examples of functions ascribed to the frequency bands

Name	Frequency range (Hz)	Exemplary function
Slow wave	<1	Sleep, Memory, Cortical excitability
Delta	1–3	Memory
Theta	4–7	Memory, Attention, Cognitive control
Alpha	8–12	Attention, Cognition, Routing of information
Beta	13–30	Attention, Cognition, Stimulus processing
Gamma	>30	Stimulus processing, Feature binding

compared with auditory or visual stimuli alone. A later EEG study extended this finding by showing that evoked gamma band power to audiovisual stimuli increases, in particular, for attended versus unattended stimuli (Senkowski et al. 2005). Interestingly, another study found increased occipital gamma band power following the presentation of incongruent audiovisual stimuli, but only if the audiovisual stimuli were integrated into a coherent perception (Bhattacharya et al. 2002). Whereas these studies demonstrate that multisensory processes or, at least, specific aspects of multisensory processes are presumably reflected in gamma band power, they did not examine the underlying cortical networks.

Traditionally, it has been assumed that multisensory integration is a higher order process that occurs after stimulus processing in unisensory cortical and subcortical areas (Driver and Noesselt 2008). However, a number of studies have challenged this idea by providing evidence for multisensory convergence in low-level sensory cortices (Schroeder and Foxe 2005; Ghazanfar and Schroeder 2006). Using intracranial recordings in monkeys, Lakatos et al. (2007) showed that somatosensory stimulation modulates activity in primary auditory areas. Interestingly, the authors found evidence for a theta band phase reset of ongoing oscillatory activity in the primary auditory cortex by concurrent somatosensory input. The authors suggested that stimulus responses are enhanced when their onset falls into a high-excitability phase and suppressed when the onset falls into a low-excitability phase. These observations are in-line with another study recording local field potentials and single-unit activity in monkeys, which highlights the role of oscillatory alpha band phase for the modulation of auditory evoked activity (Kayser et al. 2008). Analyzing ITC as a measure of transient phase synchronization in intracranial recordings from the visual cortex in humans, Mercier et al. (2013) found an influence of auditory stimulation on the processing of a concurrent visual stimulus in the theta band and low alpha band as well as in the beta band. Based on the finding of transient synchronization of delta and theta band oscillations during audiovisual stimulation in a follow-up intracranial study, Mercier et al. (2015) argued that optimally aligned low-frequency phases promote communication between cortical areas and that stimuli in one modality can reset the phase of an oscillation in a cortical area of the other modality. Taken together, these studies demonstrate cross-modal influences in primary sensory areas. Furthermore, it is likely that low-frequency oscillations mediate this cross-modal influence.

The finding that cross-modal processes influence primary sensory activity via low-frequency oscillatory activity implies a predictive process (Schroeder et al. 2008). In many natural settings, visual information precedes auditory information. For example, in audiovisual speech, the lip movements precede the articulation of phonemes (see Grant and Bernstein, Chap. 3). A mechanism that has been proposed for the transfer of information between cortical areas is neural coherence, as reflected in synchronized oscillatory activity (Fries 2005, 2015). For example, in audiovisual speech, the visual information can be transferred to the auditory cortex (Arnal et al. 2009). It has been proposed that audiovisual perception involves a network of primary visual and auditory areas as well as multisensory regions (Keil et al. 2012; Schepers et al. 2013). This network presumably reflects reentrant bottom-up and top-down interactions between primary sensory and multisensory areas (Arnal and Giraud 2012).

In summary, there is robust evidence that multisensory integration can be reflected in increased gamma band power and that cross-modal processes can modulate cortical activity in primary sensory areas (van Atteveldt 2014). Furthermore, as hypothesized (Senkowski et al. 2008; Keil and Senkowski 2018), it is likely that information transfer in a network of primary sensory, multisensory, and frontal cortical areas is instantiated through synchronized oscillatory activity.

10.3 Principles of Multisensory Integration and Oscillatory Processes

The studies described in Sect. 10.2 suggest a relationship between oscillatory activity and multisensory perception. The current section focuses on the principles of multisensory perception and how these principles relate to oscillatory activity in the auditory system. Based on findings from a wide range of studies, three *principles of multisensory integration* have been established: the spatial principle, the temporal principle, and the principle of inverse effectiveness (Stein and Meredith 1993; Stein et al., 2014). In short, the principles state that multisensory integration is the strongest when the input modalities are (1) spatially concordant, (2) temporally aligned, and (3) when the neural responses to the presented stimuli are weak. In addition to the three principles of multisensory integration, the *modality appropriateness hypothesis* has been proposed (Welch and Warren 1980). The auditory system has a relatively low spatial acuity but high temporal resolution. In contrast, the visual system has a relatively low temporal resolution but a high spatial acuity. Therefore, it has been proposed that audiovisual integration will be governed by the auditory modality in tasks requiring high temporal resolution and by the visual modality in tasks requiring high spatial acuity. The modality appropriateness hypothesis has been extended in a maximum-likelihood-estimation framework, which puts forward the idea that information from each sensory modality is weighted based on its relative reliability (Ernst and Bülthoff 2004). Therefore, it can be hypothesized that the auditory system will be especially affected by the visual system when a stimulus contains task-relevant spatial information. In turn, it can be hypothesized that the auditory system will prevail in tasks requiring high temporal resolution.

A previous EEG study examined the influence of audiovisual temporal synchrony on evoked gamma band oscillations (Senkowski et al. 2007). In line with the principle of temporal alignment, gamma band power following audiovisual stimulation was strongest when the auditory and visual inputs of an audiovisual stimulus were presented simultaneously. Interestingly, stimuli were perceived as being separated when the auditory input preceded the visual input by more than 100 ms. A later EEG study established the principle of inverse effectiveness for multisensory stimulus processing, as reflected in early event-related potentials (ERPs; Senkowski et al. 2011). In this study, ERP amplitudes were larger for bimodal audiovisual stimulation compared with combined ERPs following unimodal auditory or visual stimulation but only when the stimuli were presented at a low intensity. Moreover, a local field-potential recording study in monkeys revealed that the principle of spatial alignment and the principle of inverse effectiveness were also reflected in neural oscillations (Lakatos et al. 2007). Interestingly, these authors found that a somatosensory stimulus shifted the neural oscillations in the ipsilateral auditory cortex to a low-excitatory phase and reduced event-related responses. In contrast, a contralateral somatosensory stimulus grouped oscillations around the ideal (i.e., high-excitatory) phase and increased event-related responses. Moreover, in agreement with the principle of inverse effectiveness, the event-related response in the auditory cortex was significantly enhanced when a somatosensory stimulus was added to an auditory stimulus that elicited only a weak response in isolation. Thus, multisensory integration requires flexible neuronal processing (van Atteveldt et al. 2014).

Taken together, a number of general principles for multisensory integration and cross-modal influence have been formulated. Recent electrophysiological studies suggested that these principles are reflected in cortical network dynamics involving neural oscillations.

10.4 Influence of Visual Input on Auditory Perception in Audiovisual Speech

The auditory system, with its high temporal resolution, is very effective in the processing of temporal information. The visual system excels at spatial acuity. As described in Sect. 10.3, the auditory system might be especially affected by the visual system when a stimulus contains important spatial information. An example for the influence of visual information on auditory perception is observing the speaker's mouth movements during audiovisual speech processing. Here, the temporally complex auditory information is processed in the auditory cortex. Concurrently, the variations in the speaker's mouth movements before the utterance of a syllable are processed by the visual system. The lip movements can facilitate the processing of the auditory information. Moreover, rhythmic gestures provide coarse temporal cues for the onset of auditory stimuli (Biau et al. 2015; He et al. 2015). In an early MEG study comparing cortical activity evoked by auditory speech stimuli accompanied by either congruent or incongruent visual stimuli, Sams et al. (1991) showed that incongruent visual information from face movements influences syllable perception. Incongruent audiovisual stimuli elicited a mismatch response in the event-related field. A later MEG study used similar stimuli and found that incongruent audiovisual stimuli elicit stronger gamma band responses than congruent audiovisual stimuli (Kaiser et al. 2005). Interestingly, the study suggested a spatiotemporal hierarchy in the processing of audiovisual stimuli, which starts in posterior parietal cortical areas and spreads to occipital and frontal areas. More recently, Lange et al. (2013) compared neural oscillations to incongruent compared with congruent audiovisual speech stimuli and found increased gamma band and beta band power following congruent speech stimulation, although at a longer latency than Kaiser et al. (2005) found. Whereas the early gamma band power increase in the study by Kaiser et al. (2005) might reflect the processing of the audiovisual mismatch, the later gamma band power increase following congruent speech found by Lange et al. (2013) might be related to audiovisual integration. The idea of a processing hierarchy has recently received support by an EEG study investigating oscillatory activity during the McGurk illusion (Roa Romero et al. 2015). The McGurk illusion involves incongruent audiovisual speech stimuli, which can be fused into an integrated, subjectively congruent audiovisual percept (McGurk and MacDonald 1976). In this study, incongruent audiovisual syllables, which were perceived as an illusory novel percept, were compared with congruent audiovisual stimuli. Again, incongruent stimuli were associated with increased stimulus processing, in this case reflected in beta band power reduction. These reductions were found at two temporal stages: initially over posterior scalp regions and then over frontal scalp regions (Fig. 10.1). With respect to the cortical areas critical to this process, multistage models of audiovisual integration involving the initial processing of auditory and visual information in primary sensory areas and the subsequent integration in parietal and frontal cortical areas have been recently suggested (Peelle and Sommers 2015; Bizley et al. 2016).

Support for the notion that audiovisual speech perception involves multiple processing stages comes from a MEG study (Arnal et al. 2011). The authors investigated the neuronal signatures of valid or invalid predictions that were based on congruent or incongruent visual speech information, respectively. By correlating the ERP with time-frequency-resolved ITC, the authors found that initial processing of audiovisual speech, independent of stimulus congruence, was reflected in increased delta band and theta band ITC around 100 ms after auditory stimulus onset. Furthermore, valid cross-modal predictions in congruent audiovisual speech stimuli were reflected in increased delta band ITC around 400 ms after auditory stimulus onset. In a case of invalid predictions in incongruent audiovisual stimuli, a later beta band component around 500 ms after auditory stimulus onset was identified. This beta band component presumably reflects the error of the prediction based on the visual syllable. These findings were discussed within the predictive coding framework (Rao and Ballard 1999) to describe cortical oscillatory activity as a mechanism for multisensory integration and temporal predictions. Arnal and Giraud (2012) suggested that temporal regularities induce low-frequency oscillations to align neuronal excitability with the predicted onset of upcoming stimuli (see also Van Wassenhove and Grzeczkowski 2015). Moreover, the authors proposed that top-down signals, which are based on previously available visual information, influence oscillatory activity in neural networks. The top-down signals are primarily conveyed in beta band oscillations, whereas the bottom-up sensory signals originating in primary sensory cortical areas are conveyed in gamma band activity.

In summary, audiovisual integration of speech, as reflected in beta band and gamma band oscillations, requires early and late processing stages. The early stage could reflect sensory processing, whereas the later stages could relate to the formation of a coherent percept. Moreover, it is likely that previously available information is relayed in a feedback manner to sensory cortical areas via beta band oscillations.

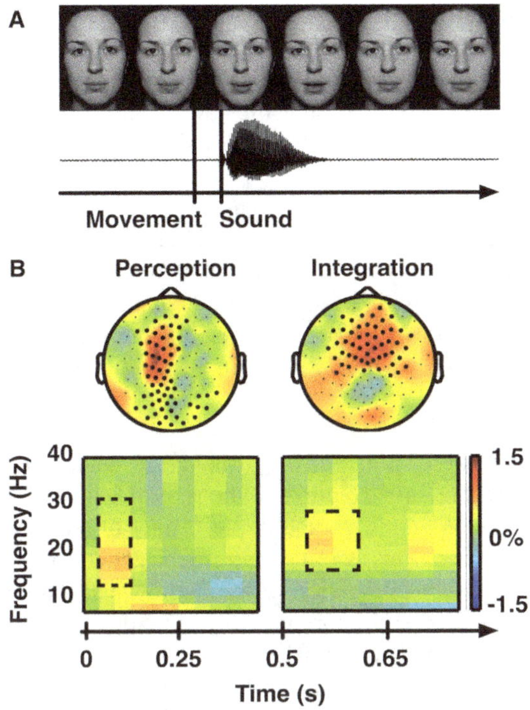

Fig. 10.1 Visual input influences auditory perception in audiovisual speech. (**A**) In the McGurk illusion, a video of an actor pronouncing a syllable is dubbed with an incongruent auditory stream. The natural mouth movement before the auditory onset provides important information about the upcoming auditory stimulus. In the case of an incongruent auditory stimulus, these predictions are violated, and the nonmatching visual and auditory information are, for specific audiovisual syllable combinations, fused to a novel percept. (**B**) Formation of a coherent percept presumably occurs in two separate stages. In a first step, auditory and visual stimuli are perceived, processed, and fed forward. In conjunction with predictions based on the mouth movements, the information obtained in the first stage are integrated into a novel, fused percept. In the topographies (*top*), *solid dots* mark significant electrodes and *shadings* represent percentage of signal change from baseline. In the time-frequency plots (*bottom*), *dashed-line boxes* mark significant effects and *shadings* represent percentage of signal change from baseline. Adapted from Roa Romero et al. (2015)

10.5 Influence of Auditory Input on Visual Perception in Audiovisual Illusions

In Sect. 10.3, it was hypothesized that the auditory system prevails in tasks requiring high temporal resolution. Examples for the influence of auditory information on visual processing are visual illusions induced by concurrently presented auditory stimuli. For example, Shams et al. (2000) have shown that a short visual flash could be perceived as multiple flashes when it is accompanied by multiple short auditory noise bursts. The perception of the sound-induced flash illusion (SIFI) is accompanied by increased ERP amplitudes over occipital electrodes (Shams et al. 2001). Notably, the SIFI only occurs if auditory and visual stimuli are presented above the detection threshold with a stimulus onset asynchrony of up to 115 ms. An interesting phenomenon in audiovisual illusions is alternating perception. In the McGurk illusion, as well as in the SIFI, individuals typically perceive the illusion in some but not all trials, even though the input is always the same. This allows for direct comparisons of the neural responses at varying perceptions but under identical audiovisual stimulation (Keil et al. 2012, 2014a). Analyzing trials in which the SIFI was perceived and trials in which no illusion occurred, Bhattacharya et al. (2002) found that the perception of the illusion is marked by a strong early gamma band power increase as well as a sustained cross-modal interaction in occipital electrodes. Mishra et al. (2007) replicated this finding in a direct comparison between oscillatory activity in illusion and nonillusion trials. Again, the perception of the illusion was marked by an increase in gamma band power in occipital electrodes. Moreover, the authors were able to distinguish an early and a late phase of audiovisual integration. A recent study replicated the role of gamma band power for the perception of the illusion and identified the left superior temporal gyrus (STG) as well as the extrastriatal cortex as putative cortical generators (Balz et al. 2016). In this study, multisensory integration of the audiovisual stimuli was marked by increased gamma band power. Importantly, the individual gamma band power was positively correlated to the SIFI rate, which represents an individual's likelihood to perceive the illusion. By additionally using magnetic resonance spectroscopy to measure neurotransmitter metabolite concentrations, it was observed that the GABA level in the STG modulated the relationship between gamma band power and the SIFI rate (Fig. 10.2). This finding points toward an influence of global cortical states on multisensory perception because the GABA concentration was recorded during rest.

Taken together, the cross-modal influence underlying the influence of auditory information on visual processing is reflected in increased induced gamma band power, which relates to the likelihood to perceive the sound induced flash illusion.

10.6 Anticipatory Activity Influences Cross-Modal Influence

In pioneering EEG research, Davis and Davis (1936) were early to suggest that the pattern and degree of cortical activity might be modified by various physiological and psychological states. In support of this idea, Lindsley (1952) demonstrated that the amplitude of auditory evoked potentials varies systematically with an underlying low-frequency phase. In the last decades, a number of studies have supported the assumption that cortical activity in response to a stimulus is influenced by the phase of ongoing oscillatory activity before the stimulus onset (Busch et al. 2009; Keil et al. 2014b). In addition to the phase, the power of oscillatory activity before stimulus onset also plays a role in perceptual processes (Van Dijk et al. 2008; Romei

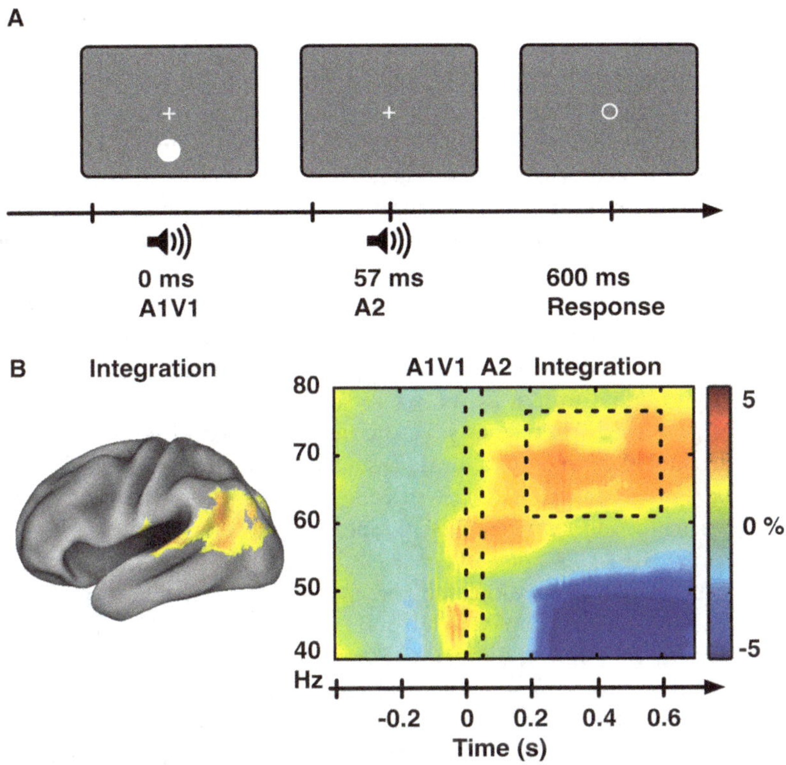

Fig. 10.2 Auditory input influences visual perception in audiovisual illusions. (**A**) In the sound-induced flash illusion, a single visual stimulus (V1) is paired with two consecutive auditory stimuli (A1 and A2). Subjects are asked to report the number of perceived visual stimuli. In approximately half of the trials, subjects reported an illusory perception of two visual stimuli. (**B**) After an initial perception of auditory and visual stimuli in primary sensory areas, incongruent information from both modalities is integrated to an illusory percept as reflected in gamma band power in the left superior temporal gyrus (STG). *Left: the shaded area* on the cortical surface represents an increase relative to baseline for poststimulus gamma band power. *Right: shadings* represent percentage of signal change from baseline; *vertical dashed lines* indicate the onset of A1, V1, and A2; *dashed-line box* indicates the time-frequency window marking increased gamma band power during multisensory integration of the audiovisual stimuli. Adapted from Balz et al. (2016)

et al. 2010). Moreover, network processes, as reflected in functional connectivity, also influence perception (Weisz et al. 2014; Leske et al. 2015). Thus far, the vast majority of studies have investigated the influence of prestimulus activity on unisensory processing.

More recently, a number of studies have started to suggest that oscillatory activity before the stimulus onset also influences the processing and perception of multisensory stimuli (Pomper et al. 2015; Keil et al. 2016). For instance, predictions based on visual information before auditory stimulus onset can modulate audiovisual integration (Arnal et al. 2011). In a similar vein, expectations based on auditory cues modulate ongoing oscillatory activity in the visual and somatosensory cortices (Pomper et al. 2015) as well as functional connectivity networks comprising frontal, parietal, and primary sensory areas (Leonardelli et al. 2015; Keil et al. 2016). Ongoing fluctuations of local cortical oscillations and functional connectivity networks have been found to also influence multisensory perception when there are no specific predictions and expectations (Lange et al. 2011; Keil et al. 2012). For example, one study compared oscillatory neural activity before stimulus onset between trials in which incongruent audiovisual speech stimuli were perceived as the McGurk illusion with trials in which either the auditory or the visual input dominated the percept (Keil et al. 2012). A main finding of this study was that prestimulus beta band power and functional connectivity influenced upcoming perception (Fig. 10.3A). More specifically, beta band power was increased in the left STG, precuneus, and middle frontal gyrus before stimulus onset, in trials in which the illusion was perceived. Interestingly, before the perception of the illusion, the left STG was decoupled from cortical areas associated with face (i.e., fusiform gyrus) or voice (i.e., Brodmann area 22) processing. Similar results were obtained in a study comparing incongruent audiovisual trials in which the SIFI was perceived and trials where the SIFI was not perceived (Keil et al. 2014a). The study revealed increased beta band power before the perception of the illusion (Fig. 10.3B). In addition, the left STG was coupled to left auditory cortical areas but decoupled from visual cortical areas before the illusion. Furthermore, the stronger the functional connectivity between the left STG and the left auditory cortex, the higher the likelihood of an illusion. These data provide strong evidence for a role of the left STG in audiovisual integration. In case of degraded bottom-up input, the formation of a fused percept is supported by strong beta band power (see also Schepers et al. 2013). In case of imbalanced reliability of the bottom-up input of various modalities, information from one modality can dominate the subjective percept.

Two recent studies using the SIFI further highlighted the role of low-frequency oscillations for audiovisual perception (Cecere et al. 2015; Keil and Senkowski 2017). Cecere et al. (2015) found a negative correlation between the participants' individual alpha band frequency (IAF) and their illusion rate, which indicates that alpha band oscillations provide a temporal window in which the cross-modal influence could induce an illusion. Underscoring the role of low-frequency oscillations for cross-modal influence, the authors also found that increasing the IAF using transcranial direct current stimulation reduces the probability of an illusion perception, where as a reduction of the IAF had the opposite effect. Recently, Keil and Senkowski (2017) corroborated the relationship between the IAF and the SIFI perception rate and localized this effect to the occipital cortex.

Taken together, local cortical activity and the information transfer between cortical network nodes critically influence the processing and perception of multisensory stimuli. Furthermore, functional connectivity networks seem to mediate how information is relayed between unisensory, multisensory, and higher order cortical areas. Hence, there is strong evidence that ongoing oscillatory activity influences unisensory and multisensory perception.

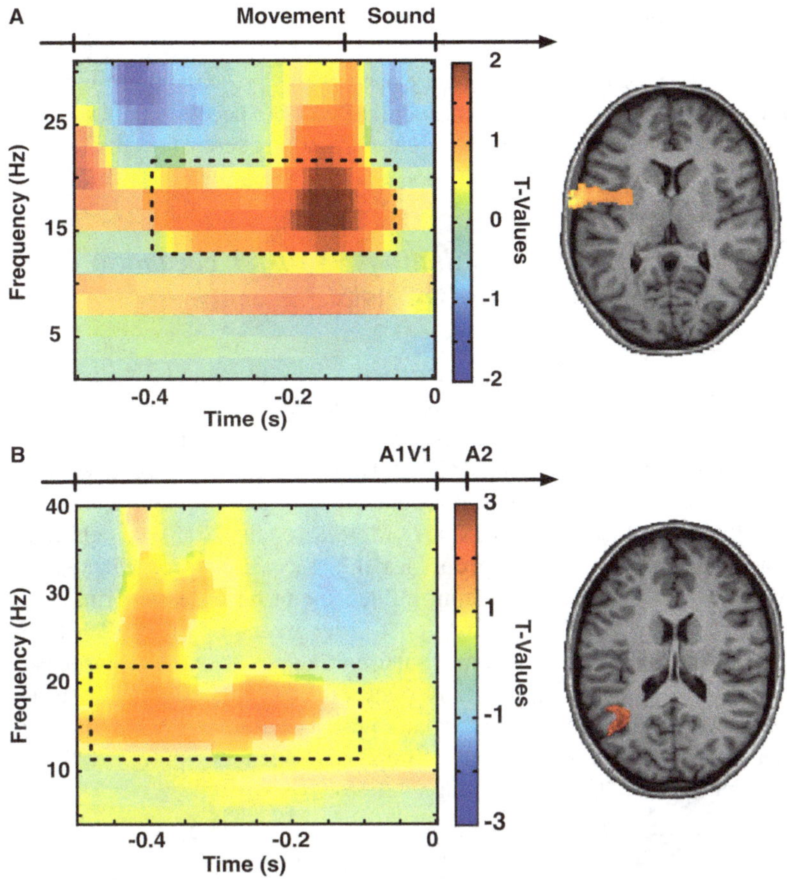

Fig. 10.3 Anticipatory activity influences cross-modal influence. Cortical activity before the onset of audiovisual stimulation influences perception of the McGurk illusion (**A**) and perception of the sound-induced flash illusion (**B**). The cross-modal influence at the behavioral level is opposite between the two illusions. However, empirical data show that similar cortical processes (i.e., increased beta band power in the left STG) influence upcoming perception in both illusions. *Left: shadings* represent results (*T* values) of the statistical comparison via *t*-tests between trials with and without the illusion; *dashed-line boxes* indicate the time-frequency window marking increased beta band power prior to multisensory integration of the audiovisual stimuli. *Right: shaded areas* in the brains represent results (*T* values) of the statistical comparison via *t*-tests between trials with and without the illusion. Adapted from Keil et al. (2012, 2014a)

10.7 Summary and Open Questions

This chapter reviewed empirical findings on the neural mechanisms underlying multisensory processing, with a focus on oscillatory activity. Based on the available findings, it can be postulated that multisensory processing and perception rely on a complex and dynamic cross-frequency interaction pattern within widespread neural networks (Keil and Senkowski 2018). Currently available evidence suggests a hierarchical interplay between low-frequency phase and high-frequency power during multisensory processing; low-frequency oscillations presumably provide temporal windows of integration for the cross-modal influence. Successful multisensory integration is subsequently reflected in increased high-frequency power.

10.7.1 Low-Frequency Oscillations Transfer Feedback Information and Cross-Modal Influence

An increasing number of studies suggest that low-frequency oscillations (delta, theta, and alpha bands), might serve as a mechanism to control local cortical activity. The phase of these oscillations has been shown to modulate stimulus evoked activity and perception (Busch et al. 2009; Keil et al. 2014b). Moreover, as demonstrated in a number of studies, low-frequency oscillations seem to reflect cross-modal influences (Lakatos et al. 2007; Mercier et al. 2015). In addition, prior information based on stimulus properties influences local cortical activity (Roa Romero et al. 2016). The modulating influence can be found in primary sensory areas (e.g., visual cortex) as well as higher order areas (e.g., frontal cortex). Thus, information that is transferred from frontal cortical areas to multisensory and unisensory cortical areas can represent abstract top-down processes, such as attention (Keil et al. 2016). Additionally, information that is transferred between these cortical areas can also represent stimulus properties, such as timing, rhythmicity, or space (Lakatos et al. 2007; Mercier et al. 2015).

10.7.2 High-Frequency Oscillations Reflect Perception and Integration

Oscillatory activity above 12 Hz (i.e., in the frequency of the beta band and gamma band) has been implied to reflect perception and stimulus integration (Senkowski et al. 2008). Furthermore, it has been shown that multisensory integration is reflected in increased gamma band power in traditional multisensory cortical areas, such as the STG (Balz et al. 2016). The analyses of beta band and gamma band power modulations during multisensory stimulus processing revealed different stages of multisensory integration (Roa Romero et al. 2015; Bizley et al. 2016). In audiovisual speech perception, stimuli are processed and different input streams are compared for congruence at a putative early stage. In a later stage, the different input streams are combined and, in case of incongruence, resolved to a subjectively congruent percept (Peelle and Sommers 2015).

10.7.3 Functional Connectivity Guides Integration

Whereas perception and multisensory integration are reflected in high-frequency power, both processes are modulated by a low-frequency oscillatory phase. Therefore, modulatory information has to be transferred within functional connectivity networks encompassing primary sensory areas, traditional multisensory areas, and higher order frontal areas (Senkowski et al. 2008; Keil and Senkowski 2018). A number of studies have shown that feedback information is conveyed in alpha band and beta band functional connectivity. Furthermore, cue-induced attention or expectations also modulate low-frequency functional connectivity (Keil et al. 2016).

10.7.4 Open Questions

In the last decade, research on the neural mechanisms underlying the integration and perception of multisensory information as well as on the role of oscillatory processes therein has made tremendous progress. It has been found that the effects in neural oscillations go along with the principles of multisensory integration, but several open questions remain to be answered. For instance, the temporal evolution of multisensory perception and integration is still not well understood. A number of studies have shown that multisensory perception, as reflected in oscillatory activity, requires multiple processing stages. However, it is so far unknown which cortical nodes are active at a given latency. Future studies could integrate recent progress in technical methods and analytical approaches to analyze time-frequency-resolved oscillatory activity on the level of cortical sources. This will help to elucidate the progression of multisensory stimulus processing. Another open question pertains to the role of attention, predictions, and expectations for multisensory perception and the underlying neural oscillatory patterns. Recent studies have highlighted the role of prior expectations and attention for prestimulus oscillations in multisensory paradigms. Yet it remains to be elucidated how cognitive processes influence multisensory processing, how they influence network architecture, and in which oscillatory signatures they are reflected. Future studies should exploit the full spectrum of information available from electrophysiological data to capture the complex network processes underlying the integration of multisensory information as well as how cognitive processes modulate neural oscillations in these networks. The functional significance of the separate network nodes for multisensory perception also has not been fully clarified. Electrophysiological as well as functional imaging studies have identified a number of cortical regions involved in multisensory perception, but these studies have mostly used correlation approaches. Future studies should therefore turn to more causal approaches in which stimulation can be used to directly test the functional role of cortical areas. For instance, transcranial magnetic stimulation could be used to apply a so-called virtual lesion to selectively interrupt activity within a cortical area to study how cortical activity, multisensory integration, and perception are influenced. In addition, entrainment of cortical networks via transcranial direct/alternating current stimulation could be used to obtain information on the role of specific oscillatory frequencies for multisensory integration and perception. In conclusion, the studies reviewed above suggest that multisensory perception relies on dynamic neural networks in which information in transferred through oscillatory activity. An important endeavor will be to more precisely study the functional roles of the different frequency bands and their interplay for multisensory integrative processing.

Compliance with Ethics Requirements Julian Keil declares that he has no conflict of interest.

Daniel Senkowski declares that he has no conflict of interest.

References

Adrian, E. D. (1942). Olfactory reactions in the brain of the hedgehog. *The Journal of Physiology, 100*(4), 459–473.

Arnal, L. H., & Giraud, A.-L. (2012). Cortical oscillations and sensory predictions. *Trends in Cognitive Sciences, 16*(7), 390–398. https://doi.org/10.1016/j.tics.2012.05.003.

Arnal, L. H., Morillon, B., Kell, C. A., & Giraud, A.-L. (2009). Dual neural routing of visual facilitation in speech processing. *The Journal of Neuroscience, 29*(43), 13445–13453. https://doi.org/10.1523/JNEUROSCI.3194-09.2009.

Arnal, L. H., Wyart, V., & Giraud, A.-L. (2011). Transitions in neural oscillations reflect prediction errors generated in audiovisual speech. *Nature Neuroscience, 14*(6), 797–801. https://doi.org/10.1038/nn.2810.

Balz, J., Keil, J., Roa Romero, Y., Mekle, R., Schubert, F., Aydin, S., Ittermann, B., Gallinat, J., & Senkowski, D. (2016). GABA concentration in superior temporal sulcus predicts gamma power and perception in the sound-induced flash illusion. *NeuroImage, 125*, 724–730. https://doi.org/10.1016/j.neuroimage.2015.10.087.

Berger, H. (1929). Über das elektroenkephalogramm des menschen. *Archiv für Psychiatrie und Nervenkrankheiten, 87*, 527–570.

Bhattacharya, J., Shams, L., & Shimojo, S. (2002). Sound-induced illusory flash perception: Role of gamma band responses. *NeuroReport, 13*(14), 1727–1730.

Biau, E., Torralba, M., Fuentemilla, L., de Diego Balaguer, R., & Soto-Faraco, S. (2015). Speaker's hand gestures modulate speech perception through phase resetting of ongoing neural oscillations. *Cortex, 68*, 76–85. https://doi.org/10.1016/j.cortex.2014.11.018.

Birbaumer, N., Elbert, T., Canavan, A. G., & Rockstroh, B. (1990). Slow potentials of the cerebral cortex and behavior. *Physiological Reviews, 70*(1), 1–41.

Bizley, J. K., Maddox, R. K., & Lee, A. K. C. (2016). Defining auditory-visual objects: Behavioral tests and physiological mechanisms. *Trends in Neurosciences, 39*(2), 74–85. https://doi.org/10.1016/j.tins.2015.12.007.

Bremer, F. (1958). Cerebral and cerebellar potentials. *Physiological Reviews, 38*(3), 357–388.

Bressler, S. L. (1990). The gamma wave: A cortical information carrier? *Trends in Neurosciences, 13*(5), 161–162.

Busch, N. A., Dubois, J., & Vanrullen, R. (2009). The phase of ongoing EEG oscillations predicts visual perception. *The Journal of Neuroscience, 29*(24), 7869–7876. https://doi.org/10.1523/JNEUROSCI.0113-09.2009.

Buzsáki, G., & Draguhn, A. (2004). Neuronal oscillations in cortical networks. *Science, 304*(5679), 1926–1929. https://doi.org/10.1126/science.1099745.

Cavanagh, J. F., & Frank, M. J. (2014). Frontal theta as a mechanism for cognitive control. *Trends in Cognitive Sciences, 18*(8), 414–421. https://doi.org/10.1016/j.tics.2014.04.012.

Cecere, R., Rees, G., & Romei, V. (2015). Individual differences in alpha frequency drive crossmodal illusory perception. *Current Biology, 25*(2), 231–235. https://doi.org/10.1016/j.cub.2014.11.034.

Chatrian, G. E., Bickford, R. G., & Uihlein, A. (1960). Depth electrographic study of a fast rhythm evoked from the human calcarine region by steady illumination. *Electroencephalography and Clinical Neurophysiology, 12*, 167–176.

Cheron, G., Cebolla, A., De Saedeleer, C., Bengoetxea, A., Leurs, F., Leroy, A., & Dan, B. (2007). Pure phase-locking of beta/gamma oscillation contributes to the N30 frontal component of somatosensory evoked potentials. *BMC Neuroscience, 8*(1), 75. https://doi.org/10.1186/1471-2202-8-75.

Davis, H., & Davis, P. A. (1936). Action potentials of the brain: In normal persons and in normal states of cerebral activity. *Archives of Neurology and Psychiatry, 36*(6), 1214–1224.

Diekelmann, S., & Born, J. (2010). The memory function of sleep. *Nature Reviews Neuroscience, 11*(2), 114–126. https://doi.org/10.1038/nrn2762.

Driver, J., & Noesselt, T. (2008). Multisensory interplay reveals crossmodal influences on "sensory-specific" brain regions, neural responses, and judgments. *Neuron, 57*(1), 11–23. https://doi.org/10.1016/j.neuron.2007.12.013.

Engel, A. K., & Fries, P. (2010). Beta-band oscillations—signalling the status quo? *Current Opinion in Neurobiology, 20*(2), 156–165. https://doi.org/10.1016/j.conb.2010.02.015.

Engel, A. K., Fries, P., & Singer, W. (2001). Dynamic predictions: Oscillations and synchrony in top-down processing. *Nature Reviews Neuroscience, 2*(10), 704–716. https://doi.org/10.1038/35094565.

Ernst, M. O., & Bülthoff, H. H. (2004). Merging the senses into a robust percept. *Trends in Cognitive Sciences, 8*(4), 162–169. https://doi.org/10.1016/j.tics.2004.02.002.

Fries, P. (2005). A mechanism for cognitive dynamics: Neuronal communication through neuronal coherence. *Trends in Cognitive Sciences, 9*(10), 474–480. https://doi.org/10.1016/j.tics.2005.08.011.

Fries, P. (2015). Rhythms for cognition: Communication through coherence. *Neuron, 88*(1), 220–235. https://doi.org/10.1016/j.neuron.2015.09.034.

Ghazanfar, A., & Schroeder, C. (2006). Is neocortex essentially multisensory? *Trends in Cognitive Sciences, 10*(6), 278–285. https://doi.org/10.1016/j.tics.2006.04.008.

He, Y., Gebhardt, H., Steines, M., Sammer, G., Kircher, T., Nagels, A., & Straube, B. (2015). The EEG and fMRI signatures of neural integration: An investigation of meaningful gestures and corresponding speech. *Neuropsychologia, 72*, 27–42. https://doi.org/10.1016/j.neuropsychologia.2015.04.018.

Herrmann, C. S., Mecklinger, A., & Pfeifer, E. (1999). Gamma responses and ERPs in a visual classification task. *Clinical Neurophysiology, 110*(4), 636–642.

Jensen, O., & Mazaheri, A. (2010). Shaping functional architecture by oscillatory alpha activity: Gating by inhibition. *Frontiers in Human Neuroscience, 4*, 186. https://doi.org/10.3389/fnhum.2010.00186.

Kaiser, J., Hertrich, I., Ackermann, H., Mathiak, K., & Lutzenberger, W. (2005). Hearing lips: Gamma-band activity during audiovisual speech perception. *Cerebral Cortex, 15*(5), 646–653. https://doi.org/10.1093/cercor/bhh166.

Kayser, C., Petkov, C. I., & Logothetis, N. K. (2008). Visual modulation of neurons in auditory cortex. *Cerebral Cortex, 18*(7), 1560–1574. https://doi.org/10.1093/cercor/bhm187.

Keil, J., & Senkowski, D. (2017). Individual alpha frequency relates to the sound-induced flash illusion. *Multisensory Research, 30*(6), 565–578. https://doi.org/10.1163/22134808-00002572.

Keil, J. & Senkowski, D. (2018). Neural oscillations orchestrate multisensory processing. *Neuroscientist, 24*(6), 609–626. https://journals.sagepub.com/doi/10.1177/10738584187553522.

Keil, J., Müller, N., Ihssen, N., & Weisz, N. (2012). On the variability of the McGurk effect: Audiovisual integration depends on prestimulus brain states. *Cerebral Cortex, 22*(1), 221–231. https://doi.org/10.1093/cercor/bhr125.

Keil, J., Müller, N., Hartmann, T., & Weisz, N. (2014a). Prestimulus beta power and phase synchrony influence the sound-induced flash illusion. *Cerebral Cortex, 24*(5), 1278–1288. https://doi.org/10.1093/cercor/bhs409.

Keil, J., Timm, J., SanMiguel, I., Schulz, H., Obleser, J., & Schönwiesner, M. (2014b). Cortical brain states and corticospinal synchronization influence TMS-evoked motor potentials. *Journal of Neurophysiology, 111*(3), 513–519. https://doi.org/10.1152/jn.00387.2013.

Keil, J., Pomper, U., & Senkowski, D. (2016). Distinct patterns of local oscillatory activity and functional connectivity underlie intersensory attention and temporal prediction. *Cortex, 74*, 277–288. https://doi.org/10.1016/j.cortex.2015.10.023.

Klimesch, W. (1999). EEG alpha and theta oscillations reflect cognitive and memory performance: A review and analysis. *Brain Research Reviews, 29*, 169–195.

Klimesch, W. (2012). Alpha-band oscillations, attention, and controlled access to stored information. *Trends in Cognitive Sciences, 16*(12), 606–617. https://doi.org/10.1016/j.tics.2012.10.007.

Lakatos, P., Chen, C.-M., O'Connell, M. N., Mills, A., & Schroeder, C. E. (2007). Neuronal oscillations and multisensory interaction in primary auditory cortex. *Neuron, 53*(2), 279–292. https://doi.org/10.1016/j.neuron.2006.12.011.

Lange, J., Oostenveld, R., & Fries, P. (2011). Perception of the touch-induced visual double-flash illusion correlates with changes of rhythmic neuronal activity in human visual and somatosensory areas. *NeuroImage, 54*(2), 1395–1405. https://doi.org/10.1016/j.neuroimage.2010.09.031.

Lange, J., Christian, N., & Schnitzler, A. (2013). Audio-visual congruency alters power and coherence of oscillatory activity within and between cortical areas. *NeuroImage, 79*, 111–120. https://doi.org/10.1016/j.neuroimage.2013.04.064.

Lange, J., Keil, J., Schnitzler, A., Van Dijk, H., & Weisz, N. (2014). The role of alpha oscillations for illusory perception. *Behavioural Brain Research, 271*, 294–301. https://doi.org/10.1016/j.bbr.2014.06.015.

Leonardelli, E., Braun, C., Weisz, N., Lithari, C., Occelli, V., & Zampini, M. (2015). Prestimulus oscillatory alpha power and connectivity patterns predispose perceptual integration of an audio and a tactile stimulus. *Human Brain Mapping, 36*(9), 3486–3498. https://doi.org/10.1002/hbm.22857.

Leske, S., Ruhnau, P., Frey, J., Lithari, C., Müller, N., Hartmann, T., & Weisz, N. (2015). Prestimulus network integration of auditory cortex predisposes near-threshold perception independently of local excitability. *Cerebral Cortex, 25*, 4898–4907. https://doi.org/10.1093/cercor/bhv212.

Lindsley, D. B. (1952). Psychological phenomena and the electroencephalogram. *Electroencephalography and Clinical Neurophysiology, 4*(4), 443–456.

Lopes da Silva, F. (1991). Neural mechanisms underlying brain waves: From neural membranes to networks. *Electroencephalography and Clinical Neurophysiology, 79*(2), 81–93.

Luck, S. J. (2014). *An introduction to the event-related potential technique*. Cambridge, MA: The MIT Press.

McGurk, H., & MacDonald, J. (1976). Hearing lips and seeing voices. *Nature, 264*(5588), 746–748.

Mercier, M. R., Foxe, J. J., Fiebelkorn, I. C., Butler, J. S., Schwartz, T. H., & Molholm, S. (2013). Auditory-driven phase reset in visual cortex: Human electrocorticography reveals mechanisms of early multisensory integration. *NeuroImage, 79*, 19–29. https://doi.org/10.1016/j.neuroimage.2013.04.060.

Mercier, M. R., Molholm, S., Fiebelkorn, I. C., Butler, J. S., Schwartz, T. H., & Foxe, J. J. (2015). Neuro-oscillatory phase alignment drives speeded multisensory response times: An electrocorticographic investigation. *The Journal of Neuroscience, 35*(22), 8546–8557. https://doi.org/10.1523/JNEUROSCI.4527-14.2015.

Michalareas, G., Vezoli, J., van Pelt, S., Schoffelen, J.-M., Kennedy, H., & Fries, P. (2016). Alpha-beta and gamma rhythms subserve feedback and feedforward influences among human visual cortical areas. *Neuron, 89*(2), 384–397. https://doi.org/10.1016/j.neuron.2015.12.018.

Mishra, J., Martinez, A., Sejnowski, T. J., & Hillyard, S. A. (2007). Early cross-modal interactions in auditory and visual cortex underlie a sound-induced visual illusion. *The Journal of Neuroscience, 27*(15), 4120–4131. https://doi.org/10.1523/JNEUROSCI.4912-06.2007.

Mitra, P. P., & Pesaran, B. (1999). Analysis of dynamic brain imaging data. *Biophysical Journal, 76*(2), 691–708. https://doi.org/10.1016/S0006-3495(99)77236-X.

Pantev, C., Makeig, S., Hoke, M., Galambos, R., Hampson, S., & Gallen, C. (1991). Human auditory evoked gamma-band magnetic fields. *Proceedings of the National Academy of Sciences of the United States of America, 88*(20), 8996–9000.

Peelle, J. E., & Sommers, M. S. (2015). Prediction and constraint in audiovisual speech perception. *Cortex, 68*, 169–181. https://doi.org/10.1016/j.cortex.2015.03.006.

Penttonen, M., & Buzsáki, G. (2003). Natural logarithmic relationship between brain oscillators. *Thalamus & Related Systems, 2*(2), 145–152. https://doi.org/10.1016/S1472-9288(03)00007-4.

Pfurtscheller, G. (1992). Event-related synchronization (ERS): An electrophysiological correlate of cortical areas at rest. *Electroencephalography and Clinical Neurophysiology, 83*(1), 62–69.

Pomper, U., Brincker, J., Harwood, J., Prikhodko, I., & Senkowski, D. (2014). Taking a call is facilitated by the multisensory processing of smartphone vibrations, sounds, and flashes. *PLoS One, 9*(8), e103238. https://doi.org/10.1371/journal.pone.0103238.s002.

Pomper, U., Keil, J., Foxe, J. J., & Senkowski, D. (2015). Intersensory selective attention and temporal orienting operate in parallel and are instantiated in spatially distinct sensory and motor cortices. *Human Brain Mapping, 36*(8), 3246–3259. https://doi.org/10.1002/hbm.22845.

Rao, R. P., & Ballard, D. H. (1999). Predictive coding in the visual cortex: A functional interpretation of some extra-classical receptive-field effects. *Nature Neuroscience, 2*(1), 79–87. https://doi.org/10.1038/4580.

Ray, W. J., & Cole, H. W. (1985). EEG alpha activity reflects attentional demands, and beta activity reflects emotional and cognitive processes. *Science, 228*(4700), 750–752.

Roa Romero, Y., Senkowski, D., & Keil, J. (2015). Early and late beta-band power reflect audiovisual perception in the McGurk illusion. *Journal of Neurophysiology, 113*(7), 2342–2350. https://doi.org/10.1152/jn.00783.2014.

Roa Romero, Y., Keil, J., Balz, J., Gallinat, J., & Senkowski, D. (2016). Reduced frontal theta oscillations indicate altered crossmodal prediction error processing in schizophrenia. *Journal of Neurophysiology, 116*(3), 1396–1407. https://doi.org/10.1152/jn.00096.2016.

Rockstroh, B., Müller, M., Cohen, R., & Elbert, T. (1992). Probing the functional brain state during P300-evocation. *Journal of Psychophysiology, 6*, 175–184.

Romei, V., Gross, J., & Thut, G. (2010). On the role of prestimulus alpha rhythms over occipito-parietal areas in visual input regulation: Correlation or causation? *The Journal of Neuroscience, 30*(25), 8692–8697. https://doi.org/10.1523/JNEUROSCI.0160-10.2010.

Sakowitz, O. W., Quiroga, R. Q., Schürmann, M., & Başar, E. (2001). Bisensory stimulation increases gamma-responses over multiple cortical regions. *Cognitive Brain Research, 11*(2), 267–279. https://doi.org/10.1016/S0926-6410(00)00081-1.

Sams, M., Aulanko, R., Hämäläinen, M., Hari, R., Lounasmaa, O. V., Lu, S. T., & Simola, J. (1991). Seeing speech: Visual information from lip movements modifies activity in the human auditory cortex. *Neuroscience Letters, 127*(1), 141–145.

Sauseng, P., Klimesch, W., Schabus, M., & Doppelmayr, M. (2005). Fronto-parietal EEG coherence in theta and upper alpha reflect central executive functions of working memory. *International Journal of Psychophysiology, 57*(2), 97–103. https://doi.org/10.1016/j.ijpsycho.2005.03.018.

Schepers, I. M., Schneider, T. R., Hipp, J. F., Engel, A. K., & Senkowski, D. (2013). Noise alters beta-band activity in superior temporal cortex during audiovisual speech processing. *NeuroImage, 70*, 101–112. https://doi.org/10.1016/j.neuroimage.2012.11.066.

Schroeder, C. E., & Foxe, J. (2005). Multisensory contributions to low-level, "unisensory" processing. *Current Opinion in Neurobiology, 15*(4), 454–458. https://doi.org/10.1016/j.conb.2005.06.008.

Schroeder, C. E., Lakatos, P., Kajikawa, Y., Partan, S., & Puce, A. (2008). Neuronal oscillations and visual amplification of speech. *Trends in Cognitive Sciences, 12*(3), 106–113. https://doi.org/10.1016/j.tics.2008.01.002.

Senkowski, D., Talsma, D., Herrmann, C. S., & Woldorff, M. G. (2005). Multisensory processing and oscillatory gamma responses: Effects of spatial selective attention. *Experimental Brain Research, 166*(3-4), 411–426. https://doi.org/10.1007/s00221-005-2381-z.

Senkowski, D., Talsma, D., Grigutsch, M., Herrmann, C. S., & Woldorff, M. G. (2007). Good times for multisensory integration: Effects of the precision of temporal synchrony as revealed by gamma-band oscillations. *Neuropsychologia, 45*(3), 561–571. https://doi.org/10.1016/j.neuropsychologia.2006.01.013.

Senkowski, D., Schneider, T., Foxe, J., & Engel, A. (2008). Crossmodal binding through neural coherence: Implications for multisensory processing. *Trends in Neurosciences, 31*(8), 401–409. https://doi.org/10.1016/j.tins.2008.05.002.

Senkowski, D., Saint-Amour, D., Höfle, M., & Foxe, J. J. (2011). Multisensory interactions in early evoked brain activity follow the principle of inverse effectiveness. *NeuroImage, 56*(4), 2200–2208. https://doi.org/10.1016/j.neuroimage.2011.03.075.

Shams, L., Kamitani, Y., & Shimojo, S. (2000). Illusions. What you see is what you hear. *Nature, 408*(6814), 788. https://doi.org/10.1038/35048669.

Shams, L., Kamitani, Y., Thompson, S., & Shimojo, S. (2001). Sound alters visual evoked potentials in humans. *NeuroReport, 12*(17), 3849–3852.

Stein, B. E., & Meredith, M. A. (1993). *The merging of the senses*. Cambridge, MA: The MIT Press.

Stein, B. E., Stanford, T. R., & Rowland, B. A. (2014). Development of multisensory integration from the perspective of the individual neuron. *Nature Reviews Neuroscience, 15*(8), 520–535. https://doi.org/10.1038/nrn3742.

Steriade, M. (2001). Impact of network activities on neuronal properties in corticothalamic systems. *Journal of Neurophysiology, 86*(1), 1–39.

Sumby, W. H., & Polack, I. (1954). Perceptual amplification of speech sounds by visual cues. *The Journal of the Acoustical Society of America, 26*, 212–215.

Tallon-Baudry, C., & Bertrand, O. (1999). Oscillatory gamma activity in humans and its role in object representation. *Trends in Cognitive Sciences, 3*(4), 151–162.

van Atteveldt, N., Murray, M. M., Thut, G., & Schroeder, C. E. (2014). Multisensory integration: Flexible use of general operations. *Neuron, 81*(6), 1240–1253. https://doi.org/10.1016/j.neuron.2014.02.044.

Van Dijk, H., Schoffelen, J.-M., Oostenveld, R., & Jensen, O. (2008). Prestimulus oscillatory activity in the alpha band predicts visual discrimination ability. *The Journal of Neuroscience, 28*(8), 1816–1823. https://doi.org/10.1523/JNEUROSCI.1853-07.2008.

van Wassenhove, V., & Grzeczkowski, L. (2015). Visual-induced expectations modulate auditory cortical responses. *Frontiers in Neuroscience, 9*, 11. https://doi.org/10.3389/fnins.2015.00011.

Walter, W. G. (1936). The location of cerebral tumours by electro-encephalography. *The Lancet, 228*(5893), 305–308.

Walter, D. O., Rhodes, J. M., Brown, D., & Adey, W. R. (1966). Comprehensive spectral analysis of human EEG generators in posterior cerebral regions. *Electroencephalography and Clinical Neurophysiology, 20*(3), 224–237.

Wang, X.-J. (2010). Neurophysiological and computational principles of cortical rhythms in cognition. *Physiological Reviews, 90*, 1195–1268. https://doi.org/10.1152/physrev.00035.2008.

Weisz, N., Wühle, A., Monittola, G., Demarchi, G., Frey, J., Popov, T., & Braun, C. (2014). Prestimulus oscillatory power and connectivity patterns predispose conscious somatosensory perception. *Proceedings of the National Academy of Sciences of the United States of America, 111*(4), E417–E425. https://doi.org/10.1073/pnas.1317267111.

Welch, R. B., & Warren, D. H. (1980). Immediate perceptual responses to intersensory discrepancy. *Psychological Bulletin, 88*(3), 638–667.

Chapter 11
Cross-Modal Learning in the Auditory System

Patrick Bruns and Brigitte Röder

Abstract Unisensory auditory representations are strongly shaped by multisensory experience, and, likewise, audition contributes to cross-modal learning in other sensory systems. This applies to lower-level sensory features like spatial and temporal processing as well as to higher-level features like speech identification. Cross-modal learning has particularly profound influences during development, but its effects on unisensory processing are ubiquitous throughout life. Moreover, influences of cross-modal learning on unisensory processing have been observed at various timescales, ranging from long-term structural changes over months to short-term plasticity of auditory representations after minutes or only seconds of cross-modal exposure. This chapter focuses particularly on cross-modal learning and its underlying neural mechanisms in the healthy adult auditory system. Recent findings suggest that cross-modal learning operates in parallel on different neural representations and at different timescales. With an increasing amount of exposure to new cross-modal associations, cross-modal learning seems to progress from higher level multisensory representations to lower level modality-specific representations, possibly even in primary auditory cortex. In addition to cortically mediated learning mechanisms, auditory representations are shaped via subcortical multisensory pathways including the superior colliculi in the midbrain. The emerging view from these findings is that auditory-guided behavior is jointly shaped by cross-modal learning in distinct neural systems. To fully understand the dynamic nature of the auditory system, it will be important to identify how short-term and long-term learning processes interact in the mature brain.

Keywords Attention · Audiovisual · Multisensory · Plasticity · Recalibration · Sensory representations · Space · Time · Ventriloquism aftereffect · Visual system

P. Bruns (✉) · B. Röder
Biological Psychology and Neuropsychology, University of Hamburg, Hamburg, Germany
e-mail: patrick.bruns@uni-hamburg.de; brigitte.roeder@uni-hamburg.de

A. K. C. Lee et al. (eds.), *Multisensory Processes*, Springer Handbook of Auditory Research 68, https://doi.org/10.1007/978-3-030-10461-0_11

11.1 Introduction

Plasticity is an inherent capacity of all sensory systems including audition and allows for a continuous adaptation of sensory representations to both environmental and bodily changes. Although neuroplasticity is most pronounced during development, sensory representations are constantly shaped by experience throughout life. Experience-dependent plasticity has been observed at various timescales, from long-term structural changes in the brain, associated, for example, with permanent damage to the sensory periphery, to short-term changes in neural-response properties and synaptic transmission, associated, for example, with transient changes in the sensory environment (Dahmen and King 2007; Tzounopoulos and Leão 2012). For instance, neurons in the inferior colliculus of the midbrain adjust their firing rate to changes in the sound level distribution in the environment and thereby improve the accuracy of neural coding within the range of the most likely occurring sound levels (Dean et al. 2005). Inferior colliculus neurons similarly adapt to the distribution of interaural level differences, which are one of the main cues for the processing of auditory space (Dahmen et al. 2010). Importantly, Dahmen et al. (2010) were able to directly relate these adaptations of neural-response properties to changes in sound localization behavior. Findings like these demonstrate that auditory processing is highly dynamic and adaptive even at subcortical levels of the auditory-processing pathway and within very short timescales.

The sensory experience in the world is usually not unisensory but typically involves input from multiple sensory modalities. Numerous studies have shown that the adult brain integrates cross-modal information in an optimal way, resulting in higher sensory accuracy and precision (Ernst and Bülthoff 2004; see also Alais and Burr, Chap. 2). Thus, it seems unsurprising that training protocols involving multisensory stimulation have often been found to be more effective for learning than unisensory training protocols (Shams and Seitz 2008). For example, the ability to localize sound sources in a monaural condition with one ear temporarily occluded was greatly improved after training with audiovisual stimuli compared with an auditory-only training condition (Strelnikov et al. 2011). This chapter starts by reviewing experimental results that demonstrate plasticity in unisensory auditory representations through cross-modal experience as well as instances in which hearing contributes to cross-modal learning in other sensory systems, with a focus on training-induced plasticity in the healthy adult auditory system (for findings in clinical populations, see Baum Miller, and Wallace, Chap. 12). In particular, this chapter further explores how cross-modal learning phenomena that operate at different timescales and on different neural representations can jointly shape auditory-guided behavior.

11.2 Influences of Cross-Modal Learning on Auditory Processing

11.2.1 Space

Although sound sources can obviously be localized in the absence of visual information (e.g., in darkness), localization accuracy improves when the potential sound sources are visible as well (Stein et al. 1988) and unisensory sound localization performance benefits from prior training with audiovisual stimuli (Passamonti et al. 2009; Strelnikov et al. 2011). Such improvements most likely reflect the fact that visual spatial information is usually more accurate and reliable than auditory spatial information. In contrast to the visual system, where space is directly represented in the sensory organ, the brain has to infer the direction of a sound source from the acoustical cues generated by the interaction of the sound waves with the head and the external ears (Recanzone and Sutter 2008). Thus, combining auditory and visual spatial information enhances the reliability of the spatial estimate (Alais and Burr 2004), thereby providing a reference for calibrating auditory spatial representations (King 2009).

Conflict situations are a useful tool to study the principles underlying cross-modal recalibration of auditory space. When auditory and visual stimuli are presented simultaneously at discrepant spatial locations, the sound is typically mislocalized toward the location of the visual input, a phenomenon referred to as the ventriloquism effect (Bertelson and Aschersleben 1998; Alais and Burr 2004). Remarkably, a number of studies have demonstrated that a brief passive exposure to audiovisual stimuli with a constant spatial disparity induces a shift in the perceived sound location in the direction of the visual stimuli that persists even when the sound is subsequently presented without the visual stimuli (see Fig. 11.1). This effect is known as the ventriloquism aftereffect and can be induced reliably both in humans (Radeau and Bertelson 1974; Recanzone 1998) and in nonhuman primates

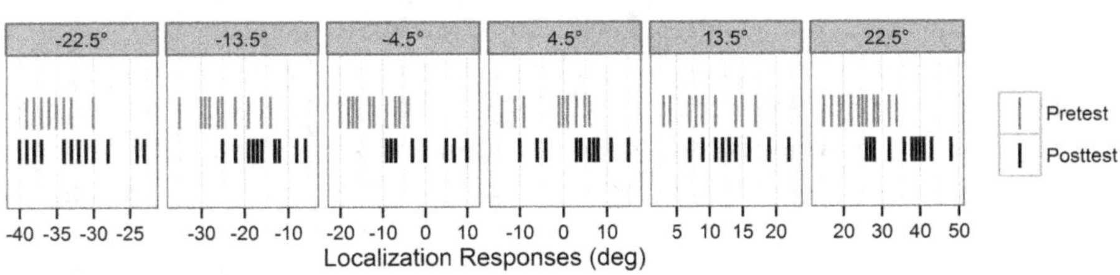

Fig. 11.1 Typical data from a participant in a ventriloquism aftereffect experiment. Single-trial localization responses are shown for six target locations (–22.5°, –13.5°, –4.5°, 4.5°, 13.5°, and 22.5°). The participant had to localize a 1,000-Hz tone before (*gray lines*) and after (*black lines*) 5 minutes of exposure to spatially conflicting audiovisual stimuli. Audiovisual stimuli were presented from varying locations, but the light was always presented 13.5° to the right of the sound source. For each of the six locations shown, unimodal sound localization responses were on average clearly shifted to the right after audiovisual spatial discrepancy training compared with the pretest

(Woods and Recanzone 2004; Kopčo et al. 2009). The ventriloquism aftereffect builds up rapidly within a few minutes of cross-modal exposure but can last for tens of minutes once the visual stimulus has been removed (Frissen et al. 2012), thus providing a compelling demonstration of cross-modally induced short-term plasticity in the auditory system (for recent reviews, see Recanzone 2009; Chen and Vroomen 2013).

Spatial recalibration in the ventriloquism aftereffect appears to be specific for the trained region of space (Bertelson et al. 2006; Kopčo et al. 2009) but does not depend on the complexity of the stimulus situation or cognitive factors, such as the plausibility of a common cause of the auditory and visual events (Radeau and Bertelson 1977, 1978). These findings suggest that cross-modal spatial recalibration primarily affects lower level auditory sensory representations. In-line with this assumption, electrophysiological results in humans have shown that the behavioral ventriloquism aftereffect is associated with a modulation of auditory event-related potentials (ERPs) as early as 100 ms poststimulus, suggesting an involvement of relatively early stages in the auditory cortical-processing stream (Bruns et al. 2011a). Moreover, some psychophysical studies have reported that the ventriloquism aftereffect does not transfer across sound frequencies when unisensory sound localization is tested with a sound differing in frequency from the sound used during the audiovisual training phase (Recanzone 1998; Lewald 2002; but see Frissen et al. 2003, 2005). This finding supports the hypothesis of a modulation of neural activity in early, tonotopically organized auditory regions. However, the possibility remains that additional or alternative neural structures within the auditory cortex or possibly higher multisensory association areas, such as the posterior parietal cortex, are involved in cross-modal spatial learning (Kopčo et al. 2009).

11.2.2 Time

Critically, multisensory integration depends on the temporal alignment of inputs from different sensory modalities. For example, audiovisual spatial integration in the ventriloquism effect (see Sect. 11.2.1) is largely reduced when the auditory and visual stimuli are not presented in temporal synchrony (Slutsky and Recanzone 2001). Although visual spatial information is normally more precise and thus dominates auditory spatial perception, the situation is reversed in temporal perception where audition is usually more precise than vision. Consequently, there is strong evidence that the perceived timing of visual stimuli can be biased toward asynchronous auditory stimuli, commonly referred to as temporal ventriloquism, in analogy to the spatial ventriloquism effect discussed in Sect. 11.2.1 (Morein-Zamir et al. 2003; Vroomen and de Gelder 2004).

The perception of cross-modal synchrony requires a certain degree of flexibility due to the different physical and neural transmission times of sound and light. A common finding is that the brain has a relatively wide window of temporal integration, meaning that humans are insensitive to small differences in the arrival time of cross-modal input (for a recent review, see Vroomen and Keetels 2010).

However, perceptual learning studies have shown that the window of temporal integration remains highly plastic even in adulthood. As little as 1 hour of training on a multisensory simultaneity judgment task resulted in a substantial narrowing of the integration window along with an increase of connectivity between multisensory areas in the posterior superior temporal sulcus and areas of the auditory and visual cortices (Powers et al. 2012). Training-induced changes in the size of the integration window were shown to be stable for at least one week after cessation of training (Powers et al. 2009). Similarly, short-term exposure to audiovisual stimuli with a consistent temporal asynchrony (e.g., sound always lagging by 100 ms) resulted in a corresponding shift of the point of subjective simultaneity (PSS) between the senses (Fujisaki et al. 2004; Vroomen et al. 2004).

One might expect that temporal recalibration depends on a temporal shift of visual perception toward audition as in the temporal ventriloquism effect. There is, however, evidence suggesting that exposure to asynchronous audiovisual input results in an adjustment of auditory rather than visual processing (see Fig. 11.2). Participants performed a speeded reaction task on unimodal visual or auditory stimuli. Compared with the baseline performance, reaction times to auditory stimuli became faster after exposure to auditory-lagging asynchronies, whereas auditory reaction times became slower after exposure to auditory-leading asynchronies (Navarra et al. 2009). Crucially, there was no effect on reaction times to visual stimuli, suggesting that auditory processing was adjusted to match the timing of the visual events. Vision might be better suited than audition to signal the timing of distal events because of the negligible physical transmission time of light and might thus serve as an anchor for recalibrating auditory temporal processing, despite the fact that visual temporal processing is usually less precise than auditory temporal processing (for a discussion of this issue, see Vroomen and Keetels 2010).

11.2.3 Speech Identification

Cross-modal recalibration is not limited to lower level sensory representations but may depend on high-level contextual correspondence between auditory and visual information such as the correspondence between voices and faces. For example, temporal recalibration can be induced by exposure to asynchronous audiovisual speech streams (Vatakis et al. 2007). Importantly, when observers were concurrently exposed to a male actor with a lagging soundtrack and a female actor with a leading soundtrack, audiovisual synchrony estimates were shifted in opposite directions for the two actors in a subsequent test phase (Roseboom and Arnold 2011). Thus, temporal recalibration can be specific for particular audiovisual pairings defined by higher level perceptual grouping mechanisms (but see Heron et al. 2012).

Speech identification does not only allow for perceptual grouping on which recalibration of lower level sensory representation can operate but can also be targeted by recalibration processes itself. Vision has a substantial influence on auditory speech perception (see Grant and Bernstein, Chap. 3; Perrodin and Petkov, Chap. 9), as illustrated by the well-known McGurk illusion: The sound of the syllable /ba/ is

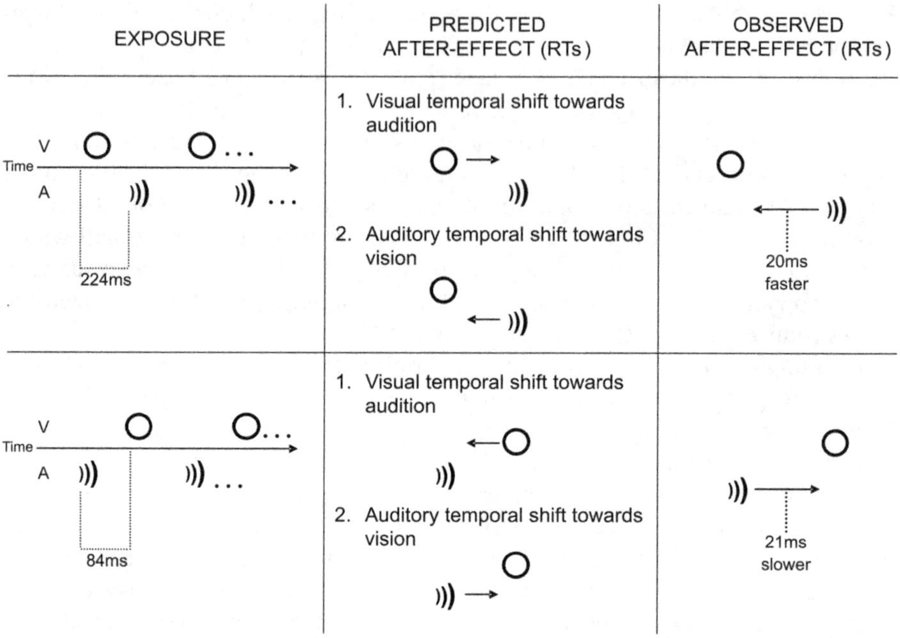

Fig. 11.2 Predicted and observed effects of audiovisual temporal recalibration on reaction times to unimodal stimuli in the study by Navarra et al. (2009). Participants were first exposed to asynchronous audiovisual stimuli (*left*). The auditory signal either always lagged the visual signal by 224 ms (VA) or always led the visual signal by 84 ms (AV). After audiovisual asynchrony training, participants performed a speeded reaction time (RT) task to unimodal auditory and visual stimuli. If audiovisual asynchrony training results in a temporal shift of vision toward audition, a specific modulation of visual RTs could be expected (Prediction 1 in *center*). In contrast, a temporal shift in audition toward vision should result in a specific modulation of auditory RTs (Prediction 2 in *center*). In their actual data and in-line with Prediction 2, Navarra et al. (2009) observed a modulation of auditory RTs of around 20 ms in the direction of the previously trained audiovisual asynchrony, but no modulation of visual RTs (*right*). Reprinted from Navarra et al. (2009), with permission

often heard as /da/ when being presented together with a face articulating the syllable /ga/ (McGurk and MacDonald 1976). Short-term exposure to such incongruent speech stimuli also leads to a recalibration of unimodal auditory speech identification so that an ambiguous sound intermediate between two syllables is more often heard as the syllable that was articulated by the concurrent face stimulus during the preceding audiovisual exposure phase (Bertelson et al. 2003). This shift in perceptual boundaries between phonemes has been associated with activity changes in early auditory cortical areas (Lüttke et al. 2016). Apart from recalibrating speech identification, multisensory training has been shown to influence voice recognition as well. Participants were better at recognizing unimodal auditory voices if they had been paired with faces before, along with an increase in functional connectivity between voice and face areas after voice-face learning (Von Kriegstein and Giraud 2006).

11.2.4 Audio-Tactile Learning

As discussed in Sects. 11.2.1 to 11.2.3, audiovisual learning can affect auditory sensory representations in a large variety of different settings and tasks. However, extrapolating these findings to other sensory combinations (in particular audio-tactile learning) is not trivial. This is particularly true for cross-modal spatial learning, considering the special role that vision might play in aligning neural representations of auditory space because of its superior spatial resolution for stimuli in external space (King 2009). Unlike visual and auditory stimuli, tactile stimuli are initially represented in anatomical (skin-based) coordinates rather than in an external spatial reference frame. It has been demonstrated, however, that auditory localization judgments can be affected by the presentation of spatially displaced tactile stimuli (Caclin et al. 2002; Bruns and Röder 2010), similar to the classic audiovisual ventriloquism effect. Crucially, auditory localization was biased toward the external spatial location of the tactile stimuli rather than toward the anatomical side of the hand that was stimulated (Bruns and Röder 2010). Moreover, exposure to auditory and tactile stimuli with a consistent spatial disparity induced a subsequent shift in unimodal sound localization in the direction of the tactile stimuli (i.e., an audio-tactile ventriloquism aftereffect) as well (Bruns et al. 2011b). This finding has important implications in demonstrating that the brain uses not only visual but also nonvisual information as a reference for calibrating the perception of auditory space. Converging evidence comes from studies on temporal recalibration, where exposure to audio-tactile asynchrony induced an aftereffect that modified subsequent temporal processing of auditory and tactile stimuli (Navarra et al. 2007), similar to that observed for audiovisual combinations.

Studies of audio-tactile recalibration have further demonstrated that cross-modal recalibration resulted in a general modulation of unisensory auditory representations rather than a modulation of specific cross-modal links between audition and touch. In the study by Bruns et al. (2011b), participants were adapted to spatially discrepant audio-tactile stimuli. Before and after audio-tactile training, they performed a relative sound localization task in which they judged whether auditory stimuli were perceived at the same or a different location as a preceding tactile stimulus (Experiment 1) or to the left or right of a preceding visual stimulus (Experiment 2). An audio-tactile ventriloquism aftereffect was observed in both versions of the task. Because only tactile stimuli, but not visual stimuli, were involved in the audio-tactile training phase, this result suggests that auditory spatial representations were modulated and not just the specific relationship between auditory and tactile spatial representations.

11.2.5 Top-Down Influences on Cross-Modal Learning

It is well-known that perceptual learning is determined not only by the sensory input but also by top-down task-dependent influences as well. For example, when exposed to auditory stimuli that varied in both frequency and intensity, rats showed improved discrimination performance along with cortical map plasticity only for the feature that was task relevant (Polley et al. 2006). In contrast, passive learning as observed in the cross-modal recalibration of space (Radeau and Bertelson 1974; Recanzone 1998), time (Fujisaki et al. 2004; Vroomen et al. 2004), and speech perception (Bertelson et al. 2003; Lüttke et al. 2016) after mere exposure to discordant audiovisual stimuli has long been assumed to be independent of top-down influences. This assumption was primarily based on findings suggesting that the (spatial) ventriloquism effect does not seem to depend on the allocation of visual attention (Bertelson et al. 2000; Vroomen et al. 2001).

More recent studies have, however, demonstrated that multisensory spatial integration in the ventriloquist situation is not fully automatic. The size of the ventriloquism effect was reduced when stimuli in the auditory modality were associated with either emotionally salient (aversive) events (Maiworm et al. 2012) or motivationally relevant monetary rewards (Bruns et al. 2014). In both studies, a reduction in the ventriloquism effect was observed, although unimodal auditory localization performance was not affected by the experimental manipulations. Thus, top-down influences from the emotion and reward systems seemed to have specifically changed the process of multisensory binding. Top-down influences have been shown to affect cross-modal spatial recalibration in the ventriloquism aftereffect as well (Eramudugolla et al. 2011). In a dual-task paradigm, participants had to detect visual targets at central fixation while they were exposed to audiovisual stimuli with a consistent spatial disparity in the background. Ventriloquism aftereffects in a subsequent auditory localization task were found to be spatially more specific than after audiovisual exposure without the attention-demanding detection task.

Attentional influences on cross-modal recalibration have been substantiated by studies on audiovisual temporal recalibration. The recalibration of temporal perception induced by exposure to asynchronous audiovisual stimuli was significantly increased when observers attended to the temporal structure of the adapting stimuli compared with when they attended to nontemporal features of the adapting stimuli (Heron et al. 2010). Moreover, it has been shown that selective attention can modulate the direction of audiovisual temporal recalibration in ambiguous situations (Ikumi and Soto-Faraco 2014). In this study, sounds were presented together with both a leading and a lagging visual stimulus that differed in color (see Fig. 11.3). After exposure to these two competing audiovisual asynchronies, the point of subjective simultaneity (PSS) differed systematically between a condition in which participants had attended to the color of the leading flash and a condition in which they had attended to the color of the lagging flash. In a control condition, in which the attended color alternated between the leading and lagging flash (i.e., in the absence of selective attention), cross-modal recalibration was driven by the leading flash similar to when participants attended the leading flash. Thus, top-down influences had a particularly strong influence when selective attention was directed to the lagging flash and countered the stimulus-driven recalibration to the leading flash (see Fig. 11.3). These findings suggest that cross-modal recalibration depends on the interplay between stimulus-driven and top-down processes (Talsma et al. 2010).

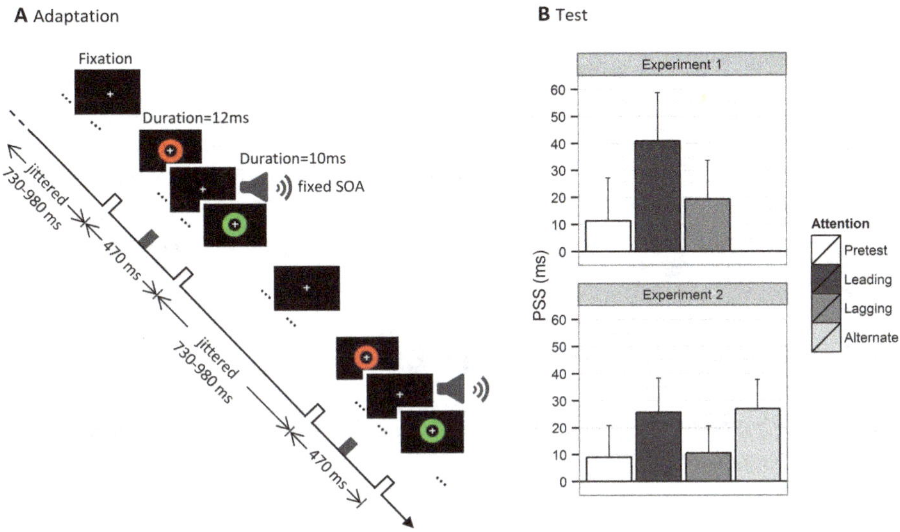

Fig. 11.3 (**A**) Schematic representation of the audiovisual asynchrony training in the study by Ikumi and Soto-Faraco (2014). Each audiovisual exposure consisted of a colored circle that was briefly flashed on a screen, followed by a brief tone and another circle with a different color. The colors of the leading and lagging circle were constant throughout the experiment. Participants had to detect rare deviant stimuli of one color and thus attended to either the leading or the lagging visual stimulus. In a control condition, the color alternated between trials. (**B**) Average point of subjective simultaneity (PSS) values (± SE) for each condition in two experiments in a test phase following the attentional manipulation, in which audiovisual stimuli (consisting of one neutrally colored visual stimulus and one sound) were presented with varying asynchronies. Compared with a pretest before the attentional manipulation (*white bars*), the flash had to be presented before the tone to perceive flash and tone as simultaneous when participants had attended either the leading flash (*dark gray bars*) or both the leading and lagging flashes (*light gray bars*). No change in PSS compared with the pretest was found when participants had attended to the lagging flash (*medium gray bars*). Adapted from Ikumi and Soto-Faraco (2014) under Creative Commons Attribution 4.0 International Public License (CC BY 4.0)

11.3 Auditory Contributions to Cross-Modal Learning in Other Sensory Systems

As demonstrated in Sect. 11.2, there are numerous examples for cross-modal learning modulating auditory processing. Auditory contributions to cross-modal learning in the visual system have been demonstrated in various paradigms as well (for a recent review, see Shams et al. 2011). A prominent example is the finding that audiovisual training can significantly facilitate visual motion-detection performance compared with a unimodal visual training (Seitz et al. 2006). In this task, randomly moving dots are presented on a screen, with a small fraction of the dots moving in a coherent direction during one of two intervals. After extensive training on this task, visual motion-detection performance improved. However, training effects were much faster and stronger when participants were trained with audiovisual stimuli in which a moving auditory stimulus was presented during the interval that contained the coherent visual motion (Seitz et al. 2006). Crucially, audiovisual training was effective only when the direction of the auditory motion was congruent with the direction of the visual motion (Kim et al. 2008) and congruent auditory motion facilitated visual motion-detection performance even when it did not indicate which of the two intervals contained visual coherent motion (Kim et al. 2012). Thus, auditory motion stimuli most likely modulated visual processing at an early perceptual stage.

11.4 Timescales of Cross-Modal Learning

11.4.1 Cross-Modal Learning During Development: Animal Studies

As with other aspects of sensory processing, auditory spatial processing is particularly sensitive to multisensory experience during development (King 2009). The important role of vision in guiding the maturation of auditory spatial representations has been demonstrated most clearly in the superior colliculus, a midbrain structure that contains topographically aligned multisensory spatial maps, and its homologue in the avian brain, the optic tectum. Laterally displacing the visual field in young barn owls (*Tyto alba*) by the use of prisms (Knudsen and Brainard 1991) or periodically exposing dark-reared cats to synchronous auditory and visual stimuli with a fixed spatial disparity (Wallace and Stein 2007) for several days resulted in a corresponding shift in the auditory receptive fields so that auditory and visual receptive fields in the superior colliculus remained aligned. These shifts constituted permanent changes in the auditory space map that were associated with a rewiring of connections in the midbrain (for a recent review, see Gutfreund and King 2012).

Early animal studies had suggested that the guiding influence of vision on the maturation of auditory spatial representations is limited to a sensitive period of development because little or no influence of altered visual input on auditory spatial maps was observed during adulthood (Knudsen and Knudsen 1990). More recent studies have, however, shown that, under certain conditions, plasticity of cross-modal spatial representation is retained throughout life. For example, in adult barn owls, adaptive changes in auditory representations and behavior in response to prism experience are greatly enhanced if the birds are allowed to hunt their prey rather than being provided with dead food (Bergan et al. 2005) or if the prismatic shift is experienced in small increments rather than in a single step (Linkenhoker and Knudsen 2002).

Even in dark-reared animals that were deprived of any multisensory experience involving vision during development, superior colliculus neurons acquired the capability for multisensory integration if the animals were exposed to cross-modal input in early adulthood (Yu et al. 2010). This finding is of particular relevance because multisensory integration in superior colliculus neurons was observed after no more than a few hours of audiovisual experience. In contrast, during typical development, multisensory integration in superior colliculus neurons emerges only after several weeks of exposure to audiovisual input (Wallace and Stein 1997). Similarly, in developing barn owls, adaptive changes in the auditory space map in response to wearing prism lenses were observed only after extensive experience for several days to weeks (Knudsen and Brainard 1991; Gutfreund and King 2012).

Taken together, animal studies of cross-modal spatial calibration and recalibration have almost exclusively focused on spatial representations in the midbrain superior colliculus. Early results had suggested that plasticity of spatial representations requires extensive exposure to cross-modal input and is largely limited to a sensitive period during development. More recent findings have, however, demonstrated that the auditory system retains the capacity to relearn the alignment of spatial representations between sensory modalities throughout life.

11.4.2 Cross-Modal Learning in the Adult System: Studies in Humans

In-line with animal studies that demonstrated visual recalibration of the auditory space map in adulthood (Linkenhoker and Knudsen 2002; Bergan et al. 2005), short-term adaptive changes in auditory localization have been reported in adult humans as well. After compressing the central part of the visual field with prism lenses for three days, a corresponding compression of auditory localization responses was observed that persisted for at least one day after removal of the prism lenses (Zwiers et al. 2003). The effects of prism adaptation in barn owls and humans resemble the ventriloquism aftereffect (see Sect. 11.2.1) in which a brief exposure to synchronous auditory and visual stimuli with a consistent spatial disparity induces a corresponding shift in unimodal auditory localization (Radeau and Bertelson 1974; Recanzone 1998).

The ventriloquism aftereffect and prism adaptation both give rise to a visually induced shift of auditory spatial perception. It is more difficult to reconcile the different timescales at which these two phenomena have been studied. Although prism adaptation usually requires days to weeks of active experience with wearing prism lenses to be effective (Zwiers et al. 2003; Bergan et al. 2005), the ventriloquism aftereffect is brought about by only a few minutes of passive exposure to spatially incongruent audiovisual stimuli (Frissen et al. 2012). Despite the much shorter timescale, however, the ventriloquism aftereffect and other comparably fast recalibration phenomena (see Sect. 11.2) typically follow accumulated evidence of a consistent and constant cross-modal mismatch similar to a typical prism adaptation experiment. Therefore, it has generally been assumed that cross-modal recalibration is initiated only in situations in which the perceptual system is confronted with a more sustained change in cross-modal correspondence. This assumption has, however, been questioned by recent studies showing robust recalibration effects even after a single exposure to a spatially (Wozny and Shams 2011) or temporally (Van der Burg et al. 2013; Noel et al. 2016) misaligned audiovisual stimulus.

It is conceivable that instantaneous cross-modal recalibration and recalibration based on accumulated evidence involve different underlying mechanisms and thus might involve distinct neural systems. Instantaneous recalibration seems to result in a reduction of temporary mismatches between sensory modalities and presumably plays an important role in achieving and maintaining a coherent multisensory representation of the external world. Enduring changes in the correspondence between cross-modal sensory cues might consecutively trigger additional mechanisms that result in more stable and longer lasting adaptation effects. Recent studies have demonstrated a dissociation between instantaneous and cumulative recalibration processes for both spatial (Bruns and Röder 2015) and temporal (Van der Burg et al. 2015) recalibration processes.

Using the spatial ventriloquism aftereffect, Bruns and Röder (2015) assessed the sound frequency specificity of instantaneous and cumulative recalibration effects to test whether both effects arise in early, tonotopically organized regions of the auditory pathway. Previous studies had shown that the cumulative ventriloquism aftereffect is specific for the sound frequency used during audiovisual adaptation and does not transfer to other frequencies (Recanzone 1998; Lewald 2002; but see Frissen et al. 2003, 2005), suggesting an involvement of early, tonotopically organized brain regions. In their study, Bruns and Röder presented two sounds, a low- and a high-frequency tone, which were paired with opposite directions of audiovisual spatial mismatch (leftward vs. rightward). In accordance with this cumulative stimulus history, localization in unimodal auditory trials was shifted in opposite directions for the two sound frequencies. On a trial-by-trial basis, however, this frequency-specific cumulative recalibration effect was reduced when the sound was preceded by an audiovisual trial featuring the other sound frequency and direction of spatial mismatch, indicative of an instantaneous recalibration effect (see Fig. 11.4). Crucially, instantaneous recalibration occurred despite the use of a different sound frequency in the audiovisual adaptation and the following auditory test trial, suggesting that instantaneous recalibration and recalibration based on accumulated evidence represent at least partially distinct processes that jointly determine sound localization behavior.

In a similar vein, Van der Burg et al. (2015) were able to show that audiovisual temporal recalibration (see Sect. 11.2.2) occurs independently at two different timescales. Participants were exposed to a constant audiovisual asynchrony for several minutes before they performed simultaneity judgments on a series of test trials with varying asynchrony. Simultaneity judgments during the test phase revealed a large but decaying cumulative recalibration effect induced by the previous exposure to a constant asynchrony. In addition, responses were modulated by the direction of asynchrony (auditory leading or lagging) in the immediately preceding test trial as well, resembling the dissociation between instantaneous and cumulative recalibration effects in the spatial domain (Bruns and Röder 2015).

Taken together, cross-modal recalibration has been shown to depend on at least two independent learning mechanisms that operate at different timescales. Given

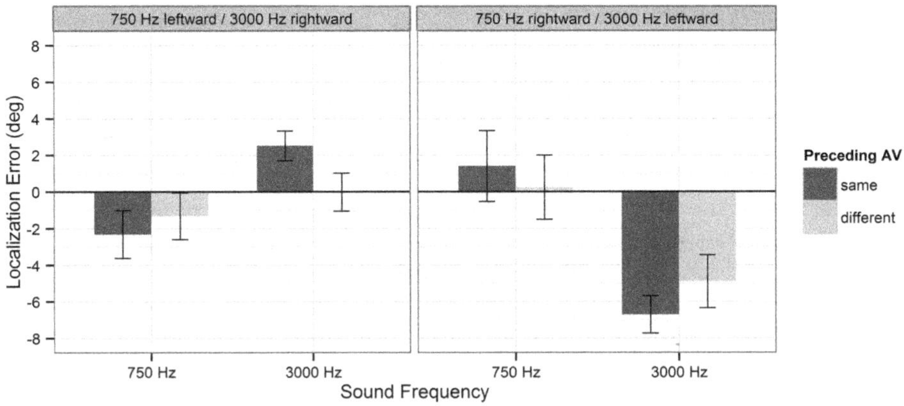

Fig. 11.4 Unimodal sound localization responses in trials following an audiovisual (AV) trial with the same or different sound frequency in the study by Bruns and Röder (2015). Mean localization errors (i.e., the mean deviation of the localization responses from the location of the sound source; ± SE) are shown, with negative values indicating localization errors to the left of the target location and positive values indicating localization errors to the right of the target location. Participants had to localize 750-Hz and 3,000-Hz tones that were presented either alone or together with a synchronous but spatially discrepant visual stimulus. In audiovisual trials, the 750-Hz tone was always paired with a visual stimulus to the left of the sound source and the 3,000-Hz tone with a visual stimulus to the right of the sound source (*left*) or vice versa (*right*). Localization responses in unimodal auditory trials were generally shifted in opposite directions for the two sound frequencies, in accordance with the sound frequency-direction pairing. However, responses were clearly modulated depending on whether the preceding AV trial featured the same sound frequency (*dark gray*) as the present auditory trial or not (*light gray*). Thus, the direction of audiovisual spatial mismatch in the preceding AV trial modulated unimodal sound localization responses irrespective of sound frequency. Adapted from Bruns and Röder (2015) under CC BY 4.0

that a dissociation between instantaneous and cumulative recalibration mechanisms was observed in both spatial (Bruns and Röder 2015) and temporal (Van der Burg et al. 2015) tasks, it seems likely that such a differentiation of learning mechanisms represents a more general strategy of the auditory and possibly other sensory systems. It remains to be shown whether cumulative recalibration mechanisms operating over minutes are dissociable from the mechanisms of longer term recalibration operating over days to weeks such as during prism adaptation.

11.5 Neural Mechanisms of Cross-Modal Learning

11.5.1 Interplay Between Cross-Modal Integration and Learning

The existence of instantaneous cross-modal recalibration effects that are induced by a single exposure to an incongruent audiovisual stimulus (Wozny and Shams 2011; Van der Burg et al. 2013) blurs the borders between cross-modal integration and learning.

Recalibration and thus learning after a single exposure suggests that sensory representations are dynamic throughout life. In this respect, each encounter of a cross-modal stimulus can be viewed as a distinct learning episode.

Several electrophysiological and neuroimaging studies have investigated the neural mechanisms of cross-modal spatial integration in the ventriloquist illusion, in which an auditory stimulus is mislocalized toward a synchronous but spatially disparate visual stimulus (see Sect. 11.2.1). It has been shown that the ventriloquist illusion is associated with a modulation of auditory cortical activity, in particular in the planum temporale. Using both ERPs and functional magnetic resonance imaging (fMRI), Bonath et al. (2007) demonstrated enhanced activity in the planum temporale of the hemisphere contralateral to the side of a visual stimulus that was presented synchronously with a central sound. However, this effect was only observed when the perceived sound location was shifted toward the side of the visual stimulus, whereas the same audiovisual stimulus elicited a bilaterally symmetrical response when the sound was correctly localized to the center. Their fMRI data demonstrated a similar response pattern in the planum temporale for unimodal auditory stimuli that were actually presented at central and lateral locations, suggesting that the modulation observed in the audiovisual trials reflected the illusory-shifted spatial percept. Due to the relatively long latency of the corresponding ERP effect (around 260 ms poststimulus), it seems likely that the visual influence on auditory cortex activity was mediated by feedback connections from multisensory association areas (for similar findings, see Bruns and Röder 2010). Multisensory association areas in the posterior parietal cortex have indeed been shown to be involved in the generation of the ventriloquist effect (Renzi et al. 2013; Rohe and Noppeney 2015). Taken together, the planum temporale seems to adjust auditory spatial input based on visual spatial cues received via feedback connections (for a broader discussion of the cortical dynamics underlying cross-modal integration, see Keil and Senkowski, Chap. 10).

Changes in unimodal sound localization behavior following exposure to spatially misaligned cross-modal stimuli, as observed in the ventriloquism aftereffect, could be mediated by the same pathway that is active during cross-modal spatial integration, namely, feedback influences from multisensory parietal regions on the secondary auditory cortex (Bonath et al. 2007). Recent evidence, however, suggests that cross-modal spatial integration and cumulative recalibration processes following repeated exposure to a consistent cross-modal spatial mismatch are mediated by dissociable mechanisms.

To identify the stage in the auditory cortical processing stream that is modulated by cumulative audiovisual spatial discrepancy training, Bruns et al. (2011a) conducted an ERP study. Participants had to report the perceived location of brief auditory stimuli before and after exposure to synchronous audiovisual stimuli with a constant spatial disparity. As expected, the behavioral responses showed a clear postadaptive shift in subjective sound localization in the same direction as the preceding visual stimuli. Importantly, this behavioral ventriloquism aftereffect was associated with a modulation of auditory ERPs in the N1 time range (around 100 ms poststimulus). Consistent with earlier findings (Bruns and Röder 2010), central sounds elicited a reduced N1 response compared with laterally presented sounds in the pretest. At posttest, the N1 was differently affected depending on the direction of audiovisual discrepancy training: the relatively lowest N1 response was observed for right sounds after leftward training but for left sounds after rightward training. Thus, a reduced N1 response, as for the central sounds in the pretest, was observed for the now centrally represented sound location, suggesting that the cortical representation of auditory space was shifted in the respective direction (see Fig. 11.5). Crucially, cumulative exposure to spatially disparate audiovisual stimuli affected earlier stages (around 100 ms poststimulus) in the auditory cortical processing stream than those initially affected by the online ventriloquist illusion during the training phase (around 260 ms poststimulus; see Bonath et al. 2007; Bruns and Röder 2010).

Psychophysical studies have provided corroborating evidence for dissociable cross-modal integration and recalibration processes. Cross-modal integration in the ventriloquist illusion is known to depend on the relative reliability of the auditory and visual stimuli (Alais and Burr 2004), whereas cross-modal recalibration does not seem to depend on cue reliability (Zaidel et al. 2011). Moreover, the occurrence

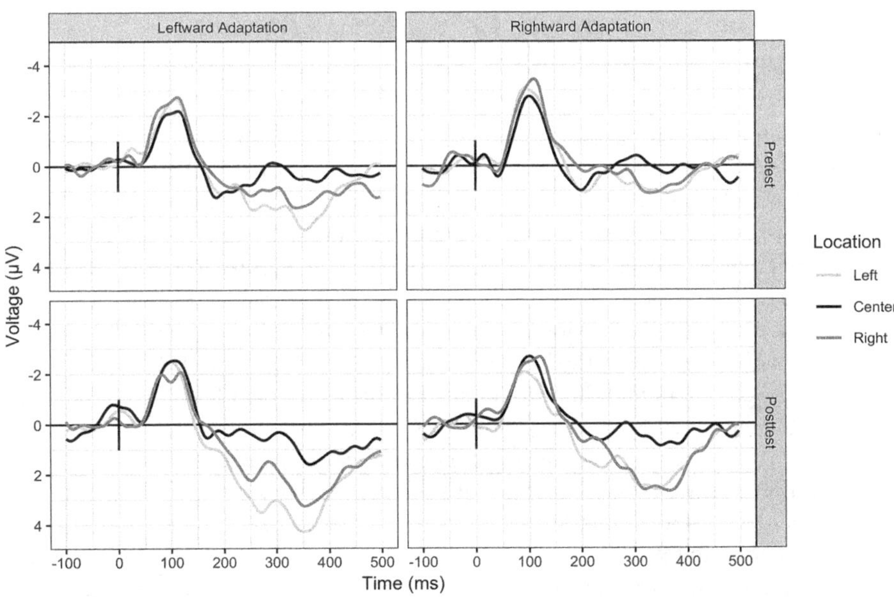

Fig. 11.5 Modulation of auditory event-related potentials (ERPs) following audiovisual spatial discrepancy training in the study by Bruns et al. (2011a). Participants had to localize brief tones that were presented at left, center, and right locations before (pretest) and after (posttest) they were exposed to audiovisual spatial discrepancy training. Audiovisual stimuli during training were presented from varying locations, but the light was either always displaced by 15° to the left of the sound source (leftward adaptation) or always displaced by 15° to the right of the sound source (rightward adaptation). At pretest, ERPs to center sounds were reduced around 100 ms poststimulus and enhanced around 260 ms poststimulus compared with both lateral locations. At posttest, ERPs were modulated depending on the direction of the preceding audiovisual spatial discrepancy training around 100 ms poststimulus but not in the later time window around 260 ms poststimulus. Adapted from Bruns et al. (2011a), with permission

of the ventriloquist illusion strongly depends on audiovisual synchrony (Slutsky and Recanzone 2001), whereas the ventriloquism aftereffect emerges independent of audiovisual synchrony, provided that the visual stimuli lag the auditory stimuli and thus allow for a spatial feedback of the auditory location (Pages and Groh 2013).

Importantly, however, recalibration mechanisms following repeated exposure to a consistent cross-modal discrepancy are dissociable from instantaneous recalibration mechanisms that operate on a trial-by-trial basis (Bruns and Röder 2015; Van der Burg et al. 2015; see Sect. 11.4.2). Possibly, these immediate adjustments of sensory representations are a result of the same feedback influences from multisensory parietal structures on the auditory cortex that are active during cross-modal spatial integration. Cumulative evidence for a consistent cross-modal mismatch might then invoke changes in unisensory auditory representations at an earlier processing stage (Bruns et al. 2011a), possibly due to a repeated and consistent activation of recalibration via multisensory parietal structures.

11.5.2 Neural Pathways for Cross-Modal Learning in the Auditory System

Studies in adult humans have suggested that short-term adaptive changes in response to mismatching cross-modal input critically depend on cortical processing (see Sect. 11.5.1). In contrast, animal studies have primarily looked at long-term changes induced by abnormal cross-modal experience during development, and these changes have mainly been investigated for subcortical circuits involving the superior colliculus (see Sect. 11.4.1). Reconciling these two strands of research is complicated by the lack of comparative studies across species. It is currently unclear whether plasticity of subcortical representations is mainly restricted to a sensitive period of development and thus would not be able to account for short-term adaptive changes during adulthood or whether the site of plasticity, cortical or subcortical, simply differs between humans and other species irrespective of age.

Some findings, however, suggest that both cortical and subcortical circuits are involved and may have complementary roles in the cross-modal recalibration of auditory representations in adults. Neuropsychological studies in hemianopic patients have shown that visual cortical processing is necessary for adapting to spatially conflicting audiovisual stimuli. When the visual stimuli were presented in the blind hemifield of the patients, no visual bias or recalibration of auditory localization was obtained (Leo et al. 2008; Passamonti et al. 2009). However, although lesions of the striate cortex prevented both the immediate ventriloquism effect and the ventriloquism aftereffect, multisensory enhancement and learning (i.e., more accurate sound localization) was still evident when the visual and auditory stimuli were presented at the same spatial location within the blind hemifield (see Fig. 11.6). This dissociation between spatially aligned and misaligned audiovisual stimulation was specific for lesions of sensory visual areas. Patients with a unilateral spatial neglect (i.e., an attentional rather than a sensory deficit) due to temporoparietal

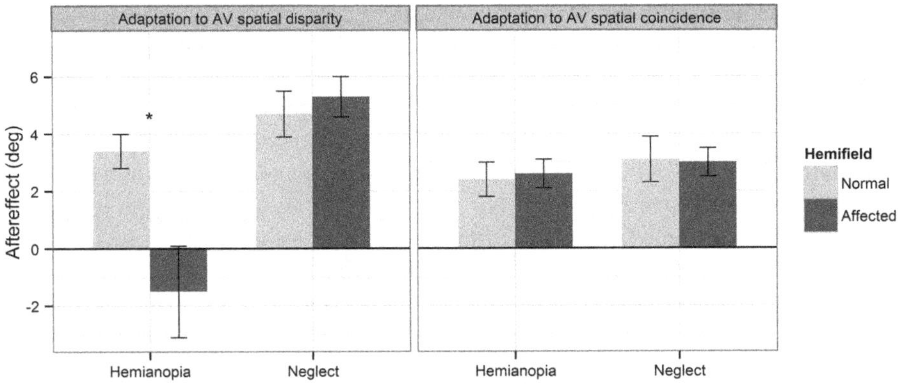

Fig. 11.6 Audiovisual spatial recalibration effects in hemianopic and neglect patients in the study by Passamonti et al. (2009). Patients performed a unimodal sound localization task before and after exposure to audiovisual spatial training. During training, audiovisual stimuli either were presented with a spatial disparity of 7.5° (adaptation to AV spatial disparity) or were spatially aligned (adaptation to AV spatial coincidence). In the AV disparity condition, sounds were always presented from straight ahead (0°), together with a light that was displaced by 7.5° either in the affected hemifield (*dark gray*) or in the normal hemifield (*light gray*). Bars show the shift in sound localization (± SE) between pre- and posttest (i.e., the ventriloquism aftereffect). Neglect patients showed a ventriloquism aftereffect in both hemifields, whereas hemianopic patients showed a ventriloquism aftereffect only in the normal hemifield, but no aftereffect when the visual stimulus was presented in their blind hemifield. In the AV coincidence condition, audiovisual stimuli were both presented at an eccentricity of 20° from the midline, either in the affected hemifield (*dark gray*) or in the normal hemifield (*light gray*). Bars show the reduction in localization errors (± SE) in the posttest compared with the pretest. Both patient groups showed a significant reduction in localization errors in both hemifields. Adapted from Passamonti et al. (2009), with permission

lesions showed a visual modulation of auditory localization irrespective of the spatial alignment of the audiovisual stimuli (Passamonti et al. 2009).

The superior colliculus has been implicated in the mediation of residual visual functions in patients with lesions of the striate cortex. Crucially, in adult animals, only cross-modal stimuli that were presented in close spatial and temporal proximity have been found to induce a response enhancement in multisensory neurons of the superior colliculus (Stein and Stanford 2008). In contrast, spatially disparate stimuli produced depression or no change in neural activity. This pattern of results has been called the spatial and temporal principles of multisensory integration (Stein and Stanford 2008). Thus, residual multisensory enhancement and learning in hemianopic patients might have been enabled by a collicular-extrastriate circuit but only for spatially aligned audiovisual stimuli that were capable of inducing a response enhancement in superior colliculus neurons. In contrast, cross-modal recalibration to mismatching input might require more flexible mechanisms operating at higher cortical processing stages (Leo et al. 2008; Passamonti et al. 2009).

Evidence obtained in cats suggests that descending inputs from the anterior ectosylvian sulcus, a subregion of multisensory association cortex in the temporal lobe, are essential for the emergence of multisensory integration in superior colliculus neurons. Deactivation of this area led to a loss of multisensory integration in superior colliculus neurons as well as a loss of behavioral benefits that are associated with multisensory integration in the superior colliculus (Stein and Stanford 2008). In-line with this observation, temporary suppression of excitability in the temporoparietal cortex induced by repetitive transcranial magnetic stimulation in healthy adult humans was found to reduce multisensory enhancement for spatially and temporally aligned audiovisual stimuli but did not modulate the ventriloquism effect with spatially misaligned stimuli (Bertini et al. 2010). The authors proposed that an inhibition of the temporoparietal cortex, which might contain the human homologue of the anterior ectosylvian sulcus in the cat, might have deteriorated multisensory integration in the superior colliculus. In contrast, inhibition of the occipital cortex suppressed the visual bias of auditory localization with spatially misaligned stimuli but did not affect multisensory enhancement for aligned stimuli (Bertini et al. 2010), a finding that nicely corroborates the results obtained in patients with permanent lesions of the visual cortex (Leo et al. 2008; Passamonti et al. 2009).

Taken together, most events in the environment convey temporally and spatially aligned information to the different senses. Under such circumstances, multisensory integration in collicular-extrastriate circuits can lead to an enhanced detection and more precise orienting responses to these events (Stein and Stanford 2008). However, the system needs to cope with situations in which inputs of independent origin overlap or inputs of the same origin differ on some dimension. This requires more flexible learning mechanisms that seem to depend on cortical processing (Passamonti et al. 2009). In particular, the brain needs to solve the causal inference problem of whether or not different sensory signals pertain to the same event and thus should be integrated and used for cross-modal recalibration. Recent evidence suggests that these computations critically depend on higher level cortical association areas (Rohe and Noppeney 2015).

11.6 Summary

The studies reviewed in this chapter provide considerable evidence that cross-modal input constantly modifies the way in which the brain processes auditory stimuli over a range of different timescales and throughout life. This applies to lower level sensory features like spatial and temporal processing as well as to higher level features like speech identification. The majority of these studies have concentrated on the visual influences on auditory perception and learning, but influences from the somatosensory system on audition and reversed influences from audition on learning in other sensory systems have been demonstrated as well. Recent findings suggest that cross-modal learning operates in parallel on different neural representations and at different timescales (Bruns and Röder 2015; Van der Burg et al. 2015). Cross-modal learning seems to progress from higher level multisensory representations to lower level modality-specific representations with an increasing amount of exposure to new cross-modal associations. For example, secondary auditory areas such as the planum temporale seem to adjust auditory spatial input to coincide with spatially disparate visual and tactile input signaled via feedback connections (Bonath et al. 2007; Bruns and Röder 2010). Such cross-modal feedback mechanisms seem to result in changes in auditory representations that affect bottom-up processing if the spatial correspondence between sensory modalities is consistently altered as in the ventriloquism aftereffect (Bruns et al. 2011a).

In addition to cortically mediated learning mechanisms, auditory representations are shaped via subcortical multisensory pathways including the superior colliculus. To fully understand the dynamic nature of the auditory system, it will be important to identify how cortically and subcortically mediated learning processes interact in the mature brain. Such studies will reveal whether there are genuine differences in the underlying neural mechanisms and behavioral properties of short-term (on the order of seconds to minutes) and long-term (on the order of days to weeks) cross-modal learning phenomena and how they are linked (see Murray et al. 2016).

Acknowledgments The authors' work was supported by German Research Foundation (DFG) Grants BR 4913/2-1 and TRR 169 Subproject A1 and the City of Hamburg Grant "Crossmodal Learning."

Compliance with Ethics Requirements Patrick Bruns declares that he has no conflict of interest.

Brigitte Röder declares that she has no conflict of interest.

References

Alais, D., & Burr, D. (2004). The ventriloquist effect results from near-optimal bimodal integration. *Current Biology, 14*, 257–262.

Bergan, J. F., Ro, P., Ro, D., & Knudsen, E. I. (2005). Hunting increases adaptive auditory map plasticity in adult barn owls. *The Journal of Neuroscience, 25*, 9816–9820.

Bertelson, P., & Aschersleben, G. (1998). Automatic visual bias of perceived auditory location. *Psychonomic Bulletin & Review, 5*, 482–489.

Bertelson, P., Vroomen, J., de Gelder, B., & Driver, J. (2000). The ventriloquist effect does not depend on the direction of deliberate visual attention. *Perception & Psychophysics, 62*, 321–332.

Bertelson, P., Vroomen, J., & de Gelder, B. (2003). Visual recalibration of auditory speech identification: A McGurk aftereffect. *Psychological Science, 14*, 592–597.

Bertelson, P., Frissen, I., Vroomen, J., & de Gelder, B. (2006). The aftereffects of ventriloquism: Patterns of spatial generalization. *Perception & Psychophysics, 68*, 428–436.

Bertini, C., Leo, F., Avenanti, A., & Làdavas, E. (2010). Independent mechanisms for ventriloquism and multisensory integration as revealed by theta-burst stimulation. *European Journal of Neuroscience, 31*, 1791–1799.

Bonath, B., Noesselt, T., Martinez, A., Mishra, J., Schwiecker, K., Heinze, H. J., & Hillyard, S. A. (2007). Neural basis of the ventriloquist illusion. *Current Biology, 17*, 1697–1703.

Bruns, P., & Röder, B. (2010). Tactile capture of auditory localization: An event-related potential study. *European Journal of Neuroscience, 31*, 1844–1857.

Bruns, P., & Röder, B. (2015). Sensory recalibration integrates information from the immediate and the cumulative past. *Scientific Reports, 5*, 12739.

Bruns, P., Liebnau, R., & Röder, B. (2011a). Cross-modal training induces changes in spatial representations early in the auditory processing pathway. *Psychological Science, 22*, 1120–1126.
Bruns, P., Spence, C., & Röder, B. (2011b). Tactile recalibration of auditory spatial representations. *Experimental Brain Research, 209*, 333–344.
Bruns, P., Maiworm, M., & Röder, B. (2014). Reward expectation influences audiovisual spatial integration. *Attention, Perception, & Psychophysics, 76*, 1815–1827.
Caclin, A., Soto-Faraco, S., Kingstone, A., & Spence, C. (2002). Tactile "capture" of audition. *Perception & Psychophysics, 64*, 616–630.
Chen, L., & Vroomen, J. (2013). Intersensory binding across space and time: A tutorial review. *Attention, Perception, & Psychophysics, 75*, 790–811.
Dahmen, J. C., & King, A. J. (2007). Learning to hear: Plasticity of auditory cortical processing. *Current Opinion in Neurobiology, 17*, 456–464.
Dahmen, J. C., Keating, P., Nodal, F. R., Schulz, A. L., & King, A. J. (2010). Adaptation to stimulus statistics in the perception and neural representation of auditory space. *Neuron, 66*, 937–948.
Dean, I., Harper, N. S., & McAlpine, D. (2005). Neural population coding of sound level adapts to stimulus statistics. *Nature Neuroscience, 8*, 1684–1689.
Eramudugolla, R., Kamke, M. R., Soto-Faraco, S., & Mattingley, J. B. (2011). Perceptual load influences auditory space perception in the ventriloquist aftereffect. *Cognition, 118*, 62–74.
Ernst, M. O., & Bülthoff, H. H. (2004). Merging the senses into a robust percept. *Trends in Cognitive Sciences, 8*, 162–169.
Frissen, I., Vroomen, J., de Gelder, B., & Bertelson, P. (2003). The aftereffects of ventriloquism: Are they sound-frequency specific? *Acta Psychologica, 113*, 315–327.
Frissen, I., Vroomen, J., de Gelder, B., & Bertelson, P. (2005). The aftereffects of ventriloquism: Generalization across sound-frequencies. *Acta Psychologica, 118*, 93–100.
Frissen, I., Vroomen, J., & de Gelder, B. (2012). The aftereffects of ventriloquism: The time course of the visual recalibration of auditory localization. *Seeing and Perceiving, 25*, 1–14.
Fujisaki, W., Shimojo, S., Kashino, M., & Nishida, S. (2004). Recalibration of audiovisual simultaneity. *Nature Neuroscience, 7*, 773–778.
Gutfreund, Y., & King, A. J. (2012). What is the role of vision in the development of the auditory space map? In B. E. Stein (Ed.), *The New Handbook of Multisensory Processing* (pp. 573–587). Cambridge, MA: The MIT Press.
Heron, J., Roach, N. W., Whitaker, D., & Hanson, J. V. M. (2010). Attention regulates the plasticity of multisensory timing. *European Journal of Neuroscience, 31*, 1755–1762.
Heron, J., Roach, N. W., Hanson, J. V. M., McGraw, P. V., & Whitaker, D. (2012). Audiovisual time perception is spatially specific. *Experimental Brain Research, 218*, 477–485.
Ikumi, N., & Soto-Faraco, S. (2014). Selective attention modulates the direction of audio-visual temporal recalibration. *PLoS One, 9*, e99311.
Kim, R. S., Seitz, A. R., & Shams, L. (2008). Benefits of stimulus congruency for multisensory facilitation of visual learning. *PLoS One, 3*, e1532.
Kim, R., Peters, M. A. K., & Shams, L. (2012). 0 + 1 > 1: How adding noninformative sound improves performance on a visual task. *Psychological Science, 23*, 6–12.
King, A. J. (2009). Visual influences on auditory spatial learning. *Philosophical Transactions of the Royal Society, B: Biological Sciences, 364*, 331–339.
Knudsen, E. I., & Brainard, M. S. (1991). Visual instruction of the neural map of auditory space in the developing optic tectum. *Science, 253*, 85–87.
Knudsen, E. I., & Knudsen, P. F. (1990). Sensitive and critical periods for visual calibration of sound localization by barn owls. *The Journal of Neuroscience, 10*, 222–232.
Kopčo, N., Lin, I.-F., Shinn-Cunningham, B. G., & Groh, J. M. (2009). Reference frame of the ventriloquism aftereffect. *The Journal of Neuroscience, 29*, 13809–13814.
Leo, F., Bolognini, N., Passamonti, C., Stein, B. E., & Làdavas, E. (2008). Cross-modal localization in hemianopia: New insights on multisensory integration. *Brain, 131*, 855–865.
Lewald, J. (2002). Rapid adaptation to auditory-visual spatial disparity. *Learning & Memory, 9*, 268–278.

Linkenhoker, B. A., & Knudsen, E. I. (2002). Incremental training increases the plasticity of the auditory space map in adult barn owls. *Nature, 419*, 293–296.
Lüttke, C. S., Ekman, M., van Gerven, M. A. J., & de Lange, F. P. (2016). McGurk illusion recalibrates subsequent auditory perception. *Scientific Reports, 6*, 32891.
Maiworm, M., Bellantoni, M., Spence, C., & Röder, B. (2012). When emotional valence modulates audiovisual integration. *Attention, Perception, & Psychophysics, 74*, 1302–1311.
McGurk, H., & MacDonald, J. (1976). Hearing lips and seeing voices. *Nature, 264*, 746–748.
Morein-Zamir, S., Soto-Faraco, S., & Kingstone, A. (2003). Auditory capture of vision: Examining temporal ventriloquism. *Cognitive Brain Research, 17*, 154–163.
Murray, M. M., Lewkowicz, D. J., Amedi, A., & Wallace, M. T. (2016). Multisensory processes: A balancing act across the lifespan. *Trends in Neurosciences, 39*, 567–579.
Navarra, J., Soto-Faraco, S., & Spence, C. (2007). Adaptation to audiotactile asynchrony. *Neuroscience Letters, 413*, 72–76.
Navarra, J., Hartcher-O'Brien, J., Piazza, E., & Spence, C. (2009). Adaptation to audiovisual asynchrony modulates the speeded detection of sound. *Proceedings of the National Academy of Sciences of the United States of America, 106*, 9169–9173.
Noel, J.-P., De Niear, M., Van der Burg, E., & Wallace, M. T. (2016). Audiovisual simultaneity judgment and rapid recalibration throughout the lifespan. *PLoS One, 11*, e0161698.
Pages, D. S., & Groh, J. M. (2013). Looking at the ventriloquist: Visual outcome of eye movements calibrates sound localization. *PLoS One, 8*, e72562.
Passamonti, C., Frissen, I., & Làdavas, E. (2009). Visual recalibration of auditory spatial perception: Two separate neural circuits for perceptual learning. *European Journal of Neuroscience, 30*, 1141–1150.
Polley, D. B., Steinberg, E. E., & Merzenich, M. M. (2006). Perceptual learning directs auditory cortical map reorganization through top-down influences. *The Journal of Neuroscience, 26*, 4970–4982.
Powers, A. R., III, Hillock, A. R., & Wallace, M. T. (2009). Perceptual training narrows the temporal window of multisensory binding. *The Journal of Neuroscience, 29*, 12265–12274.
Powers, A. R., III, Hevey, M. A., & Wallace, M. T. (2012). Neural correlates of multisensory perceptual learning. *The Journal of Neuroscience, 32*(18), 6263–6274.
Radeau, M., & Bertelson, P. (1974). The after-effects of ventriloquism. *Quarterly Journal of Experimental Psychology, 26*, 63–71.
Radeau, M., & Bertelson, P. (1977). Adaptation to auditory-visual discordance and ventriloquism in semirealistic situations. *Perception & Psychophysics, 22*, 137–146.
Radeau, M., & Bertelson, P. (1978). Cognitive factors and adaptation to auditory-visual discordance. *Perception & Psychophysics, 23*, 341–343.
Recanzone, G. H. (1998). Rapidly induced auditory plasticity: The ventriloquism aftereffect. *Proceedings of the National Academy of Sciences of the United States of America, 95*, 869–875.
Recanzone, G. H. (2009). Interactions of auditory and visual stimuli in space and time. *Hearing Research, 258*, 89–99.
Recanzone, G. H., & Sutter, M. L. (2008). The biological basis of audition. *Annual Review of Psychology, 59*, 119–142.
Renzi, C., Bruns, P., Heise, K.-F., Zimerman, M., Feldheim, J. F., Hummel, F. C., & Röder, B. (2013). Spatial remapping in the audio-tactile ventriloquism effect: A TMS investigation on the role of the ventral intraparietal area. *Journal of Cognitive Neuroscience, 25*, 790–801.
Rohe, T., & Noppeney, U. (2015). Cortical hierarchies perform Bayesian causal inference in multisensory perception. *PLoS Biology, 13*, e1002073.
Roseboom, W., & Arnold, D. H. (2011). Twice upon a time: Multiple concurrent temporal recalibrations of audiovisual speech. *Psychological Science, 22*, 872–877.
Seitz, A. R., Kim, R., & Shams, L. (2006). Sound facilitates visual learning. *Current Biology, 16*, 1422–1427.
Shams, L., & Seitz, A. R. (2008). Benefits of multisensory learning. *Trends in Cognitive Sciences, 12*, 411–417.

Shams, L., Wozny, D. R., Kim, R., & Seitz, A. (2011). Influences of multisensory experience on subsequent unisensory processing. *Frontiers in Psychology, 2*, 264.
Slutsky, D. A., & Recanzone, G. H. (2001). Temporal and spatial dependency of the ventriloquism effect. *NeuroReport, 12*, 7–10.
Stein, B. E., & Stanford, T. R. (2008). Multisensory integration: Current issues from the perspective of the single neuron. *Nature Reviews Neuroscience, 9*, 255–266.
Stein, B. E., Huneycutt, W. S., & Meredith, M. A. (1988). Neurons and behavior: The same rules of multisensory integration apply. *Brain Research, 448*, 355–358.
Strelnikov, K., Rosito, M., & Barone, P. (2011). Effect of audiovisual training on monaural spatial hearing in horizontal plane. *PLoS One, 6*, e18344.
Talsma, D., Senkowski, D., Soto-Faraco, S., & Woldorff, M. G. (2010). The multifaceted interplay between attention and multisensory integration. *Trends in Cognitive Sciences, 14*, 400–410.
Tzounopoulos, T., & Leão, R. M. (2012). Mechanisms of memory and learning in the auditory system. In L. O. Trussel, A. N. Popper, & R. R. Fay (Eds.), *Synaptic Mechanisms in the Auditory System* (pp. 203–226). New York: Springer-Verlag.
Van der Burg, E., Alais, D., & Cass, J. (2013). Rapid recalibration to audiovisual asynchrony. *The Journal of Neuroscience, 33*, 14633–14637.
Van der Burg, E., Alais, D., & Cass, J. (2015). Audiovisual temporal recalibration occurs independently at two different time scales. *Scientific Reports, 5*, 14526.
Vatakis, A., Navarra, J., Soto-Faraco, S., & Spence, C. (2007). Temporal recalibration during asynchronous audiovisual speech perception. *Experimental Brain Research, 181*, 173–181.
Von Kriegstein, K., & Giraud, A.-L. (2006). Implicit multisensory associations influence voice recognition. *PLoS Biology, 4*, e326.
Vroomen, J., & de Gelder, B. (2004). Temporal ventriloquism: Sound modulates the flash-lag effect. *Journal of Experimental Psychology: Human Perception and Performance, 30*, 513–518.
Vroomen, J., & Keetels, M. (2010). Perception of intersensory synchrony: A tutorial review. *Attention, Perception, & Psychophysics, 72*, 871–884.
Vroomen, J., Bertelson, P., & de Gelder, B. (2001). The ventriloquist effect does not depend on the direction of automatic visual attention. *Perception & Psychophysics, 63*, 651–659.
Vroomen, J., Keetels, M., de Gelder, B., & Bertelson, P. (2004). Recalibration of temporal order perception by exposure to audio-visual asynchrony. *Cognitive Brain Research, 22*, 32–35.
Wallace, M. T., & Stein, B. E. (1997). Development of multisensory neurons and multisensory integration in cat superior colliculus. *The Journal of Neuroscience, 17*, 2429–2444.
Wallace, M. T., & Stein, B. E. (2007). Early experience determines how the senses will interact. *Journal of Neurophysiology, 97*, 921–926.
Woods, T. M., & Recanzone, G. H. (2004). Visually induced plasticity of auditory spatial perception in macaques. *Current Biology, 14*, 1559–1564.
Wozny, D. R., & Shams, L. (2011). Recalibration of auditory space following milliseconds of cross-modal discrepancy. *The Journal of Neuroscience, 31*, 4607–4612.
Yu, L., Rowland, B. A., & Stein, B. E. (2010). Initiating the development of multisensory integration by manipulating sensory experience. *The Journal of Neuroscience, 30*, 4904–4913.
Zaidel, A., Turner, A. H., & Angelaki, D. E. (2011). Multisensory calibration is independent of cue reliability. *The Journal of Neuroscience, 31*, 13949–13962.
Zwiers, M. P., van Opstal, A. J., & Paige, G. D. (2003). Plasticity in human sound localization induced by compressed spatial vision. *Nature Neuroscience, 6*, 175–181.

Chapter 12
Multisensory Processing Differences in Individuals with Autism Spectrum Disorder

Sarah H. Baum Miller and Mark T. Wallace

Abstract Autism spectrum disorder (ASD) is a neurodevelopmental disorder that is characterized by a constellation of symptoms, including impairments in social communication, restricted interests, and repetitive behaviors. Although sensory issues have long been reported in clinical descriptions of ASD, only the most recent edition of the *Diagnostic and Statistical Manual of Mental Disorders* (DSM-V) has included differences in sensory processing as part of the diagnostic profile for ASD. Indeed, sensory processing differences are among the most prevalent findings in ASD, and these differences are increasingly recognized as a core component of ASD. Furthermore, characterizing ASD phenotypes on the basis of sensory processing differences has been suggested as a constructive means of creating phenotypic subgroups of ASD, which may be useful to better tailor individualized treatment strategies. Although sensory processing differences are frequently approached from the perspective of deficits in the context of ASD, there are a number of instances in which individuals with ASD outperform their neurotypical counterparts on tests of sensory function. Here, the current state of knowledge regarding sensory processing in ASD is reviewed, with a particular emphasis on auditory and multisensory (i.e., audiovisual) performance. In addition to characterizing the nature of these differences in sensory performance, the chapter focuses on the neurological correlates of these sensory processing differences and how differences in sensory function relate to the other core clinical features of ASD, with an emphasis on speech and language.

S. H. Baum Miller
Department of Psychology, Institute for Learning and Brain Sciences (I-LABS), University of Washington, Seattle, WA, USA
e-mail: shbaum@uw.edu

M. T. Wallace (✉)
Department of Hearing and Speech Sciences, Vanderbilt Brain Institute, Vanderbilt University, Nashville, TN, USA

Department of Psychiatry and Behavioral Sciences, Vanderbilt Brain Institute, Vanderbilt University, Nashville, TN, USA

Department of Psychology, Vanderbilt Brain Institute, Vanderbilt University, Nashville, TN, USA
e-mail: mark.wallace@vanderbilt.edu

A. K. C. Lee et al. (eds.), *Multisensory Processes*, Springer Handbook of Auditory Research 68, https://doi.org/10.1007/978-3-030-10461-0_12

Keywords Auditory · Binding · Causal inference · Cue combination · Integration · Neuroimaging · Perception · Psychophysics · Sensory processing · Visual

12.1 Introduction

This chapter focuses on the behavioral and neural underpinnings of altered sensory processing in autism spectrum disorder (ASD) by first looking at perception in auditory (Sect. 12.1.1), visual (Sect. 12.1.2), and audiovisual (Sect. 12.1.3) processing by comparing individuals with ASD and neurotypical (NT) development. Next, this chapter explores how these differences in perception might be tied to both structural and functional changes in the brain, first focusing on insights from magnetic resonance imaging (Sect. 12.2.1) and then moving to electrophysiology (Sect. 12.2.2). The chapter concludes by looking at the developmental trajectory of these populations and connecting sensory processing to different clinical symptoms in ASD.

12.1.1 Perceptual Differences in Auditory Processing

A given auditory stimulus can be deconstructed into a number of dimensions, such as amplitude (loudness), frequency (pitch) composition, and timbre. A great deal of human psychophysical work has focused on better understanding how these various stimulus attributes are processed by human observers. Intriguingly, in the processing of some of these primary attributes, children with ASD have been shown to exhibit superior performance (Remington and Fairnie 2017). For example, several studies have noted superior pitch perception in children with ASD compared with their NT peers (Heaton 2003; O'Riordan and Passetti 2006). This includes both the perception of simple pure tones (Fig. 12.1A; Bonnel et al. 2003) as well as pitch perception in the context of full sentences (Järvinen-Pasley et al. 2008a, b). Intriguingly, during adolescence, this advantage seems to disappear (Heaton et al. 2008; Jones et al. 2009). Somewhat paradoxically, older individuals with ASD who retain this advantage in pitch perception are more likely to have significant language difficulties (Bonnel et al. 2010).

Auditory thresholds, as measured by indices such as modulation depth discrimination, do not seem to differ in individuals with ASD compared with those who are NT (Haigh et al. 2016). However, across multiple standard audiological measures, individuals with ASD are more likely to show at least one abnormal finding, with a common finding that they are more likely to be more sensitive to sounds (lower behavioral threshold for rating a sound to be uncomfortable) than NT individuals (Demopoulos and Lewine 2016). Other findings of note include reduced otoacoustic emissions, which measure vibrations reflected backward from the cochlea in response to sound and thus the integrity of the transduction process of the ear. However, these reductions have been found only for specific frequencies but are largely not different compared with NT individuals across most frequencies

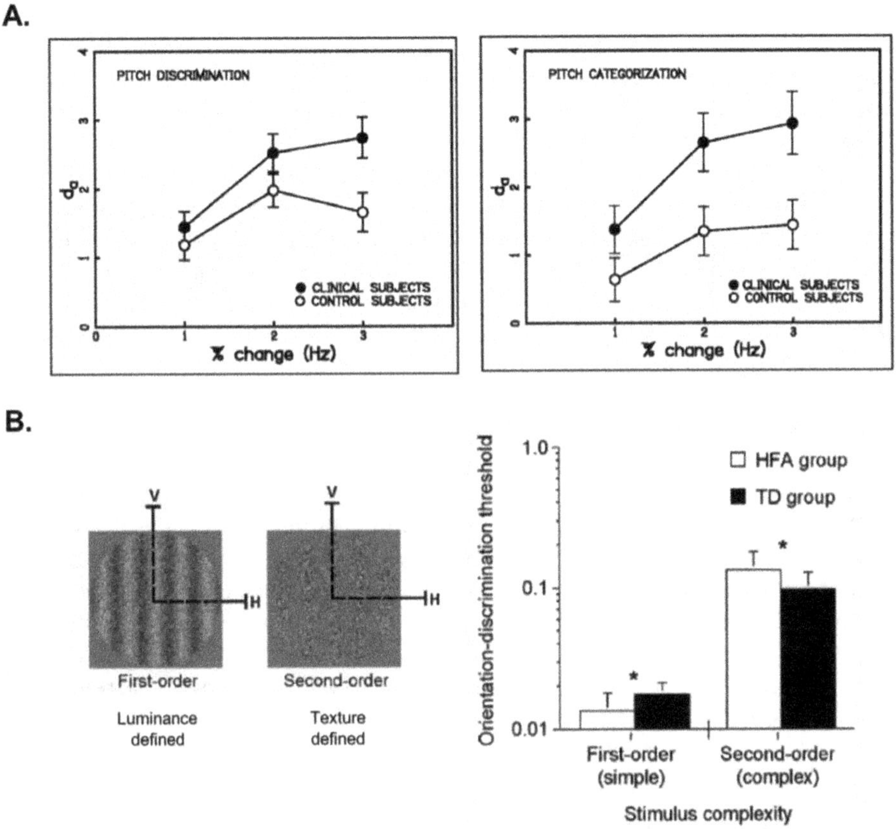

Fig. 12.1 Evidence for enhanced and impaired sensory processing in individuals with ASD. (**A**) Individuals with ASD (clinical subjects) outperform matched NT individuals (control subjects) in both the discrimination (*left*) and categorization (*right*) of differing tones. Adapted from Bonnel et al. (2003), with permission. Error bars are ±SD. (**B**) Individuals with ASD ("high-functioning" autism [HFA]) outperform NT individuals in first-order (luminance-defined) grating discrimination but show deficits relative to the NT (typically developing [TD]) group when the gratings are defined using second-order characteristics (texture). *V* and *H*, vertical and horizontal axes, respectively, of the visual stripes. Error bars are SDs. *$P < 0.05$. Adapted from Bertone et al. (2005), with permission

(Bennetto et al. 2017). Additionally, brainstem auditory evoked potentials, measuring auditory activity from the cochlea through the earliest processing stages in the brain, have longer latencies in a significant subset of children with ASD (Nagy and Loveland 2002; Ververi et al. 2015). Collectively, these findings of low-level differences suggest that auditory processing is impacted in ASD at some of the earliest stages in which auditory information is processed by the cochlea and brain.

Further along the processing hierarchy, different sound elements must be grouped into auditory objects to perceptually bind information coming from individual sources and filter background "noise" (Shinn-Cunningham et al. 2017). To do so, multiple acoustic features (spectral, temporal) must be integrated, a process that involves a number of brain regions (Christison-Lagay et al. 2015). Here, individuals with ASD have more difficulty than NT individuals in both integrating and parsing auditory (as well as visual) scenes (Lin et al. 2017). ASD individuals show a reduced ability to perceive auditory objects, particularly noted in the ability to filter out competing auditory stimuli as measured by both behavioral and electrophysiological methods (Lodhia et al. 2014). In contrast to the enhanced pitch perception for sentences, when asked instead to judge the semantic content of the sentences, children with ASD performed much worse than NT children (Järvinen-Pasley et al. 2008a). Furthermore, individuals with ASD show atypical neural responses when perceiving spatial cues like interaural time and level differences, which are used to group auditory features into auditory objects (Lodhia et al. 2014, 2018). These difficulties in utilizing auditory cues, especially for complex auditory stimuli like speech, have been hypothesized as contributing to the overwhelming nature of complex everyday sensory environments reported by many ASD individuals (Markram and Markram 2010).

When taken as a whole, the enhanced abilities on the processing of low-level auditory stimulus features contrasted with the weakness in perceptual grouping/binding represent the cornerstone of several of the more prevalent neurobiologically inspired theories of ASD. One that is strongly anchored in these data is weak central coherence, which posits that whereas local connectivity within brain circuits is either preserved or enhanced in autism, more global connectivity across brain regions is compromised (Happé 1999; Peiker et al. 2015).

12.1.2 Perceptual Differences in Visual Processing

As for audition, many measures of simple visual processing show similar or superior performance in ASD individuals compared with their NT peers. For example, individuals with ASD tend to have better performance on various visual search tasks (Simmons et al. 2009). Using very short (160-ms) display times to focus on bottom-up processing, Shirama et al. (2016) found that adults with ASD are both faster and more accurate at finding visual targets embedded within a display of distractors. However, there are some exceptions to this general improvement in the processing of low-level stimulus features, including in the ability to exploit statistical features of the stimuli over time. For example, individuals with ASD are poorer at detecting systematic biases in the location of a visual target across a series of trials (Pellicano et al. 2011) and appear to be less flexible in encoding the stimulus when it changes location (Harris et al. 2015). The ability to focus on more local features seems to result in less susceptibility to distracting features, such as surround suppression by visual stimuli in the periphery (Flevaris and Murray 2014), which appears to parallel the difference between local and global processing articulated in Sect. 12.1.1 for auditory stimuli. Extending these findings, during the perception of visual motion children with ASD show atypical patterns. Spatial suppression refers to the paradoxical finding that it is often more difficult to perceive the motion of large, high-contrast stimuli compared with smaller stimuli (Pack et al. 2005), whereas spatial facilitation refers to the relative boost in perceiving the motion direction of large stimuli of low-contrast (Tadin and Lappin 2005). Children with ASD show a reduction in spatial suppression (i.e., they show less of a performance decrement perceiving motion with large vs. small, high-contrast stimuli) as well as an enhancement of spatial facilitation (Foss-Feig et al. 2013; Sysoeva et al. 2017). One common feature in much of this work is the revelation of processing differences when stimuli become dynamic, possibly reflecting challenges in temporal integrative processes (a topic that is returned to in Sect. 12.4.2).

Although individuals with ASD are often superior at discerning distinct features of a complex visual stimulus, much like for audition, the grouping of these sensory cues into a single perceptual object seems to be weakened (Dakin and Frith 2005). In an orientation discrimination task, Bertone et al. (2005) found that ASD individuals performed better than NT individuals for discriminating simple, "first-order" (luminance-defined) orientations but worse for discerning more complex, "second-order" (texture-defined) orientations, which are likely to be processed later in the visual hierarchy (Fig. 12.1B). In a more "traits-based" approach to these questions, the association between decreased perceptual grouping and autism features was also found. Surround suppression, in which the proximity of objects to a target impedes responses to that target, was found to be reduced in individuals without a diagnosis of ASD but with higher scores on a measure of autism traits, the Autism Quotient (AQ; Flevaris and Murray 2014).

The processing of biological motion, which combines both visually complex and social information, also seems to be altered in ASD. Children with ASD spend less time looking at biological motion (Annaz et al. 2012) and have worse performance on tasks that rely on biological motion (Blake et al. 2003; Swettenham et al. 2013). Difficulties with biological motion seem to be exacerbated when the task requires the use of this motion to recognize emotions. For example, attempting to identify the emotion (e.g., happy, angry) in the body language of a walking point light display (Nackaerts et al. 2012). Such a result underscores the weaknesses in the processing of social information in ASD. Differences in performance on biological motion tasks between individuals with ASD and NT seem to diminish in adulthood (Murphy et al. 2009); however, there is evidence that this similar performance between ASD and NT adults may be mediated by distinct cortical networks (McKay et al. 2012).

As for auditory processing, the weight of the evidence in regard to visual processing in ASD points toward selective enhancements in low-level abilities coupled to processing weaknesses for more high-order stimuli. One illustrative and powerful example of this comes from work using the embedded figures test, in which subjects are asked to identify shapes that are embedded within larger illustrations (Fig. 12.2A). Children with autism show performance advantages from processing the component elements of a stimulus array but substantial deficits when asked to report on the whole image (Fig. 12.2B; Pellicano et al. 2005; Plaisted et al. 1999).

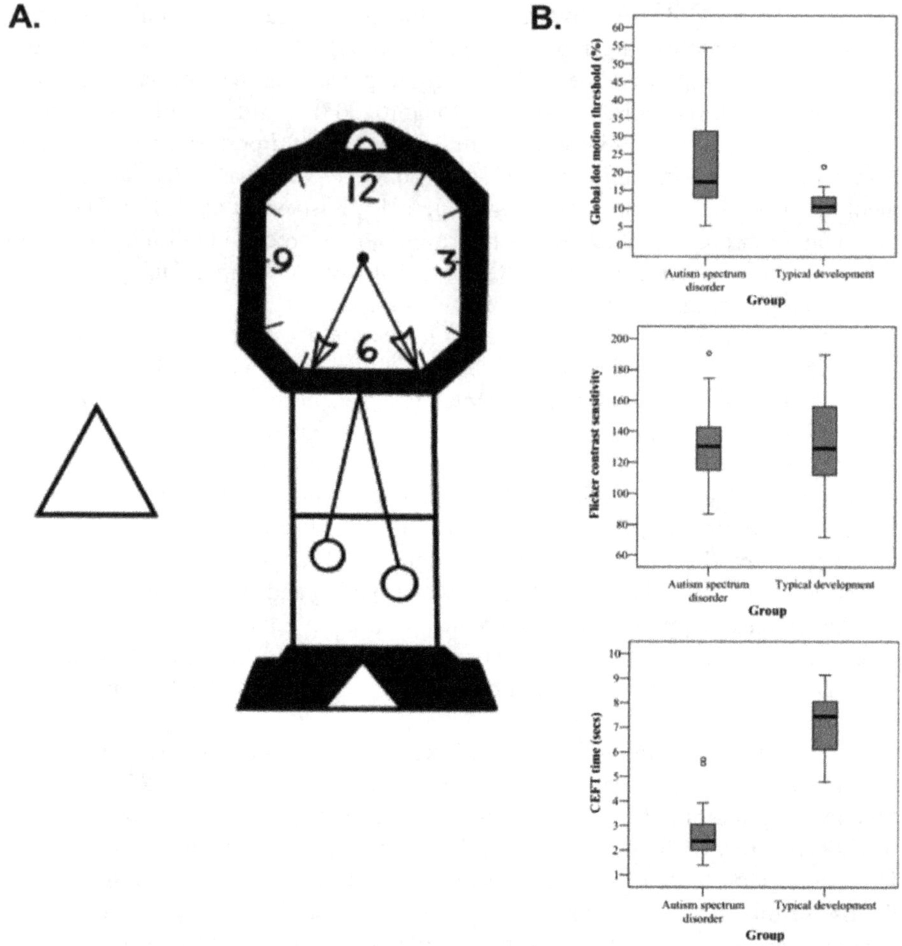

Fig. 12.2 Local and global perceptual processing in ASD. (**A**) Illustration of an image used in the children's embedded figures task (CEFT). (**B**) Performance on three different visual tasks for a group of ASD and TD participants. The global dot-motion task is a two-alternative forced choice (2AFC) task in which participants indicate the general direction of motion of a group of dots (up or down) and taps measures of visual processing later in the visual cortical hierarchy (*top*). Flicker contrast sensitivity is a two-interval forced choice (2IFC) task that quantifies the contrast at which participants can reliably identify (75% threshold) the interval with a Gaussian blob with a 10-Hz sinusoidal flicker and measures low-level visual processing (*center*). The CEFT measures how quickly participants can identify hidden features that, when assembled, result in a larger image with different meaning (*bottom*). In the example shown, the clock (**A**) is made up of a number of triangles. Note that whereas neither dot-motion thresholds nor flicker contrast sensitivity differ between groups, there is a striking difference in the CEFT, with ASD children being much faster to identify the number of components making up the whole image. Box plots are the distribution of the middle 50% of the scores. *Solid black lines,* median of each box plot. *Bars at top and bottom* of each box plot extend to include all of the data, excepting outliers, which are marked individually. Adapted from Pellicano et al. (2005), with permission

12.1.3 Perceptual Differences in the Integration of Auditory and Visual Information

In addition to the processing differences observed within individual senses like audition and vision, there is growing evidence of changes in the processing and integration of information across the different sensory modalities in individuals with ASD (for a review, see Baum et al. 2015). Similar to the differences noted in the processing of stimuli within the individual sensory modalities, changes to multisensory processing observed in individuals with ASD are manifold and differences depend on a number of features, including both the type and complexity of the stimuli that are combined.

One of the most salient cues for integrating multiple pieces of sensory information is the temporal relationship between the stimuli from the different modalities (Welch 1999; Stevenson and Wallace 2013). Stimuli that occur at the same time are likely to come from the same source and should be integrated, whereas stimuli that occur at different times should remain segregated. Overall, individuals with ASD are less able to accurately detect these multisensory temporal relationships than their NT peers (Foss-Feig et al. 2010; de Boer-Schellekens et al. 2013a), and emerging work suggests these differences may be particularly pronounced for speech stimuli (Stevenson et al. 2014a), perhaps serving as a foundation for the communication deficits that often accompany ASD (Fig. 12.3). In addition to these difficulties in the temporal processing of audiovisual speech, Foxe et al. (2015) observed that children with ASD are less able to utilize visual speech information to improve speech comprehension, and that this multisensory difference grows larger under noisy and more naturalistic conditions. This paradigm builds off of the foundational evidence that the ability to see a speaker's mouth provides a large gain in intelligibility to the spoken signal (cf. Grant and Bernstein, Chap. 3) and that these visually-mediated benefits grow larger under noisy conditions (Sumby and Pollack 1954; Ross et al. 2006).

Illusions are commonly used as one means of studying multisensory integration, where a number of audiovisual illusions have provided great insight into how auditory and visual information are synthesized (cf., Lee, Maddox, and Bizley, Chap. 4). For example, the sound-induced flash illusion consists of a single flash of light paired with two or more auditory stimuli (beeps) in quick succession. The participant is then asked to report the number of flashes while ignoring the beeps, with the task-irrelevant beeps often resulting in the illusory perception of several flashes (Shams et al. 2000). Children with ASD appear less susceptible to perceiving this illusion (Stevenson et al. 2014b), although they also seem to perceive the illusion over a wider range of temporal asynchronies (i.e., delays between the flashes and beeps) than their NT peers (Foss-Feig et al. 2010), providing further evidence for temporally based multisensory processing differences. Another common illusion, the McGurk effect, involves the presentation of an auditory syllable (e.g., /ba/) paired with an incongruent visual syllable (e.g., /ga/), which frequently results in the perception of a novel syllable (e.g., /da/), reflecting a synthesis of the auditory

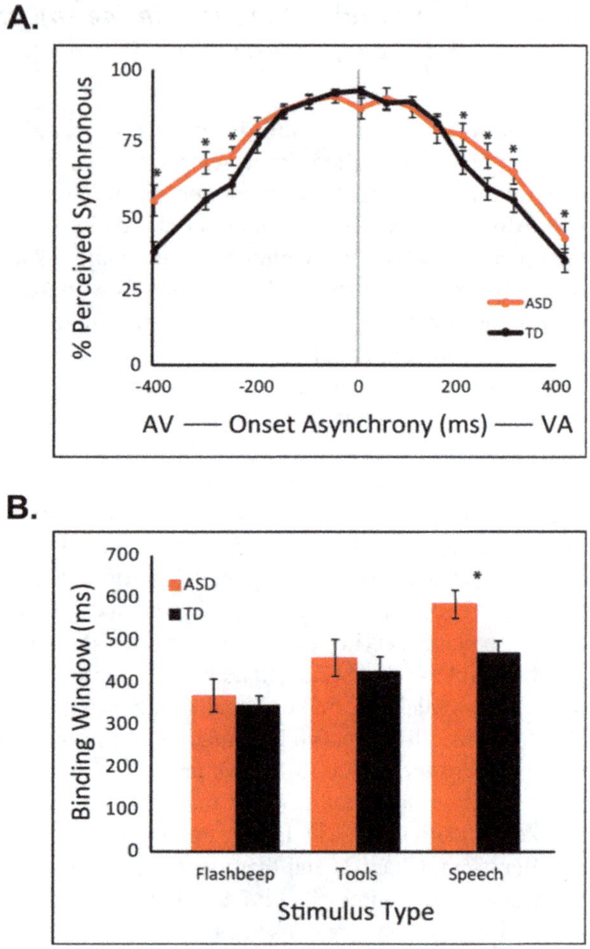

Fig. 12.3 Alterations in audiovisual temporal function in ASD. (**A**) Performance on a simultaneity judgment task reveals differences in performance between the ASD (*gray*) and NT (*black*) groups, with those with ASD showing a higher likelihood of reporting simultaneity for highly asynchronous audiovisual pairings. *AV*, auditory leads; *VA*, visual leads. *$P < 0.05$. (**B**) Group averages for the width of the audiovisual temporal binding window (TBW) as a function of stimulus complexity (flashbeep, visual flash with auditory beep; tools, video and audio of a handheld hammer hitting a table; speech, video and audio of the syllables /ba/ and /ga/) reveals preferential differences in the processing of speech-related stimuli. Error bars are ±SE of the mean. *$P < 0.05$. Adapted from Stevenson et al. (2014a), with permission

and visual cues (McGurk and MacDonald 1976). Many studies have found that individuals with ASD are less susceptible to this illusion (Irwin et al. 2011; Stevenson et al. 2014c). Recent work suggests that these differences in illusory perception may be due to differences in top-down factors (i.e., decision criterion) as opposed to differences in simple bottom-up stimulus integration (Magnotti and Beauchamp 2015).

One of the challenges in studying multisensory processing and the possible differences in ASD is teasing out the respective contributions of changes in unisensory function from changes in the integration of information across the different sensory modalities. As detailed in Sects. 12.1.1 and 12.1.2, there is substantial evidence in support of unisensory processing differences in ASD, and these differences may be responsible for many of the apparent changes in multisensory abilities. However, a number of studies have now attempted to dissociate these effects and have shown in many cases that the deficits seen in multisensory processing go beyond what is predicted from performance on unisensory tasks (Brandwein et al. 2013; Stevenson et al. 2014b).

One of the most powerful approaches to this question is through the use of computational models that endeavor to parse out the individual contributions of both the individual sensory inputs as well as the actual process of integrating the individual cues. In particular, Bayesian modeling is increasingly being applied to examine sensory processing in ASD in an attempt to disentangle possible mechanisms for these sensory and multisensory processing differences (Pellicano and Burr 2012; Van de Cruys et al. 2014). Bayesian models of sensory processing formalize perception as statistical inference, where incoming information is combined with expectations and prior experience that ultimately results in the final percept, and these perpetual experiences provide updates that inform the processing of subsequent incoming information (cf. Shams and Beierholm 2010; Alais and Burr, Chap. 2). One theory of ASD posits that internal representations of the world (so-called Bayesian "priors") are weak in ASD and thus provide a poor reference for incoming information, resulting in an overweighting of incoming sensory evidence (Pellicano and Burr 2012) and an overestimation of the volatility of this evidence (Lawson et al. 2017; Palmer et al. 2017). By formalizing possible mechanisms of perception, these models may help pinpoint how and where sensory perception diverges in ASD with testable hypotheses. For example, ongoing work suggests that internal sensory representations may actually be intact in individuals with ASD (Pell et al. 2016; Croydon et al. 2017) or are specifically impaired only in social situations (Chambon et al. 2017). Furthermore, some studies have shown that individuals with ASD can learn from and update their representation of the environment appropriately based on incoming sensory information in some contexts (Manning et al. 2017). Although more work is needed to fully characterize and understand perceptual differences in ASD, Bayesian models provide a powerful framework within which these different mechanisms may be tested.

12.2 Neural Correlates of Sensory Processing in Autism Spectrum Disorder

12.2.1 Magnetic Resonance Imaging

Functional magnetic resonance (MR) imaging (fMRI) allows for the noninvasive investigation of the neural circuitry involved in sensory and perceptual processes. Although the hemodynamic response (i.e., changes in oxygenated and deoxygenated hemoglobin) that is the basis of the fMRI signal is slow, it is highly localized, which allows for a relatively high degree of spatial resolution (Glover 2011). In addition to studying functional brain activity, MRI can also be used to investigate the underlying structural networks (i.e., anatomical connectivity) that support various neural functions (structural MRI).

As highlighted in Sect. 12.1.1 in regard to perceptual differences in auditory processing, individuals with ASD show similar behavioral performance to their NT peers in the detection of simple tones. Brain imaging done during this task shows a similar pattern of temporal (auditory) cortex activation between the ASD and NT groups, including bilateral superior and middle temporal gyri (Brodmann areas 41, 42, and 22), but also a much broader set of activity extending into right prefrontal and premotor cortices for the ASD individuals (Gomot et al. 2008). More temporally complex (frequency-modulated) sounds evoked enhanced responses in the primary auditory cortex in individuals with ASD but reduced responses in areas surrounding the auditory cortex (Samson et al. 2011). Speech processing, which involves complex and socially relevant auditory information, is an area where individuals with ASD are thought to be particularly affected. Although previous work reported a lack of voice-sensitive regions in individuals with ASD (Gervais et al. 2004), emerging work suggests that these regions do exist but show atypical activity during voice identity-recognition tasks (Schelinski et al. 2016). Intriguingly, children with ASD show a response pattern where evoked responses are reduced in response to spoken speech but look surprisingly similar to NT children when the speech is sung rather than spoken (Sharda et al. 2015), indicating that speech might be more affected in certain contexts. How these atypical neural networks and response patterns contribute to altered auditory processing in ASD is currently unclear.

Paralleling the behavioral findings of largely similar performance in tasks indexing low-level visual processing, a comprehensive study measuring visual contrast sensitivity across a range of spatial frequencies found no difference in brain activation patterns between ASD and NT individuals (Koh et al. 2010). In contrast, enhanced visual search has been a consistent finding of behavioral studies in ASD, and these enhancements seem to be accompanied by greater activity in regions of early visual cortex (Manjaly et al. 2007).

A large body of work in regard to visual processing in ASD has focused on face processing. Several studies have shown weaker activation in the fusiform face area in response to faces in individuals with ASD (Hubl et al. 2003; Corbett et al. 2009) whereas viewing pictures of inanimate items that are the focus of restricted interests (e.g., trains, cars) elicits greater activity in this same area (Foss-Feig et al. 2016). Furthermore, reductions in right fusiform face area activity in response to faces have been shown to be correlated with symptom severity in ASD (Scherf et al. 2015).

Structural and functional MRI have shown differences in the connectivity and lateralization of sensory networks in the brains of those with autism, including changes in the white matter integrity of auditory, language (Nagae et al. 2012; Berman et al. 2016), and visual (Thomas et al. 2011; Yamasaki et al. 2017) networks. In fact, these differences in network structure appear to be present well before ASD can be diagnosed. A large-scale study of infant siblings of children with ASD (and thus who are at elevated risk for being later diagnosed with autism) just found that those infants who were later diagnosed with ASD showed an enlarged cortical surface area across a range of regions of interest, including auditory and visual cortices, as early as 6–12 months old (Hazlett et al. 2017).

These differences in the structure of neural networks are also mirrored in functional connectivity findings. Reduced functional connectivity has been observed during traditional paradigms in ASD research such as theory-of-mind tasks that probe the ability to infer intentions from others' actions (Fig. 12.4A; Kana et al. 2014). Additionally, increased ASD symptom severity in regard to sensory features is correlated with reduced interhemispheric connectivity in auditory (Heschl's gyrus and superior temporal gyrus) cortices (Linke et al. 2018). Furthermore, differences in functional connectivity in voice-selective regions of the posterior superior temporal sulcus (pSTS) and other nodes in the reward system are predictive of social communication skills as indexed by both the Autism Diagnostic Observation Schedule (ADOS) and Autism Diagnostic Interview (Fig. 12.4B; Abrams et al. 2013). Although ASD has traditionally been thought of as a disorder of hypoconnectivity, ongoing work has provided evidence for both hypo- and hyperconnectivity (Hull et al. 2017). Furthermore, both short- and long-range connectivity differ across the

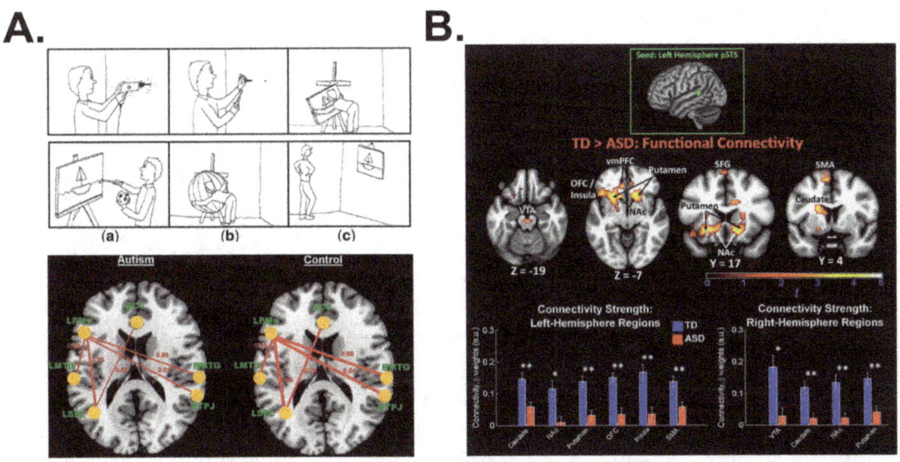

Fig. 12.4 Differences in functional connectivity between brain regions is a common feature of ASD. (**A**) Evidence for weaker functional connectivity while attributing causal explanations to the actions of others. *Top*, some of the actions that were explored in this study; *bottom*, strength of connectivity between several brain regions during the causal explanation task (both numbers and width of the lines reflect connectivity strength). Sample stimulus item from an experimental condition in the intentional causality vignette is depicted, with subjects being asked to choose the intent of the subjects after viewing the sequence depicted in *a*, *b*, and *c* (correct answer is *c*). Adapted from Kana et al. (2014). (**B**) Reduced connectivity in language areas correlates with communication subtest scores of the Autism Diagnostic Observation Schedule (ADOS) and Autism Diagnostic Interview (ADI). Figure depicts connectivity strength between a seed in the posterior superior temporal sulcus (pSTS) region and a number of brain regions in the reward network. *Yellow circles*, brain regions of interest; *red lines*, functional connectivity between them, with the numbers showing the strength of the connectivity. Adapted from Abrams et al. (2013), with permission

developmental trajectory in individuals with ASD, revealing an age by distance interaction (Long et al. 2016).

These differences in both structural and functional connectivity between individuals with ASD and NT are highly likely to impact multisensory processing, which relies on communication across both local and long-range networks. However, to date, little has been done that focuses exclusively on multisensory function and its associated brain networks. In a recent study, Ross et al. (2017) studied NT adults who had an autism risk-associated gene variant of *CNTNAP2* and explored whether this genetic variant mediated individual differences in multisensory gain as measured in a speech-in-noise task. The results showed that multisensory gain was diminished in those with the risk-associated allele, who also had overall lower fractional anisotropy (FA), a measure of the structural integrity in white matter tracts, in clusters in right precentral gyrus, and continuing into the superior longitudinal fasciculus as well as in the left corona radiata and the right superior temporal gyrus. An interesting functional corollary to this finding was that the effect of this genotype on multisensory gain was mediated by FA in the right precentral gyrus. Counterintuitively, it was found that decreased FA was linked to increased audiovisual gain. The authors suggest that although stronger structural and functional connectivity of the motor system is typically associated with improvements in speech perception (Du et al. 2014), this is usually in the case of typical development where language function is left lateralized. Reduced lateralization in language function, which has been a frequent observation in autism (Floris et al. 2016), may then explain a greater reliance on the right hemisphere and reduced performance. Much more work is needed in both the functional and structural imaging realms to clarify the differences in multisensory circuits in those living with ASD.

12.2.2 Electrophysiology

The neural underpinnings of ASD have also been investigated using electrophysiological techniques including electroencephalography (EEG), which, unlike MRI, allows for temporal resolution on a millisecond-level timescale but lacks a high degree of spatial resolution (Lee et al. 2014). EEG allows for investigations into the temporal dynamics of neural activity, which can help pinpoint when atypical brain activity emerges while processing a sensory stimulus. This information may prove useful as a biomarker for both humans and animal models of ASD (for a review, see Modi and Sahin 2017).

In the auditory domain, EEG can be particularly powerful because of the high temporal precision of the auditory system and because studies can be conducted without the loud noises of the MRI scanner as a possible confound. Paralleling the behavioral results, differences in auditory processing in ASD have been found at very early stages using EEG. For example, for the classic mismatch negativity (MMN) seen in response to an auditory oddball (an infrequently presented variant of a repeated stimulus), reduced amplitudes were seen in response to tone bursts

(Abdeltawwab and Baz 2015) in those with ASD. In an adaptation of the standard MMN paradigm, Lepistö et al. (2009) created a version of the task that either required auditory stream segregation (segregated condition) or not (integrated condition) and found a reduced MMN in individuals with ASD but only for the segregated condition. These reductions in the auditory MMN seem to be especially pronounced with speech stimuli compared with nonspeech sounds with similar spectral and temporal content (Fan and Cheng 2014) and have been interpreted as a deficit in the ability to accurately predict incoming inputs and map these on to expectations. These data can be interpreted from the perspective of weaknesses in predictive coding that have been hypothesized to play an important role in ASD (Van de Cruys et al. 2014).

Similar to the results seen in the auditory system, examining visual activation patterns using EEG in ASD reveals both similarities and differences to NT counterparts. For example, simple visual features such as retinotopic organization can be derived from EEG and have been found to be similar in the central visual field of ASD and NT individuals but to differ for the peripheral visual field (Frey et al. 2013). Other aspects of low-level visual brain activation patterns that differ in ASD include boundary detection (Vandenbroucke et al. 2008), spatial acuity (Pei et al. 2014), and visual oddball detection (Baruth et al. 2010). In addition to these differences that largely manifest as differences in response amplitude (the strength of some aspect of the EEG signal), differences in the lateralization of responses have also been noted, with the general finding of a shift toward less lateralization in ASD (Pei et al. 2014).

In addition to these changes seen in EEG markers of early visual function, there is also growing evidence for differences in more complex aspects of visual processing. For example, there is ample evidence for differences in motion processing, and, in particular, differences in the perception and processing of biological motion have been a common observation (Kröger et al. 2014). Furthermore, the ability to learn regularities in a visual scene over repeated exposure (visual statistical learning) is known to be impaired in ASD, and differences in event-related potential (ERP) amplitudes can account for these performance differences in those with ASD (Jeste et al. 2015). Additionally, neural processing of explicitly social visual stimuli like faces is also atypical in ASD. For example, differences in the lateralization of EEG responses to faces is observed in children as early as 12 months and can predict ADOS scores at 36 months of age (Keehn et al. 2015). These differences seem to continue through to adulthood, where adults with ASD show less differentiation of ERP responses to upright versus inverted faces (Webb et al. 2012) as well as a reduction in the preferential processing of one's own face compared with the face of others (Cygan et al. 2014).

EEG can also be used to investigate brain rhythms (i.e., oscillations), which appear to be critical indices of information flow through cortical circuits. In particular, abnormal oscillatory power in a variety of frequency bands has been consistently identified in ASD (Simon and Wallace 2016). For example, power in the gamma band (>30 Hz), which is thought to play a key role in perceptual integration (Keil et al. 1999), has been found to be diminished in individuals with ASD

(Snijders et al. 2013; Peiker et al. 2015). Alpha band (8- to 14-Hz) abnormalities have also been observed, with reduced power observed in individuals with ASD (Milne 2011; Murphy et al. 2014). These reductions in alpha power, which are typically thought to reflect a reduced ability to inhibit task-irrelevant information (Klimesch et al. 2007), may represent the neural correlates of deficits in sensory filtering. Furthermore, these alterations in oscillatory dynamics can be tied to ASD symptomology. For example, differences in both gamma and theta (4- to 7-Hz) activity in response to speech can predict the degree of verbal deficit and symptom severity in individuals with ASD (Jochaut et al. 2015). Differences in oscillatory power also seem to change based on task demands. Indeed, although gamma power is typically reduced in individuals with ASD during sensory processing, it is increased relative to NT individuals during the resting state, i.e., when subjects are not performing an explicit task (Cornew et al. 2012; Wang et al. 2013).

The presence of these differences in oscillatory function in ASD become increasingly important as knowledge grows concerning the central role of these oscillatory processes in information encoding and transfer. Perhaps most importantly in the context of prevailing theories of autism is the fact that these oscillations are indexing processes at various levels of the processing hierarchy, with a simple framework that the higher the frequency of the oscillation, the more local the neural process that underlies it. Thus, high-frequency (i.e., gamma) oscillations are generally thought to reflect processes within very local circuits, whereas those in the lower frequency bands are reflective of processes that are indexing communication across broad regions. Further complicating matters, these oscillatory frequencies are not independent of one another, and changes within one frequency band almost invariably results in changes across other frequency bands, through processes described as phase-phase coupling and phase-amplitude coupling (Canolty and Knight 2010). Hence, changes in one frequency band generally propagate throughout the oscillatory network.

In the context of sensory processing, an important finding has been the seeming importance of oscillations and oscillatory coupling to facilitate communication across the senses (cf. Keil and Senkowski, Chap. 10). Studies have shown that, even in the earliest regions of the sensory cortex (e.g., primary auditory cortex), input from other senses has the capacity to change the nature of the oscillations in that region (Thorne et al. 2011; Romei et al. 2012). For example, through a process known as phase reset, it has been shown that visual inputs have the ability to reset ongoing oscillations in the primary auditory cortex, thus changing the nature of information exchange at the earliest stages of auditory cortical processing. Although yet to be firmly established, such cross-modal phase resetting can provide valuable predictive information about the nature of a multisensory stimulus. For example, the articulatory movements of the mouth (i.e., visual speech) happen before the audible speech signal. If such visual information has early access to the auditory cortex, it then has the ability to provide predictive information about the auditory information that is about to arrive and thus the ability to boost (or suppress) the gain of the signal. Such findings may have important implications because there is ample evidence for oscillatory dysfunction (Simon and Wallace 2016; Larrain-Valenzuela et al. 2017), multisensory temporal deficits (Brock et al. 2002; Stevenson et al. 2014a), and weaknesses in speech comprehension (Woynaroski et al. 2013; Stevenson et al. 2017) in individuals with ASD.

12.3 Developmental Trajectory of Sensory Processing in Autism Spectrum Disorder

12.3.1 Infancy and Early Childhood

ASD cannot currently be reliably diagnosed in children younger than 2 years of age (Lord et al. 2006; Luyster et al. 2009). Therefore, younger siblings of children with ASD, who are at a higher risk of being later diagnosed with ASD, provide a useful avenue of research for assessing early signs of ASD and the development of potential biomarkers for the progression to autism (Ozonoff et al. 2011). Sensory function and the associated brain networks undergo dramatic changes in early life, and the detailing of neurotypical developmental trajectories provides an opportunity to delineate when maturation begins to deviate from the typical developmental pattern to patterns characteristic of disorders such as autism.

In the auditory domain, infants who are at high risk for ASD at 9 months show reduced habituation and sensitivity in evoked EEG responses to repeated pure tones in an auditory MMN paradigm (Guiraud et al. 2011). In a similar paradigm using consonant-vowel stimuli (used to assess the processing of more speech-related stimuli), high-risk infants showed hypersensitivity to the standard but similar responses to the deviant as infants at a low risk of being diagnosed with ASD (Seery et al. 2014). Although high-risk infants show a similar developmental progression to low-risk infants in regard to a specialization toward processing native speech sounds as they grow older, they do not show the same left-lateralized response to speech as is seen in low-risk infants between 6 and 12 months (Seery et al. 2013). This lack of left-lateralized responses to language is also observed in somewhat older children (i.e., 12–24 months old), and this pattern appears to worsen with age (Eyler et al. 2012). Another early auditory warning sign in infants who are later diagnosed with ASD is a failure to orient to their own name as early as 9 months of age (Miller et al. 2017). Indeed, this lack of response, generally captured within the domain of hyporesponsivity, is often one of the earliest concerns many parents report concerning their child.

In the visual domain, gaze-tracking studies have been conducted with infants at risk for being diagnosed with ASD as a means of assessing sensory attention and as an early marker for how infants interact with their environment (Falck-Ytter et al. 2013). A preference for looking at nonsocial images, as measured by fixation time and number of saccades, seems to emerge as early as 15 months of age in children who will progress to autism (Pierce et al. 2016). In a longitudinal study tracking gaze to faces in infants, fixation to the eye region of a face stimulus declined in infants ages 2–6 months who would later be diagnosed with ASD (Jones and Klin 2013).

As with atypical lateralization for language observed in high-risk infants, a similar pattern emerges in face perception, where high-risk infants show left hemisphere lateralization for faces while low-risk infants show more of right hemisphere lateralization (Keehn et al. 2015). In a study measuring resting-state EEG at multiple time points in the first 2 years of life, spectral power was lower in high-risk infants across the delta, theta, alpha, beta, and gamma frequency bands but eventually converged with low-risk infants in all frequency bands by 24 months. (Tierney et al. 2012). A possible confound in these studies is the relative difference in signal-to-noise ratio between low- and high-risk infants; however, these differences have not been systematically characterized.

Furthermore, as with differences in connectivity, differences in EEG power noted for individuals with ASD may be more nuanced than a mere reduction or enhancement of spectral power. For example, reduced power in the alpha band at frontal electrodes seems to predict worse expressive language (Levin et al. 2017), whereas increased power in the theta band at frontal electrodes is associated with greater sensory hyporesponsiveness (Simon et al. 2017). Clearly, much more work is needed to better understand how sensory development differs between ASD and NT infants, and to clarify how these differences relate to later differences in cognitive abilities. Additionally, little work has extended these studies into the multisensory arena to see if changes in multisensory function may serve as more sensitive indices of risk for progression to autism.

12.3.2 Adolescence and Young Adulthood

Many of the same patterns of neural processing differences between ASD and NT individuals persist into adolescence and young adulthood, particularly with regard to social stimuli such as speech and faces. Reduced left lateralization in language is present in adolescents and young adults with ASD across a wide range of tasks (Herringshaw et al. 2016), and individual differences in lateralization are tied to performance deficits in language tasks like letter fluency, which measures word knowledge and the ability to retrieve words (Kleinhans et al. 2008). Reduced specialization for native speech sounds also persists through adolescence (DePape et al. 2012). Furthermore, differences in functional connectivity between children, adolescents, and young adults with ASD and their NT peers are noted in language-processing areas across all three age groups (Lee et al. 2017).

Gaze differences observed in infancy seem to continue throughout development. School-age children with autism look at faces less in videos of social scenes (Rice et al. 2012) as well as in live social interactions (Noris et al. 2012), Furthermore, children with ASD frequently do not shift their gaze within the context of a social scene, such as following a back-and-forth conversation between two speakers (Hosozawa et al. 2012). Reduced fixation time on the eyes is also observed through adolescence and young adulthood in individuals with ASD (Sasson 2006; Frazier et al. 2017) and may instead be shifted toward the mouth

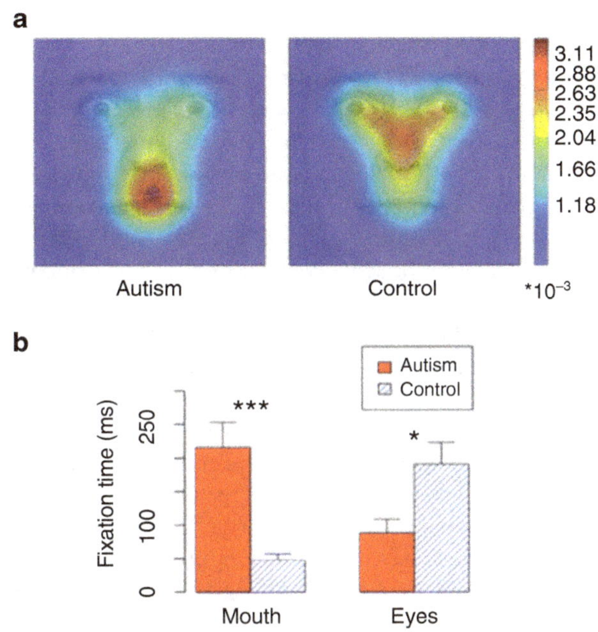

Fig. 12.5 Differences in gaze characterize in ASD. (**A**) Heat maps representing the time spent viewing various locations on the face. Note the significant differences in gaze patterns, with those with ASD seemingly avoiding looking to the eyes. (**B**) Quantification of amount of time looking at the eyes versus the mouth. Error bars are SDs. $^*P < 0.05$; $^{***}P < 0.001$. Adapted from Neumann et al. (2006), with permission

(Fig. 12.5; Neumann et al. 2006). These differences in gaze processing appear to continue into adulthood (Zalla et al. 2016).

In regard to multisensory function, much work is still needed to delineate the developmental trajectories associated with both ASD and NT development. Initial work has suggested that these trajectories show some convergence in adolescence (de Boer-Schellekens et al. 2013b; Beker et al. 2018), but it is unclear how the relationship between multisensory integration and autism symptomology changes over the lifetime. In typical development, aspects of multisensory function such as audiovisual temporal acuity mature as late as adolescence or early adulthood (Hillock-Dunn and Wallace 2012). Such a delayed developmental trajectory may be of great interventional utility because it suggests that multisensory plasticity remains quite robust well into adolescence, thus providing great opportunity to strengthen multisensory abilities through training-based approaches.

12.4 Connecting Sensory Processing to Clinical Symptoms

12.4.1 Atypical Sensory Processing Patterns

Historically, sensory processing issues in individuals with autism have been treated as unrelated to impairments in the other core domains of ASD (Rogers and Ozonoff 2005). This seems shortsighted because these domains are heavily dependent on the integrity of the incoming sensory information processing. Furthermore, a better understanding of sensory features will benefit clinical assessment of ASD, including both diagnosis and treatment strategies (Schaaf and Lane 2015).

Abnormal sensory processing in individuals with ASD is typically broken down into three broad patterns: hypersensitivity, hyposensitivity, and sensory seeking (Baranek et al. 2006; Ben-Sasson et al. 2009). It is important to note that these distinctions and divisions have historically been made on the basis of survey and observational data but that there is a growing emphasis on more rigorous empirical characterization of sensory function using psychophysical and behavioral task batteries. These patterns can be seen across multiple sensory modalities even within the same individual and have been noted not only in ASD but also in other clinical groups characterized by developmental delays such as Down syndrome (Boyd et al. 2010). Furthermore, many of these abnormalities persist into adulthood (Crane et al. 2009), although individuals with ASD seem to "catch up" by adulthood to their NT peers on a subset of tasks (Beker et al. 2018).

Recent work has sought to bridge between sensory function and the more classic domains of clinical dysfunction (i.e., social communication and restricted interests and repetitive behaviors) and strongly suggests that abnormalities across these core domains of ASD are related, at least in part, to differences in sensory function. For example, in children with ASD, sensory hyperresponsiveness is correlated with an increased presence of repetitive behaviors (like stereotypical hand flapping), whereas sensory seeking is associated with the presence of ritualistic behaviors and routines (Boyd et al. 2010). In a large-scale study by Mayer 2017, the presence of abnormal sensory processing patterns was compared with specific autism traits as measured by the AQ. Across both NT and ASD adults, greater levels of abnormal sensory processing (failure to register sensory stimuli, sensory seeking, and sensory sensitivity) were correlated with lower functioning in multiple subdomains of autism symptomology (social skills, attention switching, and communication). In a more specific example that links directly to auditory function, difficulties in focusing on an auditory stream of interest in the presence of distractors (termed auditory filtering) has been connected to cognitive problems in the classroom (Ashburner et al. 2008). The Short Sensory Profile (McIntosh et al. 1999) characterizes auditory filtering by asking the caregiver how well the child performs day-to-day activities in the presence of noise; 50% or more caregivers of children with ASD marked "always" or "frequently" to items like "doesn't respond when name is called but you know the child's hearing is OK" and "is distracted or has trouble functioning if there is a lot of noise around." More recent work has framed these perceptual differences as an increased capacity for processing sound, which carries both advantages (e.g., superior pitch perception) and disadvantages (such as a higher propsensity for sensory overload; Remington and Fairnie 2017). This study highlights an increasing awareness in sensory perception research in ASD that focuses on both the deficits and the benefits associated with differences in sensory abilities.

Atypical sensory processing has also been linked to symptom severity in the social, cognitive, and communication domains (Linke et al. 2018) and to self-injurious behavior (Duerden et al. 2012). Furthermore, these connections are found not only in individuals with ASD but also in individuals who do not have ASD but score high on measures of autistic traits (Mayer 2017). This suggests that the relationship between abnormal sensory processing and autistic symptoms occurs not just within the autism spectrum but also across the full range of clinical and subclinical autism symptomology.

12.4.2 Speech and Communication Skills

Several studies have shown that abnormal sensory processing also affects how individuals with ASD communicate and interact with their environment. For example, hyporesponsiveness is associated with increased severity on measures of social communication (Watson et al. 2011). In a study that combined both laboratory and live social interaction, facial recognition was a significant predictor of measures of playing alone and with others (Corbett et al. 2014). In particular, higher scores on a delayed recognition-of-faces task were associated with lower levels of self-play and playing on playground equipment alone and with higher levels of playing with others. Finally, a growing body of work has begun to show the presence of strong links between audiovisual temporal function and clinical measures of speech perception (Woynaroski et al. 2013; Stevenson et al. 2014b) and receptive language functioning (Patten et al. 2014). One illustration of these links was seen by examining the relationship between audiovisual temporal function and reports of the McGurk illusion (cf. Sect. 12.1.3). Here, a strong negative relationship was found between audiovisual temporal acuity (as measured by the size of the temporal window of integration) and reports of the McGurk illusion, suggesting that those with larger windows combine visual and auditory speech signals differently from those with smaller windows (Fig. 12.6; Stevenson et al. 2014a).

12.5 Summary and Future Directions of Research

Changes in auditory and multisensory (i.e., audiovisual) processing and the associated brain networks are a common feature of ASD. Although only recently added to the formal diagnostic framework for ASD, the presence of sensory features has been a long-recognized but poorly understood characteristic of the autism phenotype. Recent work has begun to establish links between sensory function and the more classic characteristics of autism, with the largest body of evidence showing strong relationships with social communication. The presence of such links makes great intuitive sense because higher order constructs such as communication skills and social interactive abilities create a scaffold on a sensory and multisensory foundation where the integrity of sensory information processing and the proper synthesis and integration across the sensory streams is key to social communicative development.

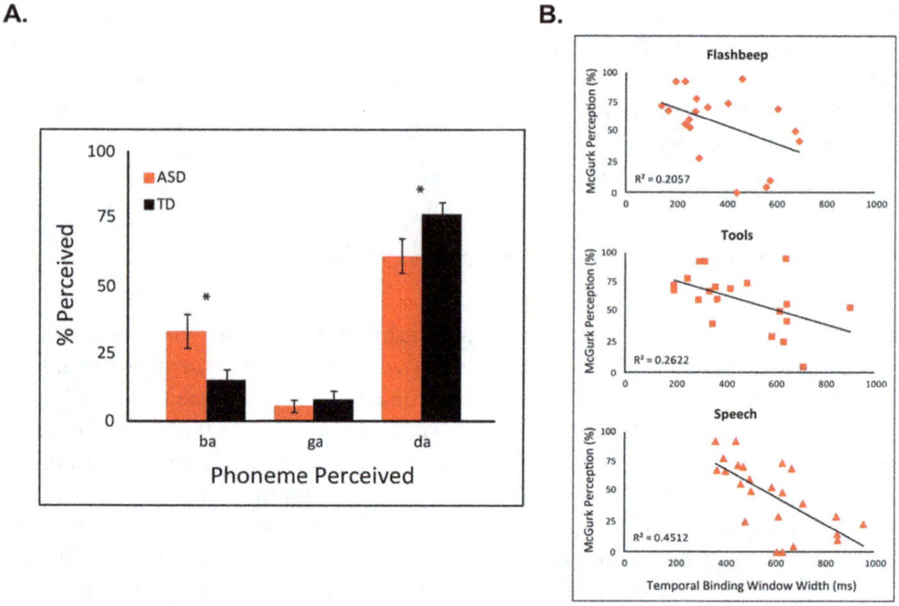

Fig. 12.6 Connections between temporal acuity and multisensory illusory perception. (**A**) Children with ASD and TD children show differences in their perception of a multisensory speech illusion (i.e., the McGurk effect). TD children report the illusory percept /da/ more often than children with ASD. Error bars are SEs of the mean. $*P < 0.05$. (**B**) Differences in susceptibility to the McGurk effect are associated with differences in multisensory temporal acuity (i.e., size of the temporal binding window [TBW]) in ASD for flashbeep ($P < 0.05$), tool ($P < 0.03$), and speech ($P < 0.001$) stimuli such that larger TBWs (worse temporal acuity) are associated with reduced susceptibility of the McGurk effect. Note the strong negative relationship for speech stimuli, in which large TBWs are associated with less reporting of the McGurk illusion. Data are from the ASD children. Adapted from Stevenson et al. (2014a), with permission

From a neurobiological perspective, a great deal of additional work is needed to better understand the circuit and network changes within sensory areas in ASD and how these changes relate to changes in brain regions supporting more cognitive abilities such as social communication. Although much of the prior neurophysiological and neuroimaging work in ASD has focused on differences in brain regions supporting these "higher order" abilities, there is a growing corpus of work oriented toward better elucidating differences in sensory regions of the brain in individuals with autism. Much like the behavioral research that has begun to establish strong associations between sensory function and social communication, these studies now need to address how processing differences in sensory regions and circuits impact the changes that are seen in brain regions responsible for more cognitively directed functions. Key nodes in this analysis will likely be multisensory regions that sit at the transition between unisensory domains of the auditory and visual cortices and regions of the parietal, temporal, and frontal cortices and that have been implicated in higher order functions including attention, executive control, and social cognition.

Two of the most intriguing of these regions are the pSTS (cf. Beauchamp, Chap. 8) and areas along the intraparietal sulcus. In addition to being strongly implicated in the integration of auditory, visual, and tactile information, these areas are also centrally involved in processes integral for speech comprehension (Venezia et al. 2017) and attentional allocation (Corbetta and Shulman 2002), respectively.

Finally, greater knowledge of sensory processing in autism is likely to play an important role in intervention and remediation. Perhaps more important than ameliorating the altered sensory characteristics seen in ASD is the potential for this work to have cascading effects on domains such as social communication. The core question here is whether sensory-based training approaches focused on strengthening aspects of sensory function will have secondary effects on processes dependent on the integrity of this sensory information and the manner in which it is integrated. There is a strong developmental emphasis to this point because the maturation of brain regions responsible for sensory processes takes place before the maturation of those regions more integral for cognitive abilities, and early intervention focused in the sensory realm may set the stage for improving the developmental trajectory of these higher order regions.

Compliance with Ethics Requirements Sarah H. Baum Miller declares that she has no conflicts of interest.

Mark T. Wallace declares that he has no conflicts of interest.

References

Abdeltawwab, M. M., & Baz, H. (2015). Automatic pre-attentive auditory responses: MMN to tone burst frequency changes in autistic school-age children. *The Journal of International Advanced Otology, 11*(1), 36–41.

Abrams, D. A., Lynch, C. J., Cheng, K. M., Phillips, J., Supekar, K., Ryali, S., Uddin, L. Q., & Menon, V. (2013). Underconnectivity between voice-selective cortex and reward circuitry in children with autism. *Proceedings of the National Academy of Sciences of the United States of America, 110*(29), 12060–12065.

Annaz, D., Campbell, R., Coleman, M., Milne, E., & Swettenham, J. (2012). Young children with autism spectrum disorder do not preferentially attend to biological motion. *Journal of Autism and Developmental Disorders, 42*(3), 401–408.

Ashburner, J., Ziviani, J., & Rodger, S. (2008). Sensory processing and classroom emotional, behavioral, and educational outcomes in children with autism spectrum disorder. *The American Journal of Occupational Therapy, 62*(5), 564–573.

Baranek, G. T., David, F. J., Poe, M. D., Stone, W. L., & Watson, L. R. (2006). Sensory Experiences Questionnaire: Discriminating sensory features in young children with autism, developmental delays, and typical development. *Journal of Child Psychology and Psychiatry, 47*(6), 591–601.

Baruth, J. M., Casanova, M. F., Sears, L., & Sokhadze, E. (2010). Early-stage visual processing abnormalities in high-functioning autism spectrum disorder (ASD). *Translational Neuroscience, 1*(2), 177–187.

Baum, S. H., Stevenson, R. A., & Wallace, M. T. (2015). Behavioral, perceptual, and neural alterations in sensory and multisensory function in autism spectrum disorder. *Progress in Neurobiology, 134*, 140–160.

Beker, S., Foxe, J. J., & Molholm, S. (2018). Ripe for solution: Delayed development of multisensory processing in autism and its remediation. *Neuroscience & Biobehavioral Reviews, 84*, 182–192.

Bennetto, L., Keith, J. M., Allen, P. D., & Luebke, A. E. (2017). Children with autism spectrum disorder have reduced otoacoustic emissions at the 1 kHz mid-frequency region. *Autism Research, 10*(2), 337–145.

Ben-Sasson, A., Hen, L., Fluss, R., Cermak, S. A., Engel-Yeger, B., & Gal, E. (2009). A meta-analysis of sensory modulation symptoms in individuals with autism spectrum disorders. *Journal of Autism and Developmental Disorders, 39*(1), 1–11.

Berman, J. I., Edgar, J. C., Blaskey, L., Kuschner, E. S., Levy, S. E., Ku, M., Dell, J., & Roberts, T. P. L. (2016). Multimodal diffusion-MRI and MEG assessment of auditory and language system development in autism spectrum disorder. *Frontiers in Neuroanatomy, 10*, 30.

Bertone, A., Mottron, L., Jelenic, P., & Faubert, J. (2005). Enhanced and diminished visuo-spatial information processing in autism depends on stimulus complexity. *Brain, 128*(10), 2430–2441.

Blake, R., Turner, L. M., Smoski, M. J., Pozdol, S. L., & Stone, W. L. (2003). Visual recognition of biological motion is impaired in children with autism. *Psychological Science, 14*(2), 151–157.

Bonnel, A., Mottron, L., Peretz, I., Trudel, M., Gallun, E., & Bonnel, A.-M. (2003). Enhanced pitch sensitivity in individuals with autism: A signal detection analysis. *Journal of Cognitive Neuroscience, 15*(2), 226–235.

Bonnel, A., McAdams, S., Smith, B., Berthiaume, C., Bertone, A., Ciocca, V., Burack, J. A., & Mottron, L. (2010). Enhanced pure-tone pitch discrimination among persons with autism but not Asperger syndrome. *Neuropsychologia, 48*(9), 2465–2475.

Boyd, B. A., Baranek, G. T., Sideris, J., Poe, M. D., Watson, L. R., Patten, E., & Miller, H. (2010). Sensory features and repetitive behaviors in children with autism and developmental delays. *Autism Research, 3*(2), 78–87.

Brandwein, A. B., Foxe, J. J., Butler, J. S., Russo, N. N., Altschuler, T. S., Gomes, H., & Molholm, S. (2013). The development of multisensory integration in high-functioning autism: High-density electrical mapping and psychophysical measures reveal impairments in the processing of audiovisual inputs. *Cerebral Cortex, 23*(6), 1329–1341.

Brock, J., Brown, C. C., Boucher, J., & Rippon, G. (2002). The temporal binding deficit hypothesis of autism. *Development and Psychopathology, 14*(2), 209–224.

Canolty, R. T., & Knight, R. T. (2010). The functional role of cross-frequency coupling. *Trends in Cognitive Sciences, 14*(11), 506–515.

Chambon, V., Farrer, C., Pacherie, E., Jacquet, P. O., Leboyer, M., & Zalla, T. (2017). Reduced sensitivity to social priors during action prediction in adults with autism spectrum disorders. *Cognition, 160*, 17–26.

Christison-Lagay, K. L., Gifford, A. M., & Cohen, Y. E. (2015). Neural correlates of auditory scene analysis and perception. *International Journal of Psychophysiology, 95*(2), 238–245.

Corbett, B. A., Carmean, V., Ravizza, S., Wendelken, C., Henry, M. L., Carter, C., & Rivera, S. M. (2009). A functional and structural study of emotion and face processing in children with autism. *Psychiatry Research: Neuroimaging, 173*(3), 196–205.

Corbett, B. A., Newsom, C., Key, A. P., Qualls, L. R., & Edmiston, E. (2014). Examining the relationship between face processing and social interaction behavior in children with and without autism spectrum disorder. *Journal of Neurodevelopmental Disorders, 6*(1), 35.

Corbetta, M., & Shulman, G. L. (2002). Control of goal-directed and stimulus-driven attention in the brain. *Nature Reviews Neuroscience, 3*(3), 215–229.

Cornew, L., Roberts, T. P. L., Blaskey, L., & Edgar, J. C. (2012). Resting-state oscillatory activity in autism spectrum disorders. *Journal of Autism and Developmental Disorders, 42*(9), 1884–1894.

Crane, L., Goddard, L., & Pring, L. (2009). Sensory processing in adults with autism spectrum disorders. *Autism, 13*(3), 215–228.
Croydon, A., Karaminis, T., Neil, L., Burr, D., & Pellicano, E. (2017). The light-from-above prior is intact in autistic children. *Journal of Experimental Child Psychology, 161*, 113–125.
Cygan, H. B., Tacikowski, P., Ostaszewski, P., Chojnicka, I., & Nowicka, A. (2014). Neural correlates of own name and own face detection in autism spectrum disorder. *PLoS One, 9*(1), e86020.
Dakin, S., & Frith, U. (2005). Vagaries of visual perception in autism. *Neuron, 48*(3), 497–507.
de Boer-Schellekens, L., Eussen, M., & Vroomen, J. (2013a). Diminished sensitivity of audiovisual temporal order in autism spectrum disorder. *Frontiers in Integrative Neuroscience, 7*, 8.
de Boer-Schellekens, L., Keetels, M., Eussen, M., & Vroomen, J. (2013b). No evidence for impaired multisensory integration of low-level audiovisual stimuli in adolescents and young adults with autism spectrum disorders. *Neuropsychologia, 51*(14), 30043013.
Demopoulos, C., & Lewine, J. D. (2016). Audiometric profiles in autism spectrum disorders: Does subclinical hearing loss impact communication? *Autism Research, 9*(1), 107–120.
DePape, A.-M. R., Hall, G. B. C., Tillmann, B., & Trainor, L. J. (2012). Auditory processing in high-functioning adolescents with autism spectrum disorder. *PLoS One, 7*(9), e44084.
Du, Y., Buchsbaum, B. R., Grady, C. L., & Alain, C. (2014). Noise differentially impacts phoneme representations in the auditory and speech motor systems. *Proceedings of the National Academy of Sciences of the United States of America, 111*(19), 7126–7131.
Duerden, E. G., Oatley, H. K., Mak-Fan, K. M., McGrath, P. A., Taylor, M. J., Szatmari, P., & Roberts, S. W. (2012). Risk factors associated with self-injurious behaviors in children and adolescents with autism spectrum disorders. *Journal of Autism and Developmental Disorders, 42*(11), 2460–2470.
Eyler, L. T., Pierce, K., & Courchesne, E. (2012). A failure of left temporal cortex to specialize for language is an early emerging and fundamental property of autism. *Brain, 135*(3), 949–960.
Falck-Ytter, T., Bölte, S., Gredebäck, G., Klin, A., Martinez-Conde, S., Pollick, F., Bolton, P., Charman, T., Baird, G., Johnson, M., Gerig, G., Hazlett, H., Schultz, R., Styner, M., Zwaigenbaum, L., & Piven, J. (2013). Eye tracking in early autism research. *Journal of Neurodevelopmental Disorders, 5*(1), 28.
Fan, Y.-T., & Cheng, Y. (2014). Atypical mismatch negativity in response to emotional voices in people with autism spectrum conditions. *PLoS One, 9*(7), e102471.
Flevaris, A. V., & Murray, S. O. (2014). Orientation-specific surround suppression in the primary visual cortex varies as a function of autistic tendency. *Frontiers in Human Neuroscience, 8*, 1017.
Floris, D. L., Barber, A. D., Nebel, M. B., Martinelli, M., Lai, M.-C., Crocetti, D., Baron-Cohen, S., Suckling, J., Pekar, J. J., & Mostofsky, S. H. (2016). Atypical lateralization of motor circuit functional connectivity in children with autism is associated with motor deficits. *Molecular Autism, 7*(1), 35.
Foss-Feig, J. H., Kwakye, L. D., Cascio, C. J., Burnette, C. P., Kadivar, H., Stone, W. L., & Wallace, M. T. (2010). An extended multisensory temporal binding window in autism spectrum disorders. *Experimental Brain Research, 203*(2), 381–389.
Foss-Feig, J. H., Tadin, D., Schauder, K. B., & Cascio, C. J. (2013). A substantial and unexpected enhancement of motion perception in autism. *The Journal of Neuroscience, 33*(19), 8243–8249.
Foss-Feig, J. H., McGugin, R. W., Gauthier, I., Mash, L. E., Ventola, P., & Cascio, C. J. (2016). A functional neuroimaging study of fusiform response to restricted interests in children and adolescents with autism spectrum disorder. *Journal of Neurodevelopmental Disorders, 8*, 15.
Foxe, J. J., Molholm, S., Del Bene, V. A., Frey, H.-P., Russo, N. N., Blanco, D., Saint-Amour, D., & Ross, L. A. (2015). Severe multisensory speech integration deficits in high-functioning school-aged children with autism spectrum disorder (ASD) and their resolution during early adolescence. *Cerebral Cortex, 25*(2), 298–312.
Frazier, T. W., Strauss, M., Klingemier, E. W., Zetzer, E. E., Hardan, A. Y., Eng, C., & Youngstrom, E. A. (2017). A meta-analysis of gaze differences to social and nonsocial information between

individuals with and without autism. *Journal of the American Academy of Child & Adolescent Psychiatry, 56*(7), 546555.

Frey, H.-P., Molholm, S., Lalor, E. C., Russo, N. N., & Foxe, J. J. (2013). Atypical cortical representation of peripheral visual space in children with an autism spectrum disorder. *European Journal of Neuroscience, 38*(1), 2125–2138.

Gervais, H., Belin, P., Boddaert, N., Leboyer, M., Coez, A., Sfaello, I., Barthélémy, C., Brunelle, F., Samson, Y., & Zilbovicius, M. (2004). Abnormal cortical voice processing in autism. *Nature Neuroscience, 7*(8), 801–802.

Glover, G. H. (2011). Overview of functional magnetic resonance imaging. *Neurosurgery Clinics of North America, 22*(2), 133–139.

Gomot, M., Belmonte, M. K., Bullmore, E. T., Bernard, F. A., & Baron-Cohen, S. (2008). Brain hyper-reactivity to auditory novel targets in children with high-functioning autism. *Brain, 131*(9), 2479–2488.

Guiraud, J. A., Kushnerenko, E., Tomalski, P., Davies, K., Ribeiro, H., Johnson, M. H., & BASIS Team. (2011). Differential habituation to repeated sounds in infants at high risk for autism. *Neuroreport, 22*(16), 845–849.

Haigh, S. M., Heeger, D. J., Heller, L. M., Gupta, A., Dinstein, I., Minshew, N. J., & Behrmann, M. (2016). No difference in cross-modal attention or sensory discrimination thresholds in autism and matched controls. *Vision Research, 121*, 85–94.

Happé, F. (1999). Autism: Cognitive deficit or cognitive style? *Trends in Cognitive Sciences, 3*(6), 216–222.

Harris, H., Israeli, D., Minshew, N., Bonneh, Y., Heeger, D. J., Behrmann, M., & Sagi, D. (2015). Perceptual learning in autism: Over-specificity and possible remedies. *Nature Neuroscience, 18*(11), 1574–1576.

Hazlett, H. C., Gu, H., Munsell, B. C., Kim, S. H., Styner, M., Wolff, J. J., Elison, J. T., Swanson, M. R., Zhu, H., Botteron, K. N., Collins, D. L., Constantino, J. N., Dager, S. R., Estes, A. M., Evans, A. C., Fonov, V. S., Gerig, G., Kostopoulos, P., McKinstry, R. C., Pandey, J., Paterson, S., Pruett, J. R., Schultz, R. T., Shaw, D. W., Zwaigenbaum, L., Piven, J., & The IBIS Network. (2017). Early brain development in infants at high risk for autism spectrum disorder. *Nature, 542*(7641), 348–351.

Heaton, P. (2003). Pitch memory, labelling and disembedding in autism. *Journal of Child Psychology and Psychiatry, 44*(4), 543–551.

Heaton, P., Williams, K., Cummins, O., & Happe, F. (2008). Autism and pitch processing splinter skills: A group and subgroup analysis. *Autism, 12*(2), 203–219.

Herringshaw, A. J., Ammons, C. J., DeRamus, T. P., & Kana, R. K. (2016). Hemispheric differences in language processing in autism spectrum disorders: A meta-analysis of neuroimaging studies. *Autism Research, 9*(10), 1046–1057.

Hillock-Dunn, A., & Wallace, M. T. (2012). Developmental changes in the multisensory temporal binding window persist into adolescence. *Developmental Science, 15*(5), 688–696.

Hosozawa, M., Tanaka, K., Shimizu, T., Nakano, T., & Kitazawa, S. (2012). How children with specific language impairment view social situations: An eye tracking study. *Pediatrics, 129*(6), e1453–e1460.

Hubl, D., Bölte, S., Feineis-Matthews, S., Lanfermann, H., Federspiel, A., Strik, W., Poustka, F., & Dierks, T. (2003). Functional imbalance of visual pathways indicates alternative face processing strategies in autism. *Neurology, 61*(9), 1232–1237.

Hull, J. V., Jacokes, Z. J., Torgerson, C. M., Irimia, A., & Van Horn, J. D. (2017). Resting-state functional connectivity in autism spectrum disorders: A review. *Frontiers in Psychiatry, 7*, 205.

Irwin, J. R., Tornatore, L. A., Brancazio, L., & Whalen, D. H. (2011). Can children with autism spectrum disorders "hear" a speaking face? *Child Development, 82*(5), 1397–1403.

Järvinen-Pasley, A., Pasley, J., & Heaton, P. (2008a). Is the linguistic content of speech less salient than its perceptual features in autism? *Journal of Autism and Developmental Disorders, 38*(2), 239–248.

Järvinen-Pasley, A., Wallace, G. L., Ramus, F., Happé, F., & Heaton, P. (2008b). Enhanced perceptual processing of speech in autism. *Developmental Science, 11*(1), 109–121.

Jeste, S. S., Kirkham, N., Senturk, D., Hasenstab, K., Sugar, C., Kupelian, C., Baker, E., Sanders, A. J., Shimizu, C., Norona, A., Paparella, T., Freeman, S. F. N., & Johnson, S. P. (2015). Electrophysiological evidence of heterogeneity in visual statistical learning in young children with ASD. *Developmental Science, 18*(1), 90–105.

Jochaut, D., Lehongre, K., Saitovitch, A., Devauchelle, A.-D., Olasagasti, I., Chabane, N., Zilbovicius, M., & Giraud, A.-L. (2015). Atypical coordination of cortical oscillations in response to speech in autism. *Frontiers in Human Neuroscience, 9*, 171.

Jones, W., & Klin, A. (2013). Attention to eyes is present but in decline in 2–6-month-old infants later diagnosed with autism. *Nature, 504*(7480), 427–431.

Jones, C. R. G., Happé, F., Baird, G., Simonoff, E., Marsden, A. J., Tregay, J., Phillips, R. J., Goswami, U., Thomson, J. M., & Charman, T. (2009). Auditory discrimination and auditory sensory behaviours in autism spectrum disorders. *Neuropsychologia, 47*(13), 2850–2858.

Kana, R. K., Libero, L. E., Hu, C. P., Deshpande, H. D., & Colburn, J. S. (2014). Functional brain networks and white matter underlying theory-of-mind in autism. *Social Cognitive and Affective Neuroscience, 9*(1), 98–105.

Keehn, B., Vogel-Farley, V., Tager-Flusberg, H., & Nelson, C. A. (2015). Atypical hemispheric specialization for faces in infants at risk for autism spectrum disorder. *Autism Research, 8*(2), 187–198.

Keil, A., Müller, M. M., Ray, W. J., Gruber, T., & Elbert, T. (1999). Human gamma band activity and perception of a gestalt. *The Journal of Neuroscience, 19*(16), 7152–7161.

Kleinhans, N. M., Müller, R.-A., Cohen, D. N., & Courchesne, E. (2008). Atypical functional lateralization of language in autism spectrum disorders. *Brain Research, 1221*, 115–125.

Klimesch, W., Sauseng, P., & Hanslmayr, S. (2007). EEG alpha oscillations: The inhibition-timing hypothesis. *Brain Research Reviews, 53*(1), 63–88.

Koh, H. C., Milne, E., & Dobkins, K. (2010). Spatial contrast sensitivity in adolescents with autism spectrum disorders. *Journal of Autism and Developmental Disorders, 40*(8), 978–987.

Kröger, A., Bletsch, A., Krick, C., Siniatchkin, M., Jarczok, T. A., Freitag, C. M., & Bender, S. (2014). Visual event-related potentials to biological motion stimuli in autism spectrum disorders. *Social Cognitive and Affective Neuroscience, 9*(8), 1214–1222.

Larrain-Valenzuela, J., Zamorano, F., Soto-Icaza, P., Carrasco, X., Herrera, C., Daiber, F., Aboitiz, F., & Billeke, P. (2017). Theta and alpha oscillation impairments in autistic spectrum disorder reflect working memory deficit. *Scientific Reports, 7*(1), 14328.

Lawson, R. P., Mathys, C., & Rees, G. (2017). Adults with autism overestimate the volatility of the sensory environment. *Nature Neuroscience, 20*(9), 1293–1299.

Lee, A. K. C., Larson, E., Maddox, R. K., & Shinn-Cunningham, B. G. (2014). Using neuroimaging to understand the cortical mechanisms of auditory selective attention. *Hearing Research, 307*, 111–120.

Lee, Y., Park, B., James, O., Kim, S.-G., & Park, H. (2017). Autism spectrum disorder related functional connectivity changes in the language network in children, adolescents and adults. *Frontiers in Human Neuroscience, 11*, 418.

Lepistö, T., Kuitunen, A., Sussman, E., Saalasti, S., Jansson-Verkasalo, E., Nieminen-von Wendt, T., & Kujala, T. (2009). Auditory stream segregation in children with Asperger syndrome. *Biological Psychology, 82*(3), 301–307.

Levin, A. R., Varcin, K. J., O'Leary, H. M., Tager-Flusberg, H., & Nelson, C. A. (2017). EEG power at 3 months in infants at high familial risk for autism. *Journal of Neurodevelopmental Disorders, 9*(1), 34.

Lin, I.-F., Shirama, A., Kato, N., & Kashino, M. (2017). The singular nature of auditory and visual scene analysis in autism. *Philosophical Transactions of the Royal Society B: Biological Sciences, 372*(1714), 20160115.

Linke, A. C., Jao Keehn, R. J., Pueschel, E. B., Fishman, I., & Müller, R. A. (2018). Children with ASD show links between aberrant sound processing, social symptoms, and atypical audi-

tory interhemispheric and thalamocortical functional connectivity. *Developmental Cognitive Neuroscience, 29*, 117–126.
Lodhia, V., Brock, J., Johnson, B. W., & Hautus, M. J. (2014). Reduced object related negativity response indicates impaired auditory scene analysis in adults with autistic spectrum disorder. *PeerJ, 2*, e261.
Lodhia, V., Hautus, M. J., Johnson, B. W., & Brock, J. (2018). Atypical brain responses to auditory spatial cues in adults with autism spectrum disorder. *European Journal of Neuroscience, 47*(6), 682–689.
Long, Z., Duan, X., Mantini, D., & Chen, H. (2016). Alteration of functional connectivity in autism spectrum disorder: Effect of age and anatomical distance. *Scientific Reports, 6*(1), 26527.
Lord, C., Risi, S., DiLavore, P. S., Shulman, C., Thurm, A., & Pickles, A. (2006). Autism from 2 to 9 years of age. *Archives of General Psychiatry, 63*(6), 694–701.
Luyster, R., Gotham, K., Guthrie, W., Coffing, M., Petrak, R., Pierce, K., Bishop, S., Esler, A., Hus, V., Oti, R., Richler, J., Risi, S., & Lord, C. (2009). The Autism Diagnostic Observation Schedule-toddler module: A new module of a standardized diagnostic measure for autism spectrum disorders. *Journal of Autism and Developmental Disorders, 39*(9), 1305–1320.
Magnotti, J. F., & Beauchamp, M. S. (2015). The noisy encoding of disparity model of the McGurk effect. *Psychonomic Bulletin & Review, 22*(3), 701–709.
Manjaly, Z. M., Bruning, N., Neufang, S., Stephan, K. E., Brieber, S., Marshall, J. C., Kamp-Becker, I., Remschmidt, H., Herpertz-Dahlmann, B., Konrad, K., & Fink, G. R. (2007). Neurophysiological correlates of relatively enhanced local visual search in autistic adolescents. *Neuroimage, 35*(1), 283–291.
Manning, C., Kilner, J., Neil, L., Karaminis, T., & Pellicano, E. (2017). Children on the autism spectrum update their behaviour in response to a volatile environment. *Developmental Science, 20*(5), e12435.
Markram, K., & Markram, H. (2010). The intense world theory-a unifying theory of the neurobiology of autism. *Frontiers in Human Neuroscience, 4*, 224.
Mayer, J. L. (2017). The relationship between autistic traits and atypical sensory functioning in neurotypical and ASD adults: A spectrum approach. *Journal of Autism and Developmental Disorders, 47*(2), 316–327.
McGurk, H., & MacDonald, J. (1976). Hearing lips and seeing voices. *Nature, 264*(5588), 746–748.
McIntosh, D., Miller, L., & Shyu, V. (1999). Development and validation of the short sensory profile. In W. Dunn (Ed.), *Sensory Profile: User's Manual* (pp. 59–73). San Antonio: The Psychological Coroporation.
McKay, L. S., Simmons, D. R., McAleer, P., Marjoram, D., Piggot, J., & Pollick, F. E. (2012). Do distinct atypical cortical networks process biological motion information in adults with autism spectrum disorders? *Neuroimage, 59*(2), 1524–1533.
Miller, M., Iosif, A.-M., Hill, M., Young, G. S., Schwichtenberg, A. J., & Ozonoff, S. (2017). Response to name in infants developing autism spectrum disorder: A prospective study. *The Journal of Pediatrics, 183*, 141–146.e1.
Milne, E. (2011). Increased intra-participant variability in children with autistic spectrum disorders: Evidence from single-trial analysis of evoked EEG. *Frontiers in Psychology, 2*, 51.
Modi, M. E., & Sahin, M. (2017). Translational use of event-related potentials to assess circuit integrity in ASD. *Nature Reviews Neurology, 13*(3), 160–170.
Murphy, P., Brady, N., Fitzgerald, M., & Troje, N. F. (2009). No evidence for impaired perception of biological motion in adults with autistic spectrum disorders. *Neuropsychologia, 47*(14), 3225–3235.
Murphy, J. W., Foxe, J. J., Peters, J. B., & Molholm, S. (2014). Susceptibility to distraction in autism spectrum disorder: Probing the integrity of oscillatory alpha-band suppression mechanisms. *Autism Research, 7*(4), 442–458.
Nackaerts, E., Wagemans, J., Helsen, W., Swinnen, S. P., Wenderoth, N., & Alaerts, K. (2012). Recognizing biological motion and emotions from point-light displays in autism spectrum disorders. *PLoS One, 7*(9), e44473.
Nagae, L. M., Zarnow, D. M., Blaskey, L., Dell, J., Khan, S. Y., Qasmieh, S., Levy, S. E., & Roberts, T. P. L. (2012). Elevated mean diffusivity in the left hemisphere superior longitudinal

fasciculus in autism spectrum disorders increases with more profound language impairment. *American Journal of Neuroradiology, 33*(9), 1720–1725.
Nagy, E., & Loveland, K. A. (2002). Prolonged brainstem auditory evoked potentials: An autism-specific or autism-nonspecific marker. *Archives of General Psychiatry, 59*(3), 288–890.
Neumann, D., Spezio, M. L., Piven, J., & Adolphs, R. (2006). Looking you in the mouth: Abnormal gaze in autism resulting from impaired top-down modulation of visual attention. *Social Cognitive and Affective Neuroscience, 1*(3), 194–202.
Noris, B., Nadel, J., Barker, M., Hadjikhani, N., & Billard, A. (2012). Investigating gaze of children with ASD in naturalistic settings. *PLoS One, 7*(9), e44144.
O'Riordan, M., & Passetti, F. (2006). Discrimination in autism within different sensory modalities. *Journal of Autism and Developmental Disorders, 36*(5), 665–675.
Ozonoff, S., Young, G. S., Carter, A., Messinger, D., Yirmiya, N., Zwaigenbaum, L., Bryson, S., Carver, L. J., Constantino, J. N., Dobkins, K., Hutman, T., Iverson, J. M., Landa, R., Rogers, S. J., Sigman, M., & Stone, W. L. (2011). Recurrence risk for autism spectrum disorders: A Baby Siblings Research Consortium study. *Pediatrics, 128*(3), e488–e495.
Pack, C. C., Hunter, J. N., & Born, R. T. (2005). Contrast dependence of suppressive influences in cortical area MT of alert macaque. *Journal of Neurophysiology, 93*(3), 1809–1815.
Palmer, C. J., Lawson, R. P., & Hohwy, J. (2017). Bayesian approaches to autism: Towards volatility, action, and behavior. *Psychological Bulletin, 143*(5), 521–542.
Patten, E., Watson, L. R., & Baranek, G. T. (2014). Temporal synchrony detection and associations with language in young children with ASD. *Autism Research and Treatment, 2014*, 1–8.
Pei, F., Baldassi, S., & Norcia, A. M. (2014). Electrophysiological measures of low-level vision reveal spatial processing deficits and hemispheric asymmetry in autism spectrum disorder. *Journal of Vision, 14*(11).
Peiker, I., David, N., Schneider, T. R., Nolte, G., Schottle, D., & Engel, A. K. (2015). Perceptual integration deficits in autism spectrum disorders are associated with reduced interhemispheric gamma-band coherence. *The Journal of Neuroscience, 35*(50), 16352–16361.
Pell, P. J., Mareschal, I., Calder, A. J., von dem Hagen, E. A. H., Clifford, C. W. G., Baron-Cohen, S., & Ewbank, M. P. (2016). Intact priors for gaze direction in adults with high-functioning autism spectrum conditions. *Molecular Autism, 7*(1), 25.
Pellicano, E., & Burr, D. (2012). When the world becomes "too real": A Bayesian explanation of autistic perception. *Trends in Cognitive Sciences, 16*(10), 504–510.
Pellicano, E., Gibson, L., Maybery, M., Durkin, K., & Badcock, D. R. (2005). Abnormal global processing along the dorsal visual pathway in autism: A possible mechanism for weak visuospatial coherence? *Neuropsychologia, 43*(7), 1044–1053.
Pellicano, E., Smith, A. D., Cristino, F., Hood, B. M., Briscoe, J., & Gilchrist, I. D. (2011). Children with autism are neither systematic nor optimal foragers. *Proceedings of the National Academy of Sciences of the United States of America, 108*(1), 421–426.
Pierce, K., Marinero, S., Hazin, R., McKenna, B., Barnes, C. C., & Malige, A. (2016). Eye tracking reveals abnormal visual preference for geometric images as an early biomarker of an autism spectrum disorder subtype associated with increased symptom severity. *Biological Psychiatry, 79*(8), 657–666.
Plaisted, K., Swettenham, J., & Rees, L. (1999). Children with autism show local precedence in a divided attention task and global precedence in a selective attention task. *Journal of Child Psychology and Psychiatry, 40*(5), 733–742.
Remington, A., & Fairnie, J. (2017). A sound advantage: Increased auditory capacity in autism. *Cognition, 166*, 459–465.
Rice, K., Moriuchi, J. M., Jones, W., & Klin, A. (2012). Parsing heterogeneity in autism spectrum disorders: Visual scanning of dynamic social scenes in school-aged children. *Journal of the American Academy of Child and Adolescent Psychiatry, 51*(3), 238–248.
Rogers, S. J., & Ozonoff, S. (2005). Annotation: What do we know about sensory dysfunction in autism? A critical review of the empirical evidence. *Journal of Child Psychology and Psychiatry, 46*(12), 1255–1268.

Romei, V., Gross, J., & Thut, G. (2012). Sounds reset rhythms of visual cortex and corresponding human visual perception. *Current Biology, 22*(9), 807–813.

Ross, L. A., Saint-Amour, D., Leavitt, V. M., Javitt, D. C., & Foxe, J. J. (2006). Do you see what I am saying? Exploring visual enhancement of speech comprehension in noisy environments. *Cerebral Cortex, 17*(5), 1147–1153.

Ross, L. A., Del Bene, V. A., Molholm, S., Jae Woo, Y., Andrade, G. N., Abrahams, B. S., & Foxe, J. J. (2017). Common variation in the autism risk gene *CNTNAP2*, brain structural connectivity and multisensory speech integration. *Brain and Language, 174*, 50–60.

Samson, F., Hyde, K. L., Bertone, A., Soulières, I., Mendrek, A., Ahad, P., Mottron, L., & Zeffiro, T. A. (2011). Atypical processing of auditory temporal complexity in autistics. *Neuropsychologia, 49*(3), 546–555.

Sasson, N. J. (2006). The development of face processing in autism. *Journal of Autism and Developmental Disorders, 36*(3), 381–394.

Schaaf, R. C., & Lane, A. E. (2015). Toward a best-practice protocol for assessment of sensory features in ASD. *Journal of Autism and Developmental Disorders, 45*(5), 1380–1395.

Schelinski, S., Borowiak, K., & von Kriegstein, K. (2016). Temporal voice areas exist in autism spectrum disorder but are dysfunctional for voice identity recognition. *Social Cognitive and Affective Neuroscience, 11*(11), 1812–1822.

Scherf, K. S., Elbich, D., Minshew, N., & Behrmann, M. (2015). Individual differences in symptom severity and behavior predict neural activation during face processing in adolescents with autism. *Neuroimage: Clinical, 7*, 53–67.

Seery, A. M., Vogel-Farley, V., Tager-Flusberg, H., & Nelson, C. A. (2013). Atypical lateralization of ERP response to native and non-native speech in infants at risk for autism spectrum disorder. *Developmental Cognitive Neuroscience, 5*, 10–24.

Seery, A. M., Tager-Flusberg, H., & Nelson, C. A. (2014). Event-related potentials to repeated speech in 9-month-old infants at risk for autism spectrum disorder. *Journal of Neurodevelopmental Disorders, 6*(1), 43.

Shams, L., & Beierholm, U. R. (2010). Causal inference in perception. *Trends in Cognitive Sciences, 14*(9), 425–432.

Shams, L., Kamitani, Y., & Shimojo, S. (2000). Illusions: What you see is what you hear. *Nature, 408*(6814), 788.

Sharda, M., Midha, R., Malik, S., Mukerji, S., & Singh, N. C. (2015). Fronto-temporal connectivity is preserved during sung but not spoken word listening, across the autism spectrum. *Autism Research, 8*(2), 174–186.

Shinn-Cunningham, B., Best, V., & Lee, A. K. C. (2017). Auditory object formation and selection. In J. C. Middlebrooks, J. Z. Simon, A. N. Popper, & R. R. Fay (Eds.), *The Auditory System at the Cocktail Party* (pp. 7–40). New York: Springer.

Shirama, A., Kato, N., & Kashino, M. (2016). When do individuals with autism spectrum disorder show superiority in visual search? *Autism, 21*(8), 942–951.

Simmons, D. R., Robertson, A. E., McKay, L. S., Toal, E., McAleer, P., & Pollick, F. E. (2009). Vision in autism spectrum disorders. *Vision Research, 49*(22), 2705–2739.

Simon, D. M., & Wallace, M. T. (2016). Dysfunction of sensory oscillations in autism spectrum disorder. *Neuroscience & Biobehavioral Reviews, 68*, 848–861.

Simon, D. M., Damiano, C. R., Woynaroski, T. G., Ibañez, L. V., Murias, M., Stone, W. L., Wallace, M. T., & Cascio, C. J. (2017). Neural correlates of sensory hyporesponsiveness in toddlers at high risk for autism spectrum disorder. *Journal of Autism and Developmental Disorders, 47*(9), 2710–2722.

Snijders, T. M., Milivojevic, B., & Kemner, C. (2013). Atypical excitation-inhibition balance in autism captured by the gamma response to contextual modulation. *Neuroimage: Clinical, 3*, 65–72.

Stevenson, R. A., & Wallace, M. T. (2013). Multisensory temporal integration: Task and stimulus dependencies. *Experimental Brain Research, 227*(2), 249–261.

Stevenson, R. A., Siemann, J. K., Schneider, B. C., Eberly, H. E., Woynaroski, T. G., Camarata, S. M., & Wallace, M. T. (2014a). Multisensory temporal integration in autism spectrum disorders. *The Journal of Neuroscience, 34*(3), 691–697.
Stevenson, R. A., Siemann, J. K., Woynaroski, T. G., Schneider, B. C., Eberly, H. E., Camarata, S. M., & Wallace, M. T. (2014b). Evidence for diminished multisensory integration in autism spectrum disorders. *Journal of Autism and Developmental Disorders, 44*(12), 3161–3167.
Stevenson, R. A., Siemann, J. K., Woynaroski, T. G., Schneider, B. C., Eberly, H. E., Camarata, S. M., & Wallace, M. T. (2014c). Brief report: Arrested development of audiovisual speech perception in autism spectrum disorders. *Journal of Autism and Developmental Disorders, 44*(6), 1470–1477.
Stevenson, R. A., Baum, S. H., Segers, M., Ferber, S., Barense, M. D., & Wallace, M. T. (2017). Multisensory speech perception in autism spectrum disorder: From phoneme to whole-word perception. *Autism Research, 10*(7), 1280–1290.
Sumby, W. H., & Pollack, I. (1954). Visual contribution to speech intelligibility in noise. *The Journal of the Acoustical Society of America, 26*(2), 212–215.
Swettenham, J., Remington, A., Laing, K., Fletcher, R., Coleman, M., & Gomez, J.-C. (2013). Perception of pointing from biological motion point-light displays in typically developing children and children with autism spectrum disorder. *Journal of Autism and Developmental Disorders, 43*(6), 1437–1446.
Sysoeva, O. V., Galuta, I. A., Davletshina, M. S., Orekhova, E. V., & Stroganova, T. A. (2017). Abnormal size-dependent modulation of motion perception in children with autism spectrum disorder (ASD). *Frontiers in Neuroscience, 11*, 164.
Tadin, D., & Lappin, J. S. (2005). Optimal size for perceiving motion decreases with contrast. *Vision Research, 45*(16), 2059–2064.
Thomas, C., Humphreys, K., Jung, K.-J., Minshew, N., & Behrmann, M. (2011). The anatomy of the callosal and visual-association pathways in high-functioning autism: A DTI tractography study. *Cortex, 47*(7), 863–873.
Thorne, J. D., De Vos, M., Viola, F. C., & Debener, S. (2011). Cross-modal phase reset predicts auditory task performance in humans. *The Journal of Neuroscience, 31*(10), 3853–3861.
Tierney, A. L., Gabard-Durnam, L., Vogel-Farley, V., Tager-Flusberg, H., & Nelson, C. A. (2012). Developmental trajectories of testing EEG power: An endophenotype of autism spectrum disorder. *PLoS One, 7*(6), e39127.
Van de Cruys, S., Evers, K., Van der Hallen, R., Van Eylen, L., Boets, B., De-Wit, L., & Wagemans, J. (2014). Precise minds in uncertain worlds: Predictive coding in autism. *Psychological Review, 121*(4), 649–675.
Vandenbroucke, M. W. G., Scholte, H. S., van Engeland, H., Lamme, V. A. F., & Kemner, C. (2008). A neural substrate for atypical low-level visual processing in autism spectrum disorder. *Brain, 131*(4), 1013–1024.
Venezia, J. H., Vaden, K. I., Rong, F., Maddox, D., Saberi, K., & Hickok, G. (2017). Auditory, visual and audiovisual speech processing streams in superior temporal sulcus. *Frontiers in Human Neuroscience, 11*, 174.
Ververi, A., Vargiami, E., Papadopoulou, V., Tryfonas, D., & Zafeiriou, D. (2015). Brainstem auditory evoked potentials in boys with autism: Still searching for the hidden truth. *Iranian Journal of Child Neurology, 9*(2), 21–28.
Wang, J., Barstein, J., Ethridge, L. E., Mosconi, M. W., Takarae, Y., & Sweeney, J. A. (2013). Resting state EEG abnormalities in autism spectrum disorders. *Journal of Neurodevelopmental Disorders, 5*(1), 24.
Watson, L. R., Patten, E., Baranek, G. T., Poe, M., Boyd, B. A., Freuler, A., & Lorenzi, J. (2011). Differential associations between sensory response patterns and language, social, and communication measures in children with autism or other developmental disabilities. *Journal of Speech Language and Hearing Research, 54*(6), 1562–1576.
Webb, S. J., Merkle, K., Murias, M., Richards, T., Aylward, E., & Dawson, G. (2012). ERP responses differentiate inverted but not upright face processing in adults with ASD. *Social Cognitive and Affective Neuroscience, 7*(5), 578–587.

Welch, R. B. (1999). Meaning, attention, and the "unity assumption" in the intersensory bias of spatial and temporal perceptions. In G. Aschersleben, T. Bachmann, & J. Musseler (Eds.), *Cognitive contributions to the perception of spatial and temporal events* (pp. 371–387). Amsterdam: Elsevier.

Woynaroski, T. G., Kwakye, L. D., Foss-Feig, J. H., Stevenson, R. A., Stone, W. L., & Wallace, M. T. (2013). Multisensory speech perception in children with autism spectrum disorders. *Journal of Autism and Developmental Disorders, 43*(12), 2891–2902.

Yamasaki, T., Maekawa, T., Fujita, T., & Tobimatsu, S. (2017). Connectopathy in autism spectrum disorders: A review of evidence from visual evoked potentials and diffusion magnetic resonance imaging. *Frontiers in Neuroscience, 11*, 627.

Zalla, T., Fernandez, L. G., Pieron, M., Seassau, M., & Leboyer, M. (2016). Reduced saccadic inhibition of return to moving eyes in autism spectrum disorders. *Vision Research, 127*, 115–121.

CPSIA information can be obtained
at www.ICGtesting.com
Printed in the USA
BVHW010822260319
543718BV00001B/2/P

9 783030 104597